The Passage to Cosmos

The Passage to Cosmos

Alexander von Humboldt and the Shaping of America

LAURA DASSOW WALLS

The University of Chicago Press
Chicago and London

The University of Chicago Press, Chicago 60637
The University of Chicago Press, Ltd., London
© 2009 by The University of Chicago
All rights reserved. Published 2009.
Paperback edition 2011
Printed in the United States of America

20 19 18 17 16 15 14 13 12 11   2 3 4 5 6

ISBN-13: 978-0-226-87182-0 (cloth)
ISBN-13: 978-0-226-87183-7 (paper)
ISBN-10: 0-226-87182-7 (cloth)
ISBN-10: 0-226-87183-5 (paper)

Frontispiece: *Natural Bridges of Icononzo*, plate 4, Alexander von Humboldt, *Vues des Cordillères* (Paris, 1810). Rare Books and Special Collections, University of South Carolina Libraries.

Library of Congress Cataloging-in-Publication Data

Walls, Laura Dassow.
    The passage to Cosmos : Alexander von Humboldt and the shaping of America /
Laura Dassow Walls.
        p. cm.
    Includes bibliographical references and index.
    ISBN-13: 978-0-226-87182-0 (cloth : alk. paper)
    ISBN-10: 0-226-87182-7 (cloth : alk. paper)
    1. Humboldt, Alexander von, 1769–1859—Influence—United States. 2. Humboldt,
Alexander von, 1769–1859. Kosmos. 3. Humboldt, Alexander von, 1769–1859—
Political and social views. 4. United States—Intellectual life—19th century.
5. Science—History—19th century.   I. Title.
    Q143.H9D388 2009
    508.092—dc22
[B]   2009003099

♾ This paper meets the requirements of ANSI/NISO Z39.48-1992 (Permanence of Paper).

Contents

Preface: Romancing the Ruby-Crowned Kinglet

Every poet has trembled on the verge of science.
HENRY DAVID THOREAU

This book was born from a very small death at a laboratory bench, more years ago than I care to remember. I grew up near Seattle on an island now home to CEOs, doctors, and Microsoft millionaires, but then a suburb of modest houses tucked into a mature second-growth forest of Douglas fir, maple, hemlock, and western cedar. I wanted to be a scientist, which to me, immersed in Jane Goodall and Niko Tinbergen, meant a natural historian. I spent endless hours learning the birds and plants of Puget Sound and weaving them into a 1960s-era ecological matrix of climax forests, disturbance, and succession, edge effects, sandy uplands, and moist ravines. All this led me to the University of Washington, where as an eager freshman I knocked on various doors until one of them opened: an ornithologist agreed to put me to work as his assistant. My first taste of heaven was setting to rights the natural history museum's musty drawers of stuffed bird skins and boxes of bird bones. Months later, that task completed, he taught me how to prepare bird skins, and I learned the delicate art of unzipping the fragile songbird body from tail to beak, folding feathers out of the way and replacing flesh with rolls of cotton. He was particularly keen to prepare as many specimens as possible of a certain sparrow for a statistical study of plumage patterns.

But one day something different came in. It was a ruby-crowned kinglet. I'd seen them in the wild, quick drab puffs of greenish-gray, but I'd never held one in my hand before. He was freshly dead, tiny, a mere tuft of down. And beautiful, so beautiful I was shocked to tears—softly shaded olive browns, elegant to the last detail, topped with that brilliant jewel-bright ruby crown. Every feather was an astonishment. I held that kinglet in my hand, bewildered, and something inside me broke. I understood my professor's need for data points, but what I beheld asked another and very different question. I didn't know then just what it was, or how to begin to search for an answer, but I did know that my path had changed. I worked a bit longer at the museum, until the day they told me my volunteer labor had proven the need for a paid assistant, a position for which I no longer qualified. After an agonized review of the university catalogue I changed my major to English. I had never taken an English course, but I loved to read and write, and of all the majors it seemed most likely to accommodate that grab bag of classes my curiosity demanded: botany, geology, animal behavior, scientific illustration, anthropology, history. So I analyzed novels and wrote poetry, and when it came time to specialize, I turned to the writings of Henry David Thoreau. I sensed that he, too, had beheld that kinglet, and asked not what statistical set this data point might fill, but that other question, Why should such things be? And what must one do about it? I felt Thoreau spoke directly to me when he wrote, "Every poet has trembled on the verge of science."

By what turns of history have we reached the point where a student must choose either to be, or not to be, a scientist? Why was my kinglet experienced as a fork in the road rather than a broadening of scientific truth? As an undergraduate, I was invited either to inhabit the conventions of science, or to turn my back on them: given the great tent of the sciences, one was either inside, or out. And similarly, transferring to English meant analyzing the novels of Dickens, never the metaphors of Darwin. Poets and artists lived in one world, scientists in another. How could I find a way to inhabit both? Thoreau offered one answer: toe that line between the written word and the natural world, just where each faces the other. Thoreau had also been tugged in two directions by what C. P. Snow called the "two cultures" of literature and science, but he had found a way to practice science as a poet, and to be a poet by means of the singular objects and discipline of science. It helped that "Science" in his day was not yet a big enclosed tent gathering insiders together, but more like a series of stalls in the great open-air agora of ideas, and that "Literature" then included writing of all sorts, science no less than poetry and fiction.

Where had Thoreau found the model for his own poetic science? How had he learned to hold together fields of intellectual inquiry that we intuitively believe to be separate? In graduate school I discovered that Thoreau's turn to science, like Charles Darwin's, was fired and shaped by the works of Alexander von Humboldt. I myself felt, on discovering Humboldt, a bit the same way Darwin did when he first stepped ashore in South America: "The mind is a chaos of delight, out of which a world of future and more quiet pleasure will arise. I am at present fit only to read Humboldt; he like another sun illumines everything I behold." In turn, the sun of Humboldt's life was his guiding idea that the physical universe and the human mind were integrated halves of a single whole, and he named that whole Cosmos.[1]

Today Humboldt is forgotten in the United States, but in Thoreau's day he was the predominant intellectual of his age, the most famous scientist in the world, and, as was widely repeated, the most famous human being after Napoleon—a pairing that Humboldt, a loyal partisan of the French Revolution, despised. He grew up nourished by Berlin's liberal Jewish intellectual circles and his friendships with Goethe and Schiller; he became the bold adventurer who had explored deep into the secret empire of Spanish America from 1799 to 1804, tracing the mysterious waterways of the Amazon and Orinoco river systems and climbing Chimborazo to the top of the world (so they thought), almost 20,000 feet, a record that stood for thirty years. He seemed to comprehend all knowledge: he founded modern plant geography (the precursor to ecology), and shaped an entire university of sciences, from biological, earth, ocean, and atmospheric sciences to physics and electrophysiology to cartography, ethnology, and linguistics, even finance and economics. He founded international cooperative science; he was the first to warn the world about the link between deforestation, catastrophic environmental change, and depopulation; his work on climate change marks the beginning of awareness of global warming. Historians today call this the age of "Humboldtian Science," for few scientists in Europe, Russia, Australia, or the Americas were not captured by the global sweep of his methods or his international network of correspondents. His most famous scientific disciples in the English-speaking world were epoch-making: Charles Lyell, who revolutionized geology, and Charles Darwin, who built on Humboldt and Lyell to found evolutionary biology. Humboldt was the catalyst for modern science. As the German physiologist Emil Du Bois-Reymond is said to have said, "Every scientist is a descendent of Humboldt. We are all his family."

In Latin America, Humboldt continues to this day to be celebrated and studied, for there he is still seen as a founding pan-national hero, something like Washington, Jefferson, and Lewis and Clark rolled together, "Liberator of peoples / And herald of Bolívar."[2] In Europe, Humboldt studies are a thriving industry, and each year sees the publication of new works on him and new editions of his writings. But oddly, the United States have forgotten him. How ironic: nineteenth-century Americans prided themselves on living in the very republic that was the unique realization of Humboldt's ideals, and Humboldt reciprocated by declaring himself "half an American" and involving himself deeply in the fortunes of the United States, corresponding with Thomas Jefferson, Albert Gallatin, and dozens more, and mentoring or inspiring writers such as Washington Irving, the historian William Hickling

Prescott, Susan Fenimore Cooper, Henry David Thoreau, Edgar Allan Poe, Walt Whitman, and John Muir, explorers and artists such as Joseph Nicollet, John C. Frémont, George Catlin and Frederic Church, and scores of scientists, including Louis Agassiz, who on Humboldt's recommendation came to Boston and stayed to revolutionize American science. More places were named after Humboldt in the United States than in any other country in the world. Yet in the twentieth century, U.S. Americans cast Humboldt onto the dustheap of antiquarian memory. Today he figures either trivially as a footnote (often confused with his brother Wilhelm, the linguist and philosopher who founded Berlin's Humboldt University), or ingloriously as yet another imperialist stooge. The myth of American exceptionalism carries a steep price. In casting away Humboldt, U.S. Americans have cut a cord connecting nineteenth-century intellectual, political, artistic, and scientific culture, unraveling a network that interlinks the antebellum United States to the rest of the globe, and severing access to his key concept, "Cosmos," the very foundation of modern environmental thought. Had Humboldt been my teacher, my youthful encounter with the kinglet would have led me toward, not away, from science.

That U.S. American literary and cultural studies have remained oblivious to Alexander von Humboldt is a scandal exactly equivalent to analyzing Romanticism without Goethe, naturalism without Darwin, modernism in ignorance of Einstein, or postmodernism without Heisenberg. Most of you who read that sentence will be skeptical. I can only enjoin you to suspend judgment and read on. There is some reason for such ignorance: since few Americanists are fluent in Spanish, German, and French, only a slender stream connects U.S. American historical, literary, and cultural scholarship with the broad currents of Humboldt scholarship flowing through Latin America and Europe. Several of Humboldt's most important works have yet to be translated into English, and the translations that do exist sound antiquated and stiff to modern ears (and are available only in rare and brittle nineteenth-century editions or expensive reprints).[3] Biographies of Humboldt in English are outdated, derivative, and frequently inaccurate.[4] Such ignorance has become self-perpetuating: every Americanist quite reasonably concludes that if Humboldt were worth bothering about, someone would have told them so.

We in the twenty-first century need to reclaim Humboldt. The goal of this book is to suggest why—by recalling Humboldt out of oblivion and asking, in a preliminary way, how our view of U.S. American literary and intellectual culture might change if seen from a Humboldtian perspective. When Humboldt looked to the horizon he saw America, and beyond, Cosmos. Recovering his cosmopolitan and multidisciplinary prospect means, first, resituating

"America" into a lively and contested global field of ideas, actions, and interests. It also means revisioning science as an intrinsic constituent of the humanities, reading beyond the "two cultures" to grasp a worldview that knew how to distinguish the natural and social sciences, the arts, and the humanities, but knew also how fully each interpenetrated all the others. Poised on the brink of modernism, Humboldt felt the winds shifting and tacked to sail before them, toward a horizon that still has not been reached. He cast his life and his books as journeys, and offered passage to Cosmos to all who could read.

If we choose to recall Humboldt's Cosmos, to recall what historical memory has lost, have we simply refined our understanding of our own past? Or have we gained a usable insight into the interrelationship of mind and nature, intellect and feeling, environmental and social justice? Before modernity could come into being it had to kill and dismember Humboldt's Cosmos. Now that modernity has fallen away into the past and the cheap thrills of postmodernity have worn thin, could remembering Cosmos help us forge a path to our uncertain future?

The book that follows makes no pretense to being definitive—an impossible task for a study that, to follow where Humboldt leads, must embrace disciplines far afield from my own specialty in American literature. Furthermore, I write as an Americanist trained to keep rigorously within the bounds of America. This is changing fast, for exciting work is being done today to globalize American literary and cultural studies, but to do this work well requires retraining ourselves to embrace scholarship in multiple disciplines and in languages other than English. That my own competence is so limited is this book's greatest limitation but also, I hope, its greatest strength: by showing that, even within this limitation, one can see so much cause to rethink the story we have been telling will, I devoutly hope, provoke others to take up where I leave off. If in a few years new writing and scholarship has rendered this book obsolete, it will have done its work.

Acknowledgments

I feel blessed, in our era of forgetting, to have found Humboldt at all, and for this I must thank Lee Sterrenburg at Indiana University, who back in 1988, in a graduate seminar on Darwin, put before us excerpts from Humboldt's works and encouraged me to believe that Humboldt really was the most exciting writer of the nineteenth century—excepting, of course, Darwin himself. Fred Churchill further encouraged my dawning obsession and in the two decades since, I have accumulated too many debts to count. A semester's fellowship from the National Endowment of the Humanities allowed me to begin researching this book in September 2001; thus this book, conceived in peace, is scored by the trauma of 9/11 and all that followed. My friends at Lafayette College were the first sounding boards for this project, and a deeply appreciated research professorship from the University of South Carolina allowed me to start writing. Strachan Donnelley, the maverick and visionary philosopher who founded the Center for Humans and Nature, believed in this project from the first. During its darkest hour, it was he who opened the passage to Cosmos by sponsoring the CHN fellowship that allowed me to finish the draft. Humboldt's spirit lived on in Strachan, and I hope that some of his generosity, boundless curiosity, and spirit of intellectual adventure lives on in these pages. To him, my most profound thanks, and to all the Fellows at CHN, that interdisciplinary "worldview boot camp"—Paul Heltne, Ron and Joan Engel, Bruce Jennings, Brooke Perry Hecht, Curt Meine, Wes Jackson, Bill Viteck, Bill Bailey, Bruce Coull, and so many others: thank you, all. You have been my best audience.

On this long path, special thanks go to Sievert Rohwer, who has never suspected the impact he and his bird studies have had on his erstwhile student assistant; to Leo Marx, for all his encouragement; to Larry Buell, who opened the doors of Harvard's libraries to me by arranging a research fellowship for 2001–2; and to the Thoreau Institute of Lincoln, Massachusetts, where I stayed in the eerie months following 9/11, walking the trails of Thoreau's woods whenever my head needed clearing. It was a tonic in those weeks to discover that the birds still sang. In the small but fervent world of U.S. American Humboldt scholarship, my great thanks go to Vera Kutzinsky, Ottmar Ette, Andreas Daum, Ingo Schwarz, Kent Mathewson, Michael Dettelbach, Rex Clark, Oliver Lubrich, Aaron Sachs, Elizabeth Millán-Zaibert, Matti Bunzl, and Ingrid Lotze and Joerg-Henner Lotze of the Humboldt Field Research Institute in Steuben, Maine. To the Humboldt Foundation in Germany, which has done so much marvelous work over the generations, thanks for the thought, and may you soon waive the limitations on funding for scholars over forty. At the University of South Carolina, my deep appreciation to the donors of the John H. Bennett, Jr., Chair of Southern Letters, who have helped support my work on this book. My deepest thanks to Joel Myerson and Greta Little, Steve Lynn, Bill Rivers, Paula Feldman, David Shields, David Miller, Leon Jackson, and Ed Madden; to my stalwart student and research assistant Jessie Bray; and to my other wonderful students, particularly Theda Wrede, Brad King, John Higgins, Matt Boehm, Harris King, Kati Kitts, and Matthew Bennett. Thanks also to the inimitable naturalist Rudy Mancke, who keeps the Humboldt tradition alive. Many in the circle of Transcendentalist scholarship have assisted my unaccountable obsession with an obscure German: of them all, special thanks to Bob Richardson, who gets it, and to Sandy Petrulionis, my sister-in-arms; to David Robinson, Len Gougeon, Phyllis Cole, Tom Potter, Ron Bosco, Barbara Packer, and Richard Kopley. In the broadening field of environmental studies, my thanks to Tom Dunlap, Rochelle Johnson, Joe Amato, Steve Holmes, David Lowenthal, and Charles Bergman. Above all I would like to honor my friend Brad Dean, whose tragic loss the Thoreau community will forever mourn, who agreed that Humboldt rocks, and who as the perfect Thoreauvian was working overtime to put the foundation under my castles in the air.

I owe a debt of gratitude to many librarians and archivists: Patrick Scott and Jeffrey Makala opened to me the exceptional and largely unknown Humboldt collection at the University of South Carolina's Cooper Library, and allowed me to photograph many of the plates for this book; the staff of the Boston Public Library Rare Book Room has been unfailingly helpful; much of the research was done at the Widener Library at Harvard, the American Antiquarian Society, and the American Philosophical Society, where I owe particular thanks to Roy Goodman, as well as the American Geographical Society Library at the University of Wisconsin–Milwaukee, where Jovanka

R. Ristič put before me many treasures. Translation assistance was rendered by Mary Beth Stein, Catharina Wuettig, Harris King, and Robert von Dassow. Several scholars read this manuscript at a late stage in its composition: Tom Dunlap, Aaron Sachs, Kent Matthewson, Sandy Petrulionis, and Robert Walls. Each of them saved me from egregious errors and helped guide me to what I hope they will agree are useful revisions, but, of course, the errors that remain are entirely my own. No academic book comes into existence without the patient concern of editors, who do more to shape and nurture the work of scholarship than anyone but a fellow editor would recognize: my great gratitude, therefore, to Mary Braun, who encouraged this project at a crucial stage; and to my editors at the University of Chicago Press, Christie Henry and Carol Saller, who have so blessed this work with their care and attention.

Finally, I have not traveled this path alone: my husband, Robert E. Walls, has walked it with me and stayed at my side through the darkness and weird afterlight of cancer. His parents, John and Louise Walls, have over the years offered boundless reserves of support and encouragement. My parents, John and Ethel Dassow, fostered the beginnings of this book though they did not live to see its end. My father, a chemist, bought me my first Humboldt, an old copy of *Cosmos*, and taught me that Humboldtian science never really died; my mother carried to the last the Humboldtian sense of the poetry of adventure, and it was she who showed me the Southern Cross. My godparents, John and Polly Dyer, were the first to lead me out onto high mountain trails, and whenever I put on my backpack and head into Washington's alpine country—the ten essentials all within reach!—I know that some of the first footprints on those trails were theirs. They awakened in me the love of wild places, and only now am I awakening to the historical importance of their long decades of hard work in protecting the wilderness lands of Washington State and beyond. Without them and their friends, that tough World War II generation of environmental activists, our Cosmos would be smaller and poorer. Johnny, Polly, this book's for you.

Prologue: Humboldt's Bridge

Once Alexander von Humboldt settled down in Paris after his monumental five-year journey through the Americas, he started to publish books. Among the first was a massive coffee-table extravaganza titled *Vues des Cordillères, et monumens des peuples indigènes de l'Amérique* (Views of the Cordilleras and monuments of the indigenous peoples of America). If you are ever so lucky as to find a copy, open it to the first plate, "Statue of an Azteck Priestess" (fig. 1). Crouched like a sphinx, stony, inscrutable, she gazes blankly out over your right shoulder, lips parted as if to speak. She looks vaguely Egyptian. Does this mean that the peoples of the Nile sent emissaries to the New World? Or did the Aztecs invent, on their own, similar sculptural forms? Humboldt thinks the latter, though he does note a case of genuine long-distance transmission: the pearls that ornament the priestess's forehead show that the cosmopolitan city of Tenochtitlan, located high in the Mexican interior plains, was in contact with the California coast, "where pearls are fished up in great numbers." Contemplating the statue leaves Humboldt with a congeries of questions: Why has she feet but no hands? Is she truly a priestess? A deity? Or simply an Aztec woman? Where did such imagery originate? Perhaps here is a reflection of the light from Asia that led to "the commencement of American civilization." But these questions cannot be answered. In the vacuum left by the wholesale Spanish destruction of her civilization, the words she speaks cannot be heard. She reigns, silenced, mysterious, alien yet familiar, over the entirety of Humboldt's works.[1]

Humboldt's next plate points directly to the cause of her silence and mystery (fig. 2). It shows the center of what was once her city—but now it is the center of Mexico City, a trim and elaborate European square built, as Humboldt points out, on the site of Tenochtitlan, "totally destroyed" by the

FIGURE 1. *Statue of an Azteck Priestess,* plate 1, Alexander von Humboldt, *Vues des Cordillères* (Paris, 1810). Rare Books and Special Collections, University of South Carolina Libraries.

Spanish in 1521. Not one stone was left on another, and out of Tenochtitlan's shattered buildings the Spanish quarried the materials for their new capital. Humboldt's image is therefore a palimpsest: where the square stands, there formerly stood the spacious temple of Mexitli. Behind the cathedral once stood the palace of the king of Axajacatl, where Montezuma lodged his guests, the Spaniards. To the right of the cathedral now stands the palace of the viceroy of New Spain; to the left once stood Montezuma's own palace. These specifics may, Humboldt drily remarks, interest those who study the conquest of Mexico. Here, in its opulent presence, ready to vie with the finest cities of Europe, rises an erasure of all the past, the heart of the American civilization Humboldt is attempting to resurrect. Where are its people? Humboldt admires the great equestrian statue of his patron Carlos IV in the center of the square but notes that the Indians call it "the great horse" (not "the great king"); he also notes that the four ornamental gates to the

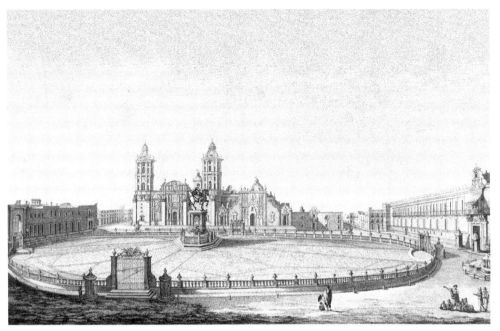

FIGURE 2. *View of the Great Square of Mexico*, plate 3, Alexander von Humboldt, *Vues des Cordillères* (Paris, 1810). Rare Books and Special Collections, University of South Carolina Libraries.

statue's raised enclosure are kept "closed, to the great discontent of the natives." Indeed, the square itself is strikingly vacant, the inhabitants evacuated. Humboldt peoples the margins of this apparently European city not with elegant Spaniards or bustling Creoles but with a handful of "Guachinangoes," the mixed race "lower class of the Mexican people": their places destroyed, they yet remain, the erased and the excluded, strolling through the space of their ancestors, gazing on the emptiness that was their history.[2]

Next in this sequence comes the first of Humboldt's natural scenes (see frontispiece). The German scholar Ottmar Ette has remarked that in Humboldt, the ocean becomes "not the separating element but the one that engages everything into worldwide communication and connects everything."[3] Here, in *Natural Bridges of Icononzo*, is an image of separation and connection iconic of all Humboldt's work, itself a bridge between natural and human. His text tells us that in the Andean high plains it is not the mountains but the valleys that stagger the European imagination, for they carve depth into the landscape, a wild aspect that fills the soul with "astonishment and terror." Icononzo is the name of the ancient Muysco Indian village at the southern end of the valley of Pandi in the kingdom of New Granada (now known as Colombia), a valley sliced in two by this impassible gulf which yet is bridged, twice, by the hand of nature. Down its length thunders the Rio de la Summa Paz. Humboldt believes this chasm to have been created by an earthquake, which a single stratum of compact quartzose resisted, forming a natural bridge nearly a hundred meters above the torrent below. The second, lower bridge was formed when masses of rock tumbled into the gulf in such a way as to support each other, with the middle rock forming a keystone—a fact which in time, he speculates, might have given the Indians the concept of the arch, unknown to the Americas as it was to ancient Egypt. Here in this image are natural history and human purpose united: travelers use the upper bridge to pass between the valley's northern and southern halves, and there on its flank the Indians have erected for the safety of travelers "a small balustrade of reeds."

That these bridges are natural is important: they were already inscribed in nature, principles of architecture embodied in unworked stone. But humans have discovered them and use them—and so there is that frail reed fence: Don't fall off! it cries. The artist, down in the riverbed looking up, turns utility into aesthetics, a lived experiential reality into a "view." Yet this view can exist only in imagination. True, it is based on Humboldt's own sketch, made in the field—but Humboldt sketched the bridges from the northern valley above, in a side view. The Paris artist has accomplished the impossible, vaulted us to the midst of the river far below, a place so dark and inaccessible that the only way Humboldt can see the thousands of cave-dwelling birds that live and fly in its depths is to bounce rockets off the canyon sides and glimpse them flying in the flare of artificial light.

In this view not just two but many worlds cross: the plate tectonics that split open the rock, the waters that plunge down the crevasse, the vegetation that clothes the rock walls, the birds who call this chasm home and fill it with their "lugubrious cries," the Muysco Indians who name the birds—*cacas*, Humboldt tells us—and who travel to market across this bridge, the European scientist who ventures to place the Cordilleran birds in their Linnaean genus, *Caprimulgus*, even as he records the native's speech and sketches the site in his notebook, the artist, M. Bouquet of Paris, who reimagined Humboldt's field sketch as a sublime view, the viewer who sees in this one image the braiding together of fact and beauty, science and poetry, nature and society, history poised on the present instant.

The discipline I work within is conventionally named "literature," understood to include that subset of written texts encompassing fiction and poetry and understood as "aesthetic"—expressive and emotive—rather than "scientific," objective and factual, or "political," social and contested. But Humboldt resisted the tectonic shift that was splitting the world of knowl-

edge under his feet even as he wrote and lived, and in his writings he created
his own natural bridge. More: this is not a light structure thrown across like
a plank to provide perilous passage between two great realms. It is pulled
into place by gravity and by its weight supports the whole, becoming the
keystone to Humboldt's cosmic architecture. Humboldt chose the title for
his late work, *Cosmos*, with care; though it daunted him a little, he stood
by it, for it allowed him to articulate both landings of his bridge: first, what
Thoreau called "hard matter and rocks in place," the physical universe as it
exists apart from human purpose; and second, the beauty and order of that
universe, the very *idea* of the whole. The physical universe exists without us,
no doubt, beyond us and other than us; but the Cosmos needs us. Only in
the dance of world and mind, subject and object, does Humboldt's Cosmos
come into being. It is the act of human art—as in this very image—to represent the real, not to copy or replicate it but to make it "present" to our minds
and hearts and souls, an image or trace of the Cosmos realized, renewed, and
revitalized in mind.

Vues des Cordillères was a gorgeous production, but it priced itself right out of
the market; only a few copies exist in the United States today. So Humboldt's
friend, the English radical poet and publisher Helen Maria Williams, persuaded him to select just a few of the illustrations and republish the book in a
smaller and cheaper London edition. Her translation, published in 1814 under
the rather gusty title (complete with exclamation point) *Researches concerning the Institutions and Monuments of the Ancient Inhabitants of America, with
Descriptions and Views of Some of the Most Striking Scenes in the Cordilleras!*,
circulated Humboldt's ideas much more widely among English-speaking
audiences. In any language, this was a very odd book, a jumble tossing together
sculpture, costumes, mountains, buildings, hieroglyphs, waterfalls, calendars,
pyramids, and geological wonders. Was there a method to this joyous, exuberant madness? Yes: each illustration is supported by an essay, and each
essay references an illustration. As text and picture, theory and illustration,
complete each other, so do science and art, nature and mind. The landscape
cannot be understood without its people, nor can the various cultures and
civilizations Humboldt documents be understood without seeing the landscape that shaped them and that they, in turn, shaped. Thus the experimental,
hybrid form of this book puts on display the complex ways in which nature
and culture produce each other within the historical context of Spanish colonialism. Humboldt provided a fascinating glimpse into the exotic and secret kingdom of Spain's New World colonies, and his illustrations not only

influenced the development of Latin American art, but modeled for North American scientist-explorers a new way to represent the landscapes of the American West.[4]

Humboldt periodically flashes into anger at the stupidity, barbarism, and fanaticism of the Spanish conquerors who destroyed all they touched, leveling the cities, burning the libraries, and abandoning what little was left to decay and misfortune. From the shards and fragments he attempts to piece together the moral, aesthetic, political, cultural, and religious life of pre-Columbian American Indians, centering on Mexico and the Andes and borrowing from the languages and historical traditions of their descendants. His approach is deliberately comparative. How much, he asks, do New World nations resemble those of the Old World? From whence came these peoples, with their hundreds of languages, their legends and sophisticated calendars, their technologies and distinctive religions? Most likely from East Asia, he decides, pointing for evidence to linguistic, cultural, and technological roots across the Pacific among the Tibetans and the Japanese. Yet so long ago had they migrated that American nations developed wholly new and independent civilizations, in isolation from the rest of the world. The implications were startling: civilization did not dawn once in the Old World, spreading outward from a single cradle; it had multiple centers of origin, many cradles, and one of them was in the New World. Given the affinity of the hundreds of American languages and the similarities in "the cosmogonies, the monuments, the hieroglyphics, and institutions," Humboldt is convinced that a single people entered North America from Asia to spread and diversify across two continents.[5]

Many cradles, but one people. Everywhere Humboldt looks he sees one great truth verified: "The Caucasian, Mongul, American, Malay, and Negro races" are not "insulated" from one another but form one "great family of the human race, one single organic type, modified by circumstances which perhaps will ever remain unknown." If all humanity forms one "great family," then all human works, even those that do not meet our European standards of beauty, are worthy of respect and attention, for they all tell a part of the greater human story. As Humboldt says, different nations have followed many "different roads in their progress toward social perfection," and their progress is helped or hindered not by internal, biological limitations of race but by external, or environmental, circumstances and accidents. The Americas, for instance, lacked milk-giving ruminants capable of domestication. Instead of cattle, the Asian immigrants encountered untamable musk oxen and bison, and so their road from hunting to agriculture necessarily skipped over the supposedly required "pastoral" stage, derailing the traditional doctrine that human cultures must advance through the three "stages" of hunting, herding, and agriculture. By another trick of geography, American peoples were cut off from communication with the rest of mankind, left to struggle with a "savage and disordered nature" with no resources other than their own ingenuity. Who can wonder, then, at the apparently "rude" style or "incorrect" expression of the arts of native America? or at their slower progress? How many nations, asks Humboldt, can boast the mild climate of the gentle Mediterranean?[6]

Humboldt argues, then, that all races form one great human family, and their great diversity comes from adapting to their many and various environments. Since environment is crucial to understanding how human unity flowers into such diverse societies, the only way to understand a people is to become immersed in their landscape. Cultures cannot be judged from afar. Scholars who "never quitted Europe" say foolish things—for instance, that America is a marshy country with few animals, overrun by savage hordes. Travel allows us to overcome such prejudices, to explore the ways human and natural history shed mutual light on each other, and so Humboldt's method emerges in his book's mad scramble of nature and culture. As he declares, "An accurate knowledge of the origin of the arts can be acquired only from studying the nature of the site where they arose." Archaeological research must take into account climate and soil, the presence or absence of animals, the physiognomy of plants and of landforms, for they all influence the progress and style of human arts. Thus his landscape views are not intended as decorative but to drive forward his argument: the cultures of the mountain peoples of the Cordilleras are stamped by the massive and wild nature of their high peaks and hanging valleys. After apologizing for the book's lack of order, Humboldt ends his introduction with a frustrated but hopeful note: may his "feeble sketches" lead other travelers to visit these regions and "retrace accurately" the "stupendous scenes, to which the Old Continent offers no resemblance."[7]

Travelers came, of course, not all of them so idealistic as Humboldt. In his time, the face of nature he so loved was being remade by colonial imperialism, global capitalism, and the industrial revolution. He himself was a knowing participant, both in creating and circulating new regimes of knowledge and in helping to construct a global economy that would, he perhaps naively believed, advance the cause of freedom through free trade and the open exchange of ideas. Though his ideas and methods were co-opted by the imperialist projects of Europe and the United States, Humboldt consistently protested against the evils of colonial exploitation, particularly slavery and the oppression of indigenous peoples, and he deliberately incorporated the voices and knowledges of ethnic and colonial peoples into his planetary project.

Humboldt attempted, in short, to create a counternarrative to the drumbeat of imperial progress, and in this attempt he effectively created what we would now call an environmental discourse. His foundational assumption was that neither humans nor nature can be understood in isolation. In his social writings, nature was never merely background but played an essential role in the development of human societies; in his natural writings, the ways various societies construct their views of nature were crucial to understanding their physical environment. For Humboldt that environment was overwhelmingly historical and spatial: though he worked some in physics and chemistry, his interests always centered on the earth and the processes that generate its forms and surfaces. His scientific methods were relentlessly inductive, for he sought to identify patterns in nature by combining and collating hundreds of measurements—myriads of data points—until out of what the Russian physical geographer Peter Kropotkin called "the bewildering chaos of scattered observations" flashed a new vision of the "harmonious whole" they described, "like an Alpine chain suddenly emerging in all its grandeur from the mists which concealed it the moment before, glittering under the rays of the sun in all its simplicity and variety, all its mightiness and beauty." Far out on the horizon, "the eye detects the outlines of new and still wider generalizations." As Kropotkin's lyrical description suggests, such quantitative work was not the enemy but the ally to poetic insight. For the Humboldtian scientist, the doing of science combined rigorous and exacting labor with the joy of poetic creation and an almost spiritual sense of revelation, as if nature borrowed the mind and hand of the scientist to describe its own most beautiful laws and structures.[8]

Generating an environmental discourse was only half the journey. To complete it meant enrolling others to the cause: in letters and mentoring relationships with young scientists and artists, salon conversations, political negotiations, the organization of scientific societies and international scientific research, and lectures and publications both popular and technical, Humboldt sought to create institutions and practices that would spread his particular way of thinking about humans and nature. In one sense he succeeded spectacularly. He virtually invented modern international science and seeded so many fields with productive new ideas that historians of science call the era "Humboldtian." However, the scientific results of his initiating efforts eventually passed him by, and by his death his prediscplinary insistence that the physical and natural sciences, economics, politics, cultural history, ethnology, linguistics, and aesthetics all be practiced together in an environmental network of interacting discourses was resisted as heroic but impossible, fractured by the rise of specialization and standards of scientific objectivity, and suppressed as useless and old-fashioned. The old Baron, the most loquacious man of his time, was effectively silenced.

This silencing has done real damage to environmental studies, for Humboldt stands at the head of environmental and ecological thinking today. Recovering Humboldt does more than deepen our knowledge of the long foreground of such iconic figures as Henry David Thoreau, George Perkins Marsh, and John Muir—Humboldtians all. First, it locates the first global wave of environmental studies just prior to the era of scientific specialization, when scientific discourses were fluid and a single mind could still innovate in multiple disciplines across the humanities and social and natural sciences, allowing each to inform the others. Second, and more importantly, it places at the head of environmental studies an alternative narrative that closes the gap between mind and nature by demonstrating how each creates or constructs the other, a concept that, thanks to modernism's persistent dualisms, still seems novel today.

Finally, recovering Humboldt positions the first wave of environmental thinking not within a nationalistic debate over resource exploitation but within a global debate over capitalism and imperial power. The apparent roots of Anglo-American nature writing in imperial discourses of exploration have made it too easy to dismiss such writing as ideologically complicit. But the story is deeper and more complex than this. Humboldt grew up knowing that nature was the site of deeply political conflict, and his popular works deplored the tragic destruction of the civilizations of the Americas and tried to reconnect them with the global human community, by showing that colonial exploitation of the land went hand in hand with the destruction and continuing oppression of its ethnic peoples. Building on the insights of eighteenth-century colonial scientists, Humboldt became what Ramachandra Guha calls "a pioneering analyst of global deforestation," arguing as early as 1805 that cutting down forests causes climate change, and in later works attributing the alarming and inexplicable fall in water levels in both Mexico's Lake Tetzcoco and Venezuela's Lake Valencia to deforestation by the Spanish, which desiccated the landscape and caused periodic destructive floods. This cycle was made even worse in Mexico by imperial Spain's ill-informed and catastrophic attempt to reengineer Mexico City's water system, in which colonial rulers forced natives to build a massive canal to drain the upland lakes. As Humboldt documents, the folly of the Europeans ruined once-plentiful water resources and poisoned once-fertile agricultural land, and their abusive labor practices killed untold numbers of workers and plunged the entire Indian population into poverty.[9]

Richard Grove argues that the nineteenth-century growth of natural knowledge and resulting new ecological concepts of nature were—and are—inseparable from European colonialism. Humboldt added the necessary conclusion that environmental destruction was also socially devastating: natural ecology entails social ecology. In Humboldt the two were forged together, humans and nature forming two sides of the same coin. Humboldt's work is refracted in the writings of Henry David Thoreau, who also joined the causes of natural and social justice; their separation into noncommunicating fields came later. As Aaron Sachs has recently argued, today, with social justice forming an exciting new front in ecocriticism, recovering Humboldt would open a radical environmental tradition that would link with the social justice orientation of postcolonialism.[10] Splitting humans from nature has other consequences, too: the separation of scientific from literary knowledge is now so total that leading ecocritics like Ursula Heise can see no viable connection. Academic environmental studies have "to date not established any significant links between literary and scientific approaches to the environment." Instead, literature and art have become "bulwarks" against the encroachments of science and technology rather than "sites of encounter between different types of knowledge and discourse." Even works sympathetic to science show no "conceptual bridging between scientific description and aesthetic valuation."[11]

We are back, standing, once again, on Humboldt's bridge. Two hundred years later, does it still hold? Or has it, too, broken into ruins, one more monument to a lost civilization? Let us imagine it holding. In bridging peoples, disciplines, places, and historical eras, Humboldt sought to create a zone of exchange rather than domination, a pluriform and multivocalic world that would allow humans and natures to speak through a range of representations, from scientific to social to aesthetic, augmenting the stripped-down world of scientific fact by presenting those facts as elements of a renewed and revitalized Cosmos. Humboldt tried to intervene at the discursive level—which is to say he collected, he wrote, published, and lectured, voluminously and persistently, letter by letter, essay by essay, book by book, person by person, building over seventy years a massive global network of scientific and cultural knowledge and artistic expression by which he hoped to bring natural knowledge into the public sphere as a form of liberation. In an era of bloodshed, revolution, imperial warfare, and Malthusian struggle against a nature "red in tooth and claw," Humboldt found in nature not limitation and conflict but freedom, justice, and harmony. His was a quixotic vision, aimed at changing the course of history, and it failed.

But not entirely. Humboldt's subversive vision of science for the people lived on in Europe (most provocatively in the work of Peter Kropotkin), and in the United States he succeeded in bringing into being a discourse, a way of speaking, about nature that we now call "environmental": namely, a planetary interactive causal network operating across multiple scale levels, temporal and spatial, individual to social to natural, scientific to aesthetic to spiritual. Darwin, one of Humboldt's closest readers, would envision an interactive network of chance and inheritance working across time and space to evolve new life forms. Thoreau, another of Humboldt's closest readers, would recast Humboldt's methods into the idiom of American Transcendentalism, thereby, with John Muir, the "American Humboldt," and George Perkins Marsh, another Humboldt convert, founding North American environmental thought. Soon this new discourse of nature would receive a name: ecology.[12] The new name designated a science, one more subspecialty in the widening panorama of natural knowledges. But before it was a science, before it could be a science, "ecology" was a discourse. It was, in fact, Humboldt's discourse. It had first to be imagined, thence to be represented, circulated, and reimagined in works of great beauty and power, from Humboldt forward, among thinkers, poets, and painters. His writings and ideas are like a rhizome, the root connecting a ramifying community: Coleridge, Darwin, Emerson, Susan Cooper, Thoreau, Whitman, and Poe; George Catlin and Frederic Church; John Muir and George Perkins Marsh; Franz Boas and Lewis Mumford. Each was moved by Humboldt's words and pictures to imagine a new way of envisioning nature, a way that stamped its mark on a distinctive American literature and art and that remains alive in American culture today.

Confluences

How novel and original must be each new mans view of the universe—for though
the world is so old—& so many books have been written—each object appears wholly
undescribed to our experience—each field of thought wholly unexplored—The whole
world is an America—a *New World.*

HENRY DAVID THOREAU

Humboldt's America

John Locke tells us that "in the beginning all the world was America." In the
end, thought the young Humboldt, all the world would be again—an America transformed from place to prophecy, universal freedom restored to humanity through enlightened Republican politics and the spread of science,
art, and culture. He was born in the right place and at the right time to im-
bibe such ideas—Berlin in 1769, to a family of minor Prussian aristocracy
loosely attached to the court of King Frederick the Great. Frederick's rule was
in its waning years, and accounts make eighteenth-century Berlin sound un-
prepossessing enough, yet still, it was here that the king had for some years
given refuge to Voltaire, the Enlightenment French philosopher-poet who inspired the rise of European liberal thought. While Enlightenment thinking
was hardly mainstream, with the blessings of such a king, liberal ideas circulated widely among the city's intellectual elite, including the Humboldts.
By the time young Alexander was a college student, he made sure to study in
Hamburg, the center of *Amerikunde* (or American studies), where the political events of 1776 were being turned into an ideology that would become a
pillar of nineteenth-century German liberalism. During his five years in prerevolutionary Latin America, he often measured the political discourse he encountered against the standard set by Washington and Jefferson, and before
he returned to Europe he made a pilgrimage to the United States to meet the
heroes of the American Revolution in person. Thus America was on his horizon from the start.

Late in life he returned to the importance of America to the development
of the Cosmos: as he wrote, it was the discovery of America that planted the
seeds of the Cosmos, for the land Humboldt liked to call the "new continent"
opened a new sense "for the appreciation of the grand and the boundless,"

making possible "higher views" that would show humanity the interconnections of all phenomena. Columbus himself, wrote Humboldt, understood this, and "on his arrival in a new world and under a new heaven, he examined with care the form of continental masses, the physiognomy of vegetation, the habits of animals, and the distribution of heat and the variations in terrestrial magnetism"—sounding remarkably like Humboldt himself. In their turn, the Spanish writers who followed Columbus opened up important questions still unanswered: the unity of the human race amidst so many variations; the affinities of America's many languages; the migrations of plants, animals, and nations; the causes of trade winds and ocean currents, volcanoes and earthquakes. Never before, Humboldt wrote, had the sphere of ideas been "so wonderfully enlarged." Even in his own day three centuries later, such questions could still enlarge the sphere of ideas by embracing the dazzling diversity of humans, animals, plants, and natural phenomena in a single—today one wants to say "ecological"—vision.[1]

Books, journals, and newspapers across the New World hailed Humboldt as "the *second* Columbus," the scientific discoverer of America. Partly this was the appeal of coincidence: as his first biographer wondered, who better than Humboldt to write a history of Europe's fifteenth-century discoveries? "Had *he* not also gone to sea from Spain as the second discoverer of America, and had he not stood on the same spot where Columbus had landed and taken possession of the new continent?" But there were ideological reasons as well: as Alfred Stillé told the graduating class of Pennsylvania College in 1859 (just before it was renamed Gettysburg College after the Civil War battle), Columbus entering Barcelona in triumph with baskets of gold and jewels and surrounded by captive Indians did not bring gifts nearly so precious as Humboldt: "The one opened to Spain the gates of a new empire, the other revealed to the world the secrets of nature and the laws of the universe." While the one caused whole nations to be reduced to servitude, the other "paved the way for the revolutions which rendered the nations of South America once more independent." If Columbus stood for the discovery of riches leading to servitude, Humboldt stood for the discovery of knowledge leading to liberation: even as he had been inspired by the Revolution of 1776, so the *next* American revolutions were inspired, it was widely agreed, by Humboldt.[2]

Celebrating Humboldt as a "second Columbus" carried darker undertones which the celebrants worked hard to subdue, for as Stillé recalls, the transcendent achievement of Columbus was tainted by the enslavement and genocide of America's indigenous peoples. Though Stillé followed Washington Irving's popular biography (and indeed Humboldt himself) in defending the innocence of the Genovese navigator from the crimes unleashed by his discovery, all Anglo-America rose up to condemn the Spanish conquistadors who came afterward. Indeed, the vehemence of the "Black Legend" that had grown up around Hernando Cortés and Francisco Pizarro—conquerors and destroyers of the Aztec and Inca civilizations respectively—was fanned by the guilt of those who spread it. The more bestial was the violence of the Spaniards and the more cruel their monomaniacal demands for gold, the more easily Anglo-Americans could portray themselves by contrast as agents of humanity and reason. Yet it was not an argument that stood up to close scrutiny. Whereas the Spanish government had made at least some attempt to limit and mitigate the enslavement of both Indians and Africans, the British had introduced slavery to their colonies and the Americans were perpetuating it even as they fought their war of "liberation." And whereas the Spanish had incorporated Indian populations into their colonial administration (and the French had befriended and allied with them), the English had swept them off the map and the U.S. Americans were exiling the remnants to bleak western desert lands. Colonial imperialism had much to answer for, no matter which European nation bore the weaponry.

Humboldt as the "second Columbus" seemed, in an age vexed by imperial anxiety, to redeem all this. He was the "enlightened" discoverer, the anticonquistador, hailing from a weak and fractured nation with no imperial ambitions and celebrated as the center of European learning. He traveled not with armies and weapons but unarmed and alone but for a companion or two, a guide or two, and mules laden with scientific instruments. He took not gold and silver but notes and samples—pebbles and bones, a few flowers and leaves, sketches and astronomical measurements. Of this new and innocent Columbus, all Europe could be proud. As Mary Louise Pratt observes, the naturalist as traveler could both invoke the heroism of the Conquest and provide safe distance from its depredations.[3]

Humboldt was also renowned as the most famous man after Napoleon. The two were exact contemporaries, born the same year, a coincidence that linked them at every birthday memorial. In this pairing, Humboldt continued to represent the antitype to the empire of force and bloodshed. In a poem celebrating "the Napoleon of Science" (written for the Boston Humboldt centennial in 1869), Oliver Wendell Holmes invoked Humboldt's "bloodless triumphs" that "cost no sufferer's tear! / Hero of knowledge, be our tribute thine!" Two anecdotes were widely circulated to confirm this ideology of peaceful conquest. In one, Humboldt, laden with awards and adulation after his return from America, was presented at Napoleon's court. "You collect plants?" asked the emperor. "Yes," answered Humboldt. "So does my wife," sneered Napoleon. In another brush with royalty, the young brothers

Humboldt were honored in their Berlin home with a visit by Frederick the Great. Of Wilhelm, the elder, the king is said to have asked, "Do you not wish to become a soldier?" "No, Sire," answered the boy, "I wish to have my career in literature." Turning to Alexander, the king reminded the eight-year-old of his great namesake, the "earth-conqueror." "Do you wish to be a conqueror too?" "Yes, Sire," answered Alexander, "but with my head."[4]

How was it that, three long centuries after Columbus, America still needed to be conquered by knowledge? Even after so many generations, the New World continents were still largely unknown and unassimilated into Western learning, still seen as a problem and a mystery. As J. H. Elliott pointed out in his classic study, America's very existence "constituted a challenge to a whole body of traditional assumptions, beliefs, and attitudes." The newness of the American lands, their flora and fauna and peoples, was so overwhelming that "the mental shutters came down" and Europeans retreated to "the half-light of their traditional mental world." It did not help that the Spanish colonial government refused to publish the reports and observations that came flooding back across the Atlantic, but buried them in archives, forgotten by their own administrators, while forbidding travelers of other nations from entering Spanish territory. The exception was the first scientific expedition in South America, sponsored by France and led by La Condamine from 1735 to 1744, but he and his men traveled with officials who controlled its every movement. Humboldt was faced with a wall of ignorance in Europe and North America alike of the most basic realities of Spanish America, its peoples, and cities no less than its geology and geography, flora and fauna. Much of his writing is directed against the eighteenth-century French naturalist Buffon, who (from an armchair in Paris) proclaimed authoritatively that New World life was degenerate, its climate hostile, its creatures, including its human creatures, diminished in size and potency. One of Humboldt's goals in South America was to confirm the existence of the Casiquiare Canal connecting the Orinoco river system with the continental system of the Amazon, a claim made by La Condamine on the basis of South American reports and disputed ever since. On the very eve of Humboldt's departure, the existence of the Casiquiare was finally and decisively repudiated—on paper—by learned European geographers, even as missionaries and Indians were navigating its waters, as they had been, Humboldt pointed out, for generations.[5]

When Europeans did look at the New World, they tended to see it as the mirror image of themselves, normalizing its alien beings to fit familiar patterns. Anne Bradstreet, the Puritan poet, wrote movingly of hearing nightingales (a British bird) sing in New England; Columbus, facing the Orinoco, located it on the eastern coast of Asia and decoded its meaning in Biblical terms. As Humboldt wrote, the cool evening air, the clarity of the stars, and "the balmy fragrance of flowers, wafted to him by the land breeze—all led him to suppose . . . that he was approaching the garden of Eden, the sacred abode of our first parents."[6] The inability to see the New World on its own terms, the need to translate it into the familiar categories of European custom and religion, had serious consequences beyond the irony of Venezuelans barging groceries and mail along an officially nonexistent canal. Indigenous hierarchies were translated into European-style monarchies, so that early Virginians in Jamestown hailed Powhatan as an "emperor," even arranging a royal state marriage to ally English and Algonquin nations through the wedding of his daughter, "Princess" Pocahontas, to the adventurer John Rolfe (a union from which, in a genealogical fable like that of the Mayflower, untold millions of Americans are descended.) The Christian narrative cast Indians as minions of Satan, or else God's lost people (perhaps the descendants of a wandering tribe of Israel), justifying on the one hand genocide, on the other the missionary zeal of Christopher "Christ-bearer" Columbus and the Spanish mission system that so successfully "tamed" and clothed South American Indians, teaching them their catechism while denying them their culture. Secular narratives cast the Indians as "barbarians" (using the Greek word for uncouth outsiders, whose language sounded like "bar-bar"), or "savages" more animal than human, to be eliminated by genocide when assimilation failed. Humboldt thus broke with long tradition when he advised that Indian artifacts, however uncouth to European taste, must not be judged by the standards of classical Greece, and that Indian architecture (what little was left) ought to be valued and preserved, not treated as convenient quarries of precut stone ready for assembly into European buildings. Finally, leveling New World forests to recreate Iberian plains and English meadows had had the unintended side-effects of desiccating the climate and eroding the soil needed to grow crops. Was it not possible, Humboldt argued, to imagine America as *America*, not a diminished Europe? Did it not have its own identity, and should not its peoples be allowed to seek their own destiny?

In pursuing this argument, Humboldt as "second Columbus" discovered for Europe an America to be seen on its own terms, not as an artifact of Europe's making or an appendage to its power. As he traveled from mission to mission, he sorrowed at the vacant and beaten look of the missionized Indians, and pointed to their unchristianized fellows not as heathen or "savages," a word he repeatedly rejects, but as "independent" peoples with their own distinctive character, dignity, language, and contribution to the great human story. The Creoles he visited and worked with—the American-born peoples of European descent—were, as he reported, restless and angry under colonial rule, and would soon claim their own independence, their own self-governed political, republican future. If African slaves and the "copper-coloured" races were prevented from full realization of their human rights, they too would rise up and throw off their oppressors, as Tupac Amaru had tried to do in Peru and the slaves had succeeded in doing in Haiti. Everywhere Humboldt went he took the temperature of the social as well as the natural climate, and he found it near the boiling point. The laws of nature, just as Jefferson had said in the Declaration of Independence, would soon assert themselves and right the injustices and imbalances in the political realm. And as for American physical nature, Humboldt found it incalculably grander and more sublime than anything Europe had to offer. Here, man had not everywhere dominated and subdued the wild, nor could he, for the destructive forces of volcanoes and earthquakes and the creative power of tropical heat and light would always make American nature an equal, if not a dominant, partner with human enterprise. As later generations would say of the United States, all America was "Nature's Nation."[7]

Thus Humboldt did far more than unlock the closed gates of the Spanish empire; he showed Americans how to imagine themselves as something more than offshoots of European ambition. This is why Humboldt became a culture hero to both Latin and North Americans, from the masses to the intelligentsia. He literally put America on the global map, positioning its history, nations, and resources in relation to the rest of the world, and drawing the detailed and extensive maps by which Americans could find, and know, themselves. He even traced the origin of the very *word* "America," hitherto a puzzle, to its source in a German mapmaker in 1507, giving it a genealogy not in Columbus's tainted legacy but in the relatively innocent explorations of Amerigo Vespucci. It was widely said (though the story may be apocryphal) that when a young Creole named Simón Bolívar met the triumphant Humboldt at a Paris salon in 1804, Bolívar remarked on "the glittering destiny of a South America freed from the yoke of oppression." Yes, agreed Humboldt—if only someone could be found capable of leading its war for liberation. The rest is, as they say, history: in 1810 Bolívar led the Venezuelan revolution, starting a movement that he carried over the next fifteen years to Colombia, Ecuador, Peru, and Bolivia (named in his honor), and that spread to Mexico in 1821.[8]

For their part, U.S. Americans also felt uniquely bonded to Humboldt. When he called himself "half an American," U.S. Americans were pleased to think he was claiming fellow citizenship in their republic, an interpretation made all the easier by their success in appropriating the term "America," which covered two continents, for their own nation. As Edward Everett declared, eliding the difference between North and South, "His American voyage—was performed on the soil of *this* continent." There was some justice here, for Humboldt did praise the United States as the pattern for humanity's future and befriended dozens of influential U.S. Americans, starting with President Jefferson. Once the mail steamers from New York to Bremen made travel to Germany routine, it seemed every U.S. traveler on his grand tour stopped in Berlin to visit the aging Humboldt, and of all the foreign students in Berlin at the time of Humboldt's death, only the U.S. Americans marched in his funeral procession. By then much of their country had been named for Humboldt—towns, counties, rivers, lakes, parks, marshes, caves, forests, eventually a university, and very nearly the entire state of Nevada. Humboldt had aided their "manifest destiny" through his maps and advice, and when the United States invaded Mexico, President Polk's secretary of war, the historian George Bancroft, was anxious to secure Humboldt's blessing.[9]

It could even be said that Humboldt was the father of modern America. Mary Louise Pratt, in the single most often cited treatment of Humboldt by a U.S. American, is fascinated to discover that his journey and the monumental volumes of print it produced "laid down the lines for the ideological reinvention of South America" on both sides of the Atlantic. On the European side, Humboldt opened prospects "of vast expansionist possibilities for European capital, technologies, commodities, and systems of knowledge." On the American side, newly independent Creole elites found in his writings the resources to reinvent themselves, in relation both to Europe and to the "non-European masses they sought to govern." As Pratt argues, Humboldt's impact on the public imagination was made less through his technical writings than his popular books, widely read, reviewed, and discussed in journals and periodicals starting in the 1810s: *Ansichten der Natur* (or *Views/Aspects of Nature,* first published in 1808 but not translated to English until the third edition of 1849); *Vues des Cordillères, et monumens . . . (Views)* (1810, translated 1814); and *Relation Historique,* the narrative of his American travels (1814–25, translated as the *Personal Narrative* between 1818 and 1829 and again in 1852). She could have added his *Political Essay on the Kingdom of New Spain* (1808–11, trans. 1811), the book that first made Humboldt's name a household word and that had particular impact on the United States; also the American edition of *Political Essay on the Island of Cuba* published in 1856 that started a public controversy and played a role in that year's presidential election, and of course *Cosmos,* whose multiple volumes flooded the market starting in 1845. In short, from the 1810s through the 1860s a veritable tidal wave of publications made Humboldt a celebrity across Europe and the Americas. As Pratt observes, in these "bold discursive experiments," Humboldt sought not only to reinvent popular imaginings of America but "of the planet itself," heading off the emerging split between objective and subjective knowledge, "science and sentiment, information and experience."[10]

Here Pratt is exactly right: what made Humboldt so enormously attractive, apart from the sheer romance of his travels, was the newness of his approach to narrating those travels. Humboldt blended an Enlightenment-derived certainty in the agency of reason, factuality, and precision with a Romantic's enthusiasm for feeling and poetry. His views of nature responded deeply to the emotions awakened by each region's unique natural features, from the brilliant skies of Italy to sublime equatorial mountains, endless barren plains, or the gentle meadows of northern Europe. Yet his richly particularized descriptions vaulted seamlessly from the unique to the generalized, locating individual features in the grand pattern of the planetary whole, and linking the powerful emotions of awe or wonder or delight they evoked to an ever-widening sphere of knowledge. Humboldt's science had heart. And because in his philosophy humans were an essential part of the Cosmos, his description of America as "Nature's Nation" never excluded the human, whether the indigenous peoples who were so deeply shaped by their landscape, the Europeans who so variously aided or defeated the land's potential, or himself, seeing and feeling it all for us, our representative mind and heart. Wherever Humboldt goes in the world he looks for traces of the human: New World nature was exhilarating in its vastness and power, but also, more than once, deeply terrifying, or deadly in its monotony, creating a challenge to his philosophy that Humboldt had to work to overcome. He did not call for mere daredevil explorers who saw wild nature as a stage for their exploits, nor for calculating capitalists who would cut, dig, profit, and run. Such men came, of course—how could he have stopped them?—but what Humboldt did call for were dwellers who would weave the land into their dreams, and artists who would write it and paint it, bringing their experience to those who lived in distant and far different lands.

Thus the New World that Humboldt reinvented for the nineteenth century was indeed "America as Nature," as Pratt says, but Pratt goes on to insist that Humboldt invented America as "primal" nature, emptied of human history in order, in the classic imperialist mode, to repopulate it with white European systems and goals. Her Humboldt becomes one more "imperial eye," handmaiden to colonial domination, blind to the Other and full of himself, omniscient and godlike lord of all he surveys. Her highly selective interpretation, driven by a need to privilege binary oppositions rather than pluralistic differentiations, has become canonical in postcolonial studies. The effect has been to silence Humboldt all over again. True, Humboldt cannot step very

far outside the networks of colonial power; even his attempts to do so—if
one is willing to grant they were more than mere self-delusion—ultimately
made it stronger, as the following chapters will show. However, to deny him
the agency to recognize, protest, and on occasion even subvert those networks is to deny the moral reach of his arguments—worse, of anyone's arguments. All argument becomes complicit with merely strategic interest, all
agency the passive reproduction of ideology. Unlike Pratt, I do wish to grant
active moral agency to Humboldt, and by extension to anyone who, like him,
becomes aware they are struggling within, and penetrated by, structures of
power. As Edward Said has observed, "American identity is too varied to be
a unitary and homogenous thing"—it is, in fact, split "between advocates of
a unitary identity and those who see the whole as a complex but not reduc-
tively unified one." In this opposition of "two historiographies, one linear
and subsuming, the other contrapuntal and often nomadic," I would place
Humboldt in the latter: for him, intellectual nomad as he was, every situation
was the outcome of complex, even contrapuntal, historical experience. His
writing anticipates Said's further insight that "partly because of empire, all
cultures are involved in one another; none is single and pure, all are hybrid,
heterogenous, extraordinarily differentiated, and unmonolithic." Indeed, I
take this to be the lesson of Humboldt's works, starting with the provocative
jumble in *Vues de Cordillères* of empire's variously refracted and suppressed
natural and cultural artifacts.[11]

There were, to be sure, plenty of "imperial eyes" stalking the New World,
but Humboldt's project was different, and that difference matters. To begin
with, he lacked imperial sponsorship. While the Spanish passport gave him
freedom of access, he repaid this generosity by depositing his collections in
Paris and London and Berlin, publishing the results in French and German,
disseminating his maps to Spain's enemies, and inciting its colonial peoples to
rebel or at least reform Spain's endlessly inept and destructive colonial policies. Humboldt was determinedly self-sponsored and independent, grateful
for aid but beholden to no one, and he expended every penny of his personal
fortune to keep it that way. But perhaps most important, he felt (and that
is precisely the right word) that nature without humanity lacked meaning.
Far from emptying the landscape of its human presence, it grieved him that
Indians, victimized by whites, had fled their homes on the banks of the Orinoco, leaving the river empty of canoes and the jungle canopy unbroken by
villages. When he did find Indian villages he delighted to report on their inhabitants: their appearance, thoughts, language, culture, manufactures, history and, most of all, their ingenious deployments of native plants. For every
region he visited he described the myriad historical, cultural, and environ-

mental forces that might account for the movements of native peoples, from individuals to entire populations, and his political essays included shrewd analyses of the conflicts of interest between natives, Creoles, and Europeans. For Humboldt, a secular philosopher looking for material causes, discovery of the New World had catalyzed modernity by turning all the globe into a contact zone. From Columbus on, all histories were mingled, all worlds interlinked, all peoples cosmopolitan. For him, "America as Nature" meant nature as an equal partner with human purpose, expressed through science, art, technology, and commerce in a cosmic exchange. In short, Humboldt was a dissident who spoke out, loudly and persistently, against European imperialism and American slavery, and he was both honored and condemned as a dangerous man. Popular adulation, professional reputation, and his dense network of high-placed friends protected him to some extent from Napoleon's charges of espionage and, later, the insinuations of his enemies at the Berlin court, but his outspokenness against the Spanish government cost him dearly. Never had he thought his travels finished, and his dearest dream was to journey across Asia through the Himalayas to India, whose literature had helped form his philosophy. Year after year he laid plans to open the British Empire to his searching scrutiny, only to be thwarted by the British East India Company, who had no desire to see their own colonial policies made the butt of his next wave of books. Finally, his money spent, all he could manage was a state-sponsored and tightly controlled expedition across Russia, in 1829. "Unfortunately, we are scarce a moment alone," he complained; "we cannot take a step without being led by the arm like an invalid."[12] Rendered by then dependent on the king of Prussia for his income, he was increasingly muzzled during the reactionary years which saw European monarchies put down the Revolutions of 1830 and 1848. He withdrew into science, philosophy, and poetry, repressing and even destroying his writings of social protest.

Humboldt was not alone in his outspoken antiimperialist politics. His teachers, friends, and readings came out of what Sankar Muthu has recently identified as a "historically anomalous and understudied episode" in political thinking, namely, a tradition of late-eighteenth-century European intellectuals who attacked not merely the evils of imperialism, but its fundamental assumption of the right to subjugate other peoples. According to Muthu, key leaders of this tradition (Rousseau, Diderot, Kant, Herder—all of critical importance to Humboldt) based their critique on their view of *"humanity as cultural agency"* rather than an unchanging universal essence. That is, they saw all human beings, including indigenous peoples, as active, independent cultural agents, freely and creatively interacting with their diverse natural environments to generate "a wide plurality of individual and collective ways of life." Cultural differences were not pathologies or departures from a true way of life but creative adaptations which pointed to "the dignity of a universal, shared humanity as fundamentally intertwined ethical and political commitments." As Muthu concludes, scholarly views of "the" Enlightenment need to be pluralized in order to do justice to this group of thinkers, whose views were not allowed to enter the mainstream of nineteenth-century political thought. Instead they were "ridiculed and defeated," and by the early nineteenth century, virtually absent.[13]

Humboldt is clearly part of this dissident and repressed tradition, which seems to have held on far longer in Germany than in western Europe, and Muthu's attempt to recapture and foreground it helps place Humboldt in relationship to the Enlightenment thought he inherited, and helps account for the peculiar distortions of his work by his followers, who with few exceptions appropriated from him what was ideologically useful for their own projects and ignored or repressed what they found inconvenient. As Aaron Sachs observes repeatedly in his recent book *The Humboldt Current*, there was a strong "current" of Humboldtian thought in the United States that can be followed throughout the course of the nineteenth century, but the radical social dimension of that thought was seldom assimilated.[14]

Thus in the United States Humboldt becomes variously a colorful explorer, a romantic adventurer, a positivist scientist, an apologist for manifest destiny, a crusader against slavery, an inspiration to the Hudson River school of painters—but less often the radical social reformer who defended the rights of the oppressed. Humboldt had learned to weave society, nature, and culture into a single complex and seamless tapestry, but in his own lifetime that tapestry was unraveled and rewoven into smaller units. He proved too big to swallow whole; a century devoted to dissecting nature had to dissect him too, diminishing the qualities that made him unique, and uniquely productive, among intellectuals. After his death, those who recalled his vision of the whole called it "old-fashioned," the product of a bygone age when the world was so small that one capacious intellect could still see it end to end, all round. What they missed was the secret of Humboldt's success: not merely encyclopedic range, but a vision, a method, and a philosophy so generative that it marked not the end of one era but the beginning of another—one not yet born.

Humboldt's Europe

It is impossible to write a biography of Humboldt, says his latest biographer: he is too expansive and many-sided, has been appropriated by too many di-

vergent groups and viewpoints, and holds too many secrets. Nicolaas Rupke's
solution is to write a "metabiography," juxtaposing "clusters of characteristic
features" as they have been represented by various biographers, each embedded "in the remembrance culture of any one period of political history,"
from the liberal democrats of 1848 who claimed Humboldt as a fellow revolutionary, through the Nazis who ignored him, to competing appropriations by
the variously gay, green, global, postmodern, and postcolonial Humboldts of
today.[15] This present book does not aspire to be a biography of Rupke's one-
man pantheon. Yet readers will find a little biographical information helpful
in placing Humboldt in relation to his time. When he packed his trunks for
his American voyage, what, figuratively speaking, was in them? How did he
find himself, on the fifth of June, 1799, boarding the *Pizarro* in the Spanish
port of La Coruña with his companion and colleague Aimé Bonpland, shadowed by British warships just offshore? Why, and how, America?

Baron Friedrich Wilhelm Karl Heinrich Alexander von Humboldt was
born in Berlin on 14 September 1769. He hated the "von," which went against
his principles and which was apparently only a courtesy title for the family
anyway; he signed himself simply "Alexander Humboldt." His father, Alexander George von Humboldt, had after many years in the Prussian military
served as chamberlain in the king's court and developed a special, confiden-
tial relationship with him. He was in line for the position of minister upon the
ascension of Crown Prince Friedrich Wilhelm II, but the family's hopes were
dashed by his premature death. Alexander and his older brother Wilhelm
were only nine and eleven years old. Biographers call their father an intelligent, cheerful, and charitable man. To their mother they are less kind, referring often to her emotional coldness. It was she, Maria Elisabeth von Colomb
von Hollwege (descended from French Huguenots who had sought refuge in
Germany), who had brought money and property into the family, and under
that chill exterior she burned with ambition for her two sons. When Alexander was sixteen and still at home, a visiting friend reported to her sister that
everything in the Humboldt household "is just as it was"—the same old dog
asleep on the sofa, the widowed mother wearing the same "smooth, neat, and
simple" hairstyle, "the same pale countenance and delicately formed features,
upon which no trace of emotion is ever visible, the same soft voice, the same
cold though sincere greeting, and the same unalterable faithfulness towards
all with whom she is connected." For her sons, only the best education was
good enough, and after her husband's death she mortgaged her property to
continue to provide them with private tutors.[16]

The first of those tutors was Joachim Heinrich Campe, the German translator (in 1780) of the only book Rousseau thought appropriate for children:

Daniel Defoe's *Robinson Crusoe*, which described a man's reliance on his own ingenuity and resourcefulness during a retreat to elemental nature. Wilhelm recalled Campe's "happy knack" for stimulating the mind of a child, although Alexander was barely more than an infant when Campe departed, and it was Wilhelm, not Alexander, whom Campe took with him to Paris in 1789 to celebrate the French Revolution. But the educational principles laid out by Rousseau continued to shape the Humboldt brothers, particularly when the household was joined in 1777 by Gottlob J. Christian Kunth, a cultured, liberal, and impecunious young man who quickly became a substitute father, manager of the boys' education and of the Humboldt family estate. Rousseau had written in *Émile* that "Nature, not man" is the boy's schoolmaster, "and he learns all the quicker because he is not aware that he has any lesson to learn. So mind and body work together," each developing the other until the man achieves "the reason of the philosopher and the vigour of the athlete." Young Alexander apparently had neither. While his brother Wilhelm flourished, Alexander was "slow" at his studies and often sickly. He confessed later to a friend that his tutors despaired of his reaching even normal intelligence, and only in "quite later boyhood" did he show any evidence of "mental vigour."[17]

It was in these years that he turned to nature, dragging home so many rocks and bits of plants that his family began to call him, perhaps unkindly, "the Little Apothecary." He remembered these as unhappy years of terrible loneliness, living "among people who loved me and showed me kindness, but with whom I had not the least sympathy." Nature was both escape and redemption: cultivating the forests, meadows, and hills of Tegel, the family estate where he grew up, had been the love and continuing project of his departed father. The grounds were more than merely decorative. Tegel was a botanical garden, used as a nursery for foreign trees and shrubs, mostly from North America, and it supplied over five hundred varieties of plants to the Royal Gardens at Potsdam and elsewhere. Here young Alexander also developed his talent for art. He was promising enough that the director of the Berlin Academy of the Arts took him on as a pupil, and in 1786 his pencil copy of a Rembrandt painting hung at a Berlin art show. The same family friend who reported on the eternal sameness of life at Tegel remarked that Alexander, a "petit esprit malin" (or "clever little devil") was a remarkably talented artist, drawing heads and landscapes: "The walls of his mother's bedroom are hung round with these productions." Did she approve of his other ambition, to travel? In later years Humboldt often remarked on how he longed as a boy to follow his favorite writers to the earth's far corners: to see the great Dragon Tree of Orotava, on the Canary Islands; to behold the

Southern Cross, the constellation that dominates equatorial skies; to stand, like Balboa in Panama, on the crest of the mountains and descry the Pacific. One sees in Humboldt's reminiscences a melancholy and lonely child, walking in Rousseauian forest reveries, attempting to capture the mysteries of nature and people in drawings, dreaming of distant lands. In twenty years he would be famous across three continents for his exotic adventures and richly evocative descriptions of nature. How many lonely children grow up to realize their dreams?[18]

Alexander's world began to open up when, in 1785, he attended the popular lectures of the Jewish physician Marcus Herz on physics and Kantian philosophy. There in Herz's Berlin home Alexander witnessed the groundbreaking electrical experiments first performed by Benjamin Franklin and Count Volta. The brothers Humboldt quickly became intimate friends of the Herz family, taken up especially by Herz's young and famously beautiful wife Henriette, who taught Alexander Hebrew. The teenaged Alexander wrote long wistful letters in Hebrew to her, signing them off from "Schloss Langweil," "the Castle of Boredom." As Henriette said in her autobiography, Alexander was not merely practicing his language skills: "It was not to be thought of, that a young nobleman should confess in letters which could be read by everyone, that the society of Jewish ladies was more entertaining to him than a visit to the mansion of his ancestors." The Christian intellectuals of Berlin gathered in the homes of Marcus Herz and his friend Moses Mendelssohn (grandfather of the composer Felix) to discuss the philosophy of Kant and the poetry of Goethe, Schiller, and Lessing. Here was the center of Berlin's intellectual life, where stuffy convention could be defied in the world that the Jewish community, excluded from the common rights of European citizens, had built for themselves. Moses Mendelssohn, who had arisen from poverty and had to educate himself in secret, kept open house on Friday evenings for Jews and Gentiles alike. The Humboldt brothers were introduced to him by their mathematics tutor, who also taught Mendelssohn's two sons, Joseph and Nathan, with whom Alexander formed a lifelong friendship. It was his Jewish friends who made possible Humboldt's American travels, for it was the Mendelssohns' Berlin bank that extended Humboldt a letter of credit when another Berlin banking house withdrew it, just as he was about to board the *Pizarro*. And in turn Humboldt fought, during his later years in Berlin, for Jewish emancipation and liberalization of the laws restricting their participation in German society. In the brilliant salons of Berlin's Jewish intellectuals and socialites, Humboldt first witnessed cultural prejudice from the perspective of the oppressed and came to identify with their cause.[19]

In September 1787, the Humboldt brothers were sent to school in nearby Frankfurt, ushered by the faithful Kunth. The university's only recommendation seems to have been its proximity to mother's apron strings. It had no facilities for teaching science—no museum, botanical garden, observatory, hall of anatomy, or decent library or bookstore. Scoffed Alexander, "The Goddess of Science . . . certainly has no temple here." But that didn't matter, for his mother intended for him to study finance, not science, and in any case Alexander kept busy at his studies and discovering a new circle of friends. After a few months, Wilhelm went on to Göttingen, then the foremost university in Germany, and Alexander returned home for another year of studying Greek, drawing landscapes, and posting Hebrew letters to Henriette. His great discovery that year was botany, through his new friend Karl Ludwig Willdenow, a young man already famous for his *Flora of Berlin* and his innovative research into plant forms and their geographic distribution. Humboldt would bring him plants, and Willdenow would classify them, opening up for Humboldt a new passion—especially when plants came in from such exotic places as Japan. In a letter he gushed over the pleasures of collecting mosses, lichens, and fungi: "How sad to wander about alone! And yet there is something attractive in this solitude when occupied with nature." Yet he was not always alone; sometimes he and Willdenow wandered together, "hand in hand, through the vast temple of nature. Would you believe," asks Humboldt aghast, that out of 145,000 Berliners, there were scarcely four others who cultivated botany? Europeans assumed they already knew everything worth knowing.[20]

Finally, in the spring of 1789—the year of the French Revolution—Alexander was allowed to join Wilhelm at Göttingen, where he moved in with his roommate Count Metternich (future prime minister of Austria and another lifelong friend) and immediately joined the elite circle of intellectuals drawn to Wilhelm's extraordinary talents. Here he studied with the renowned Professor Blumenbach, still remembered today as a founder of modern anthropology and a pioneer of comparative anatomy, particularly of human racial differences. When Humboldt succeeded in shipping back a human skull from an Indian burial site in Venezuela, he made sure it ended up in Blumenbach's famous collection. He also became a student of Christian Gottlob Heyne, famous for his researches in classics and the history of civilization. More important to him, though, were the professor's family connections: it was at Heyne's home where Humboldt met Heyne's famous son-in-law, Georg Forster. This relatively brief friendship would be one of the most important of Humboldt's life.

It was almost as though Humboldt had met his older self. As a biographer notes, Forster was virtually a "prototype" of the mature Humboldt, who gave his vague childhood dreams force, direction, and a model. Forster was celebrated for having circumnavigated the globe on Cook's second voyage, from 1772 to 1775, with his father Reinhold; they were the first Germans involved in voyages of exploration, which Prussia was far too small and poor to mount. The irascible Reinhold, restrained for legal reasons from writing the narrative of the voyage, passed the task to his son, then just twenty-one years old. Georg Forster's unofficial (and uncensored) account was published in London in 1777 as *A Voyage Round the World*, and it created a sensation; a German edition soon followed. Forster emphasized not the dry technical details of navigation or the natural history specimens they collected, but the human stories of their adventure, including the Europeans' impact on native peoples. Humboldt memorialized him in *Cosmos*, attributing to Forster's delineations of the South Sea Islands much of his "early and fixed desire" to visit the tropics, and honoring his "celebrated teacher and friend" as the German writer who did more than any other to encourage the scientific study of nature: "Through him began a new era of scientific voyages, the aim of which was to arrive at a knowledge of the comparative history and geography of different countries. . . . George Forster was the first to depict in pleasing colors the changing stages of vegetation, the relations of climate and of articles of food in their influence on the civilization of mankind. . . . All that can give truth, individuality, and distinctiveness to the delineation of exotic nature is united in his works."[21]

Forster had nurtured the romance of the South Sea Islands first ignited by Bougainville's accounts and rendered by Diderot's commentary on them into an ideological battle over the nature of humanity. Diderot challenged Captain Bougainville's right to claim Tahiti for France, causing the destruction of a free people with their own cultural achievements, whose social felicity and sexual freedom gave the lie to European hierarchy and oppression. Suddenly world travels were the forum for cultural critique and philosophical reflection, and in this climate the young Forster felt free to express his doubts over Europe's civilizing mission. He was on a dangerous path. Professor Heyne, in an effort to keep his daughter and her husband safe and close, arranged a position for Forster at the University of Mainz. But Forster's progressive politics soon isolated him from the mainstream of German intellectuals. It was in this period, from March through July in 1790, that Forster and Humboldt journeyed up the Rhine to the Netherlands, France, and England, Forster teaching his young friend how to see and feel as a truly scientific traveler. (Humboldt always remembered that it was Forster who first showed him the ocean.) In Paris they celebrated the first anniversary of the French Revolution, and Humboldt loaded sand in a wheelbarrow to help construct the Temple of Liberty. The trip completed, Humboldt went on to the School of Commerce in Hamburg; back in Mainz, Forster became a leader of the German Jacobins. After France captured Mainz he went to Paris to petition for the annexation of his homeland to France. In his absence, the Prussian army recaptured Mainz, marooning him in Paris. Abandoned by his family, reviled in Germany as a traitor, Forster died in Paris, of illness, early in 1794. As Humboldt said in his memorial to his old friend, "But alas! even to his noble, sensitive, and ever-hopeful spirit, life yielded no happiness."[22]

Humboldt's term in the German seaport of Hamburg extended his new cosmopolitanism, for nearly half the students there were from other countries, and to improve his English he roomed with a young Englishman, John Gill (who went on to make a fortune in commerce). When not attending lectures on accounting, or on the history, economy, and culture of the United States (imbibing through his liberal professors the principles of democracy and religious tolerance), or teaching himself geology and botany, he took up the study of Danish and Swedish (bringing him up to seven languages and counting).[23] He had yet one more stop to make before he considered his formal education complete, and that was at the Freiberg School of Mines, led by the celebrated geologist Abraham Gottlob Werner. His application included a book he had recently completed on the basalt formations of the Rhine in which he advanced Werner's own "Neptunian" theories, and he was duly enrolled and warmly received by the great professor. Here for eight months Humboldt studied geology not in books but on the ground, as well as mineralogy, mining, surveying, and the law. He drove himself hard: mornings in the mines, afternoons in lectures, "moss hunting" (as he called his botanical walks) in the evenings, studying chemistry in his free time, and striking up another of his lavish friendships, this time with Karl Freieslaben, a fellow mining student at whose home Humboldt was rooming. Their subterranean explorations took them the length and breadth of Freiberg's labyrinthine mines looking for fossils, and out of these ramblings came Humboldt's next book, *Florae Fribergensis*, a botany of dark caverns. How is it, Humboldt wondered, that underground plants were so oddly tinged with green? His novel undertaking, combining chemistry, physiology, and geology, was a brilliant success, winning him election in 1793 to the Imperial Academy of Sciences, an honorary doctoral degree, and a gold medal presented by the elector of Saxony.

In these years the energies of the young man once described as slow and sickly were exploding in every direction. If anything, he exhausted himself with overwork, worrying his friends and family. His biographers see a psychosomatic element in all this: a passionate young man frustrated in love, displacing his energies to work. Humboldt was aware of the pattern, including his tendency to keep even his closest friends at a certain distance. In a letter to his classmate Wilhelm Gabriel Wegener he supposed that "no strong passion will ever sway me with an overwhelming power. Serious occupation and the calm induced by an absorbing study of nature will preserve me from the temptations of life." Indeed, no single individual ever swept Humboldt off his feet, but he seems always to have been in love, or longing for it. As his letter continues, "How happy, how inexpressibly happy should I be, if I had a friend like you by my side!" Yet his ambition was becoming boundless, crowding out all who did not contribute to what he was beginning to see as his life's work. A few months later he wrote again to Wegener from Hamburg with a revealing self-observation. "There is an eager impulse within me which often carries me, I fear, over the bounds of reason; and yet such impetuosity is always necessary to ensure success." As he noted in 1794 to Freieslaben, "I am quite mad enough to be engaged upon three books at once." No wonder his friends were worried.[24]

Humboldt, however, had barely begun to tap unsuspected reservoirs of energy, will, and adaptability. The harder he worked the happier he seemed, so long as he could surround himself with what Walt Whitman would call the love of comrades. He never married, and while he and his family kept his sexuality veiled, it was and remains an open secret that his deepest and most passionate attachments were with men. Same-sex desire was still unnamed then, even as laws governing private sexual desire were being liberalized, and although the question is still hotly debated, it is clear that Humboldt participated in a Romantic culture where intimate and intense relationships with men were, as Joan Steigerwald observes, "an important mode of sexual and self-expression"—certainly they were important for Humboldt's own self-figuring. It was also clear that he relished the salon culture of the day and was capable of forming passionate friendships with women as well, starting with Henriette Herz, so long as they had minds of their own—and used them.[25]

What was the goal of Humboldt's increasingly frenzied schedule? To all appearances, to become a mining engineer. Three days after finishing school in 1792 he went to work as an assessor of mines; six months later he was writing joyously to Freieslaben of his promotion to superintendent of all the mines in Franconia. He plunged into the work, descending into the mines by day and by night digging through trunkfuls of neglected medieval documents for his latest project on the history of mining. He was proud that he had gained the "universal confidence" of his miners, for daily he was out among them, in the "unbearable" heat and "enervating" atmosphere of the mines. It all seems improbable—why mining? It was the one profession that could satisfy his mother's expectations while giving him practical experience in natural history. He must have looked like the model Prussian civil servant— Douglas Botting calls him "a quite exceptionally talented eighteenth-century whiz kid"—but the fact was, he didn't need the job and had no intention of keeping it. This gave him the fearlessness of independence. Confronted with the primitive and dangerous working conditions of the miners and their staggering ignorance, he did not agitate for reform but set about building devices and institutions to help. Appalled that his miners couldn't identify even common minerals, he set up a free school, paying the teacher out of his own pocket. His official report persuaded the bureaucrats to expand the concept, and a second school was opened, initiating the movement for better education among the working classes. To mitigate the hazards the miners were exposed to, he designed a respirator to help them breathe and a variety of lamps to help them see, lamps that would not ignite the flammable gases they worked in. (Sir Humphry Davy, British chemist and friend of Coleridge, would later improve on Humboldt's designs.) He tested his inventions on himself, once nearly dying when, carried away by joy at seeing his new lamp burn brightly in gases that had extinguished every other, the fumes overcame him and he was nearly extinguished himself. The foreman pulled the unconscious Humboldt out by the feet to fresher air.[26]

But then his boss promoted Humboldt to director of mines, and he had to come clean. In turning down the promotion he admitted that all his work in the mines was really just preparation for a scientific expedition. His boss renewed the offer: travel all you want, he told his rising star; we will keep paying your salary and await your return. Humboldt did not budge. He donated his salary back for distribution to the miners, and set off to explore Italy with his new companion, Reinhard von Haeften, then Switzerland with Freieslaben, investigating the relationship between flora and the stratification of mountains. An anecdote from this period captures Humboldt's position in Prussian society: invited to a high-level diplomatic dinner, he rushed in late and sweaty, "in boots and travelling dress," fresh off the mountains. The reigning count put the shocked assemblage at ease "by a shrug of the shoulders and the whispered apology, 'A philosopher.'"[27]

Humboldt didn't spend all his time in the mines and mountains. These were also the years when he discovered himself as an intellectual. In 1794

Wilhelm moved to Jena, where he was welcomed into the circle of Jena and Weimar intellectuals. Wilhelm had long been friends with Schiller, chair of history at Jena, and Goethe's residence at the court of Karl August in nearby Weimar had made it a center for the arts and sciences. Alexander joined his brother, and Goethe, who had long been friends with the Humboldt family—his visit to Tegel in 1778 is memorialized in *Faust*—was attracted immediately to the budding scientist. Years before, Goethe had made a name for himself in science—comparative anatomy, the metamorphosis of plants, and optics—and the Humboldt brothers reinvigorated his interests. Alexander plunged into experimental researches on how muscles were excited by electricity, and Goethe joined in, expounding his own theories of comparative anatomy at such tiresome length that the Humboldts grew "impatient" and insisted he write them down. For his part, Schiller was so excited by Alexander's ideas that in 1794 he invited him to contribute to his journal *Die Horen*. Humboldt wrote up for him an odd little allegory, "The Rhodian Genius," that inquired into nothing less than the secret of life. At this point, Humboldt thought life was a vital principle that kept certain chemical affinities at bay, temporarily forestalling dissolution and death.[28]

Schiller's enthusiasm cooled when Humboldt later published his research and retracted his belief in a vital principle: life is rather "the joint action of well-known substances in their material forces," not reducible to chemistry and physics—more like meteorology, Humboldt thought, in its complications of phenomena and multitude of forces. Today this seems like a breathtakingly modern view, but all Schiller saw in it was crude materialism, and he let loose with a long and devastating critique of Humboldt's heartless and shallow intellect and "keen cold reason" that so "shamelessly" exposed to scrutiny mysteries of nature that should be kept secret. By contrast, Goethe was effusive in his praise for Humboldt, at least when he wasn't disagreeing with him. (Goethe hated Humboldt's theory that volcanic forces lifted mountains, which he mocked in *Faust*.) But even then, the force of Humboldt's reasoning took Goethe's breath away. When Humboldt visited him in 1826, Goethe cheered him on: "He resembles a living fountain, whence flow many streams, yielding to all comers a quickening and freshening draught." A few hours with Humboldt taught him more, Goethe thought, than years in the company of other men. As for Humboldt, he dedicated his most innovative work, *Essay on the Geography of Plants*, to Goethe (who in turn produced a delightfully cartoonish watercolor illustration of Humboldt's ideas). Goethe's attention to the wholeness of nature, to the transformation of its parts, and to the aesthetic impression made by nature on human beings, resonated deeply with Humboldt, who blended the fact-based positivism Schiller condemned with the aesthetic holism Goethe honored, a combination unique in its time and one that Goethe was better able than most to appreciate. Humboldt repaid the favor in *Cosmos* when he asked who had done more than Goethe "to renew the bond which in the dawn of mankind united together philosophy, physics, and poetry?" In Humboldt's Goethean synthesis, details took their meaning from the whole, and the whole could not be understood apart from its details.[29]

That this reciprocity between part and whole took place within a larger environmental context was one of the key ideas Humboldt gained from his familiarity with the work of Johann Gottfried Herder, another student of Kant's, who was publishing his *Outlines of a Philosophy of the History of Man* between 1784 and 1791. Early in this book Herder issued a call for "a new world of knowledge" that would gather together discussions of heat and cold, electricity, chemistry, and the chemical composition of the atmosphere, and by showing their influence on "the mineral and vegetable kingdoms, and on men and animals" collect all "into one natural system"—a "geographical aerology" that would explain the formation of mankind in both mind and body. Men, insofar as they, like plants, are formed by nature, ought to be studied in their reciprocal connections with "the elevation and quality of the land, air, water, and temperature," for "every race of men, in its proper region, is organized in the manner most natural to it." Humboldt would take Herder's concept into the field; he would also echo Herder's thinking on the universality of *Bildung*, the cultivation of self and society. In Herder's view, the joint processes of cultivation, the culture of the ground, and enlightenment, the culture of the creative mind, were not at all restricted to Europe or the West: "The chain of light and cultivation reaches to the end of the earth," no less to the New World inhabitants of California or Tierra del Fuego. All have "language and ideas, practices and arts, which they learned, as we learn them." While some peoples may be less enlightened or cultivated than others, all are striving for happiness and a better economy of life—so-called "savages" perhaps most actively of all. Thus, for Herder, the development of human culture and civilization was inseparable from the wider history of the earth, and all peoples were caught up in a dynamic interchange with their environment—ideas Humboldt activated when he traveled through the widely different landscapes of Spanish America.[30]

It's not clear whether Humboldt actually met the man who influenced him most deeply of all: Immanuel Kant, the transcendental philosopher of Königsberg. However, Humboldt grew up reading Kant and discussing him with friends, beginning with the Jewish circle around Kant's friends Marcus Herz and Moses Mendelssohn. In Kant lies the key to Humboldt's thinking, which presents a bit of a puzzle. Too positivist to please German idealists like Schiller, he was also too idealist to please the French positivists. He doesn't quite seem to fit in. Later writers almost universally emphasize one side over the other, turning Humboldt into either, in Julius Löwenberg's words, "one of those scrupulous empirics who observe and collect nothing but facts," or else a poetic mystic who romances the whole at the expense of hard science. Humboldt fell into neither camp, because he was forging something new, a truly Kantian synthesis, a dynamic concept of nature that would correct the excesses and blindnesses of both sides. Failure to grasp this has led to Humboldt being perhaps the most deeply and perennially misunderstood major intellectual of his time.[31]

Kant himself set out to create a synthesis of "rationalism" (knowing the world through pure thought) and "empiricism" (knowing through experience only). He had been unnerved by the skepticism of Hume, who undercut the certainty of scientific knowledge by showing that since we know only what our senses tell us, we really can know nothing at all. In response, Kant tried to redefine the conditions of knowledge, creating what soon was called a "Copernican revolution" in philosophy. In a nutshell, Kant argued that experience provides the content of our knowledge while reason provides the form; neither can exist without the other, so knowledge needs both, acting together. Hume was in one sense right: we cannot know with absolute certainty anything outside our perceptions or experience. The *ding an sich*, the thing in itself apart from our perceptions and conceptions, remains hidden from us. But we can know phenomenal reality, the world as it presents itself to us through our experience. As Robert J. Richards says, "In our experience of this realm, we discover both unity and order, whose ultimate foundations could only be the mind." Thus the human mind irradiates nature as the sun the solar system, illuminating it into meaning. To take a Humboldtian example: volcanoes exist, whether we know about them or not. While we cannot know volcanoes in an absolute sense, their character is given to us by the point of view that can experience the nature of the rocks that compose them, the degree of heat, the smell of fumes, their destructive power, their manner of growth. "Hence," says Roger Scruton in his illuminating introduction to Kant, "in describing my experience I am referring to an ordered perspective on an independent world." In other words, an ordering intelligence uses experience in the world to draw forth the order of its inner self, and of the outer world, together. To know nature better is thus to know ourselves better, for knowledge is a deeply human project.[32]

Humboldt saw it something like this: the idealists are wrong because they cannot see the universe; the positivists are wrong because they cannot see the

Cosmos. Humboldt would provide a third way, a Kantian-derived synthesis intended to open a new phase in intellectual life. He shared with his German idealist friends the Goethean desire to grasp the whole. But while the rational holist starts with a concept of the whole and thinks it down to the necessary parts, the empirical holist starts with the pieces and particulars as they present themselves to her ordering intelligence and works upward and outward, seeking connections and drawing them into patterns. As a sense of the whole emerges, it guides a deepening understanding of the interrelationship of the parts, in a reciprocal spiral of ever-deeper and wider knowledge. Kant's outline of the conditions of knowledge set the ground rules for empirical holism, and "Cosmos," Humboldt's term for the ultimate whole, is its fruition. Humboldt in Berlin worked under the gaze of a bust of Kant, and he told at least one American visitor not to leave Germany without purchasing one of his own.[33]

In February 1796, in the midst of this ferment of philosophy, poetry, and science, Humboldt was recalled to the bedside of his mother, who was losing her long battle with breast cancer. It was a lingering and torturous death, and ten months later, when it was finally over, he wrote to a friend of his mixed reactions: "We have always been strangers more or less to one another; but who could have remained unmoved at the sight of her unremitting sufferings!"[34] This was, as all his biographers recognize, the turning point of Humboldt's life. Kunth divided the estate between the two brothers, and when it was settled Alexander was at last a free man. Henceforth he would be wholly devoted to science. For the next year and a half he bounced across Europe, buying up the finest scientific instruments available and learning how to use them until they became, however strange the environment, his old familiar friends. He was in readiness for his life's work, a scientific expedition, and when the next one sailed he meant to be on it.

His timing was terrible. He wished he had been born forty years sooner, or forty years later. The French Revolution he honored and defended had devolved into the Napoleonic wars. Borders were closing, armed ships were blockading ports, travel was desperately dangerous. He had planned to go to Italy to study its volcanoes, but Napoleon got there first; for a spell Humboldt hung out in Austria waiting for the war to stop. He'd gotten an intriguing offer: the English Lord Bristol was putting together a junket to Upper Egypt, and he wanted Humboldt along, all expenses paid. Humboldt accepted and plunged into the study of ancient Egypt (a useful investment, it turned out, when it came time to examine the monuments of the Incas and Aztecs.) They were about to embark when Napoleon's forces invaded Egypt, rendering it no place for an Englishman, and the mad old genius canceled his plans. By now Humboldt was in Paris, the imperial center of European science, loving it and meeting everyone of any importance. Here he learned of a major French expedition to be led by Captain Nicolas Baudin: they would explore South America, the South Pacific, Madagascar, and beyond—a voyage round the world, Humboldt's dream. What must he have felt when he was picked to go along? While waiting to get started he struck up what would be a lifelong friendship with the young French physician Aimé Bonpland, selected as the expedition's botanist; the two were rooming at the Hotel Boston, where they bumped into each other on the stairs. But the theater of war was expanding across Italy and Germany, and the French government could afford either a war or else a scientific expedition, but not both. War won, and Baudin's expedition was postponed indefinitely. (It finally sailed in 1801.)

Humboldt was crushed, but at least in Bonpland he had a companion who shared his dream, and his disappointment. Together they left Paris for Marseilles, where yet another opportunity had materialized: the Swedish government would transport them to Egypt, where they could continue on their own. In a fever of anticipation, the two friends several times daily climbed the hill above the port, straining to spy the sail of the Swedish frigate that was to bear them off to see the world. Two months went by before they learned the frigate had been shipwrecked, and would not sail until repaired sometime the next year.

Now they were getting desperate. On a side trip to Toulon they happened across Captain Bougainville's old ship—Bougainville himself, the very captain who had discovered Tahiti to the world and been so kind and encouraging to Humboldt in Paris. They went aboard, and Humboldt sought out Philibert Commerson's old cabin—Commerson himself, the very naturalist who had first stepped onto Tahiti, whose writings had sensationalized it as the paradigmatic tropical island paradise. Humboldt could practically smell the flowers, hear the breeze in the palms. There in Commerson's cabin he had a small meltdown: "I lay for full ten minutes in the window, contemplating the bright vision. At length they came to seek me; I could almost have shed tears as I thought of my shattered prospects."[35] Would they never get out of Europe? The two booked passage on a small ship bound for Tunisia, but on the verge of sailing, they learned that the Tunisian government was throwing everyone arriving from a French port into prison. So Humboldt and Bonpland walked across the Pyrenees from France to Spain. Perhaps they would have better luck sailing from a Spanish port.

They did. In Madrid Humboldt's personal connections finally came through. The ambassador from Saxony was none other than the Baron von Forell, brother of an acquaintance in Dresden, "well versed in mineralogy" and a patron of the sciences. Forell had a hunch that the Spanish minister Don Mariano Luis de Urquillo, an enlightened man of liberal views who had the ear of the queen, could successfully pitch Humboldt's project at court. In March 1799, Humboldt was presented to the king and queen of Spain, to whom he outlined his desire to visit, at his own expense, the interior of Spanish America. King Carlos IV, a man no one would call enlightened (Botting calls him "so stupid he was nearly imbecile"), nevertheless was persuaded. Perhaps he just went along with the queen. In any case, Urquillo smoothed the way, and Humboldt was granted two passports, one from the secretary of state and another from the colonial administration. It was quite a coup: "Never before," he said, "has a permission so unlimited been granted to any traveller, and never before has a foreigner been honoured by such marks of confidence from the Spanish Government." After cramming for weeks in Madrid's New World collections and archives, Humboldt and Bonpland traveled on to La Coruña, where they booked passage on the *Pizarro*, a mail boat bound for Cuba. She was reported to be a slow ship but a lucky one, good at evading British warships. This time there would be no last-minute cancelation. On 5 June 1799, at 2 p.m., they set sail for the New World.[36]

A New Earth and a New Heaven

The morning of the day they weighed anchor, Humboldt posted last-minute goodbye letters to his friends. To Freieslaben, back home rising steadily in the ministry of mining, he wrote that he was dizzy with joy: "Man must strive for the good and the great! Within a few hours we sail around Cape Finisterre. I shall collect plants and fossils and make astronomic observations. But that's not the main purpose of my expedition. I shall try to find out how the forces of nature interreact upon one another and how the geographic environment influences plant and animal life. In other words, I must find out about the unity of nature." He was not driven by personal glory or the collector's zeal to add new entries to the catalogue of nature. As he noted in the introduction to his *Personal Narrative*, "The discovery of an unknown genus seemed to me far less interesting than an observation on the geographical relations of the vegetable world." His vision, shaped as it was by the grand geographic syntheses of Kant and Herder, could encompass potentially everything he was to experience. Where other scientists separated out their individual objects of study and examined them from a narrow point of view, he wished to

> comprise in one view the climate, and its influence on organized beings, the aspect of the country, varied according to the nature of the soil and its veg-

etable covering, the direction of the mountains, and the rivers which separate the races of men as well as the tribes of vegetables; and finally, those modifications, which the state of nations, placed in different latitudes, and in circumstances more or less favorable to the display of their faculties undergoes.

What must come from his travels was not a heap of fragments but a new vision of the total connectivity of nature. As Helen Maria Williams exclaimed, Humboldt will show us not the New World but "a new earth, and even new skies!" The starting point was enchantment. The ordinary would be made marvelous, and the marvelous a step in the passage to Cosmos.[37]

Humboldt wrestled with what to call his new science. "Cosmos" would come much later, after all his Latin American materials had been published and he was ready to make a conclusive statement. Meanwhile he ventured a few names: "natural history of the world, theory of the earth, or physical geography." "Natural history," which sought to name and tame the diversity of the physical and social worlds, did not reach for their underlying causal connections, while "theory of the earth" slighted the material basis for that theory, sounding too rationalist, not sufficiently empirical. He settled, when he had to settle, on "physical geography," linking his name to this science as a founding father. In the United States today, geographers are the specialists most likely to know Humboldt's name and place in intellectual history.[38]

Humboldt's use of the term "physical geography" derived directly from Kant, who taught the subject for four decades but did not himself travel or do geographical research. Although the British geographical tradition excluded human beings, the Germans did not. Kant's 1757 description of the field inserts humans into the midst of a capacious sequence: "Physical geography considers only the natural condition of the earth and what is contained on it: seas, continents, mountains, rivers, the atmosphere, man, animals, plants, and minerals." According to Margarita Bowen's history of geography, it was Humboldt who developed Kant's unfinished suggestions into a science, using Kant as a "fruitful source of inspiration." Humboldt was particularly provoked by Kant's distinction, in the *Metaphysics of Natural Science*, between "natural science," in which the whole is rationally ordered by a generative principle, and empirical or "historical" science, which since it lacked an organizing internal principle, was not really a "science" at all but merely a systematic ordering of facts, as in taxonomy or, as Coleridge said contemptuously, "a dictionary." Humboldt's goal in *Cosmos* would be to create a form of "physical geography" that would order facts not descriptively but by their causal connections, turning physical geography into a true science. But never a complete science: as Anne Macpherson observes, Humboldt asserted that "we cannot now, or perhaps ever, derive the unity of nature from a single concept. . . . We must be satisfied with partial explanation, based on empirical observation." Unlike Kant, Humboldt would wrestle with these problems on the ground, making the whole world, rather than a room in Königsberg, his study.[39]

Down on the ground things tended to look pretty messy. Humboldt was not part of the increasingly streamlined machine of the government-sponsored scientific expedition. That had been his first choice, but it had been denied him. Instead he and his invaluable companion Bonpland arrived on the American coast alone and unsponsored, with two pieces of paper granting them permission to be there, a line of credit on a Jewish bank, and several crates of scientific instruments. He had the advantage of independence, which would, here as in the mines of Silesia, give him a certain fearlessness; as he wrote to Willdenow in 1801, his independence was so valuable that to keep it he refused all government money. The tradeoff he thus made quite literally shadowed his sailing from La Coruña: as the ship tacked against the wind, his eyes and Bonpland's were "fixed on the castle of St. Anthony, where the unfortunate Malaspina was then a captive in a state prison." As the ship tacked back and forth they had plenty of time to reflect on Malaspina's fate: after leading a brilliant government-sponsored Spanish expedition around the world, he had returned in 1794 brimful of plans to turn the Spanish empire into a commonwealth. For a year the Spanish government let the great explorer agitate and intrigue, until officials lost patience, arrested, and imprisoned him—and there he was still, his fate looming over them like a warning: Stick with science. Beyond that, keep your mouth shut. By the time they returned, Napoleon had intervened and ordered Malaspina's release, but the great explorer was broken and silenced, and his narrative of the journey was never published. Sighed Humboldt, "I could have wished to have fixed my thoughts on some object less affecting."[40]

Packed away in boxes were the instruments they had managed to bring from Spain. Some had stayed behind at Marseilles, never reaching them in time. It was a hard lesson Humboldt never forgot: "Never," he advised his readers, "lose sight of instruments, manuscripts, or collections." In a time of warfare, they might not be seen again. Everywhere he and Bonpland went, they had to take everything with them, all their instruments, their growing collections, their precious manuscripts. Humboldt's list of instruments alone takes up seven full pages: chronometers, telescopes and microscopes, sextants and theodolites and quadrants, compasses and dipping needles to measure the earth's magnetic field, barometers (so delicate, and so important in measuring altitude, that for five years Humboldt assigned one guide to carry his one best barometer), hygrometers to measure humidity, electrometers, a "cyanometer" to measure the blue of the sky. Not to be left out were "a great number of small tools necessary for travellers to repair such instruments as might be deranged from the frequent falls of beasts of burden."[41]

Humboldt marveled at the sagacity and surefootedness of mules, but they, too, were only human; river crossings (as many as twenty-seven in one day) were ordeals, as he and Bonpland watched from the riverbank in agony lest a mule go under. The moments when they did lose a mule were long and terrible: was it carrying food or camp gear, meaning only temporary hardship? Or instruments, meaning scientific results lost forever? Or part of their collections, weeks or months of hard labor vanished? Or worst of all, their manuscripts? They never had time to make copies; lose the manuscripts and they might just as well have stayed home. In seas "infested with privateers," most of their letters and shipments home never made it, and they went for two years without receiving any letters from Europe. Humboldt didn't know whether a journal of astronomical measurements and altitudes had made it safely to Paris until he read the news of its arrival—in a Philadelphia lending library. "I could scarcely suppress an exclamation of joy." At strategic points they divided their collections and papers, amounting to dozens of boxes: a third stowed safely in port and a third shipped to each of the warring sides: France via Spain, or England via Germany or the United States. They learned the hard way the sad wisdom of their precautions. One particularly precious shipment, consisting of duplicate plants, all of Bonpland's Orinoco insects, and bones from the extinct tribe of Atures Indians, was entrusted to the personal care of Juan Gonzales, a "gay, intelligent, and obliging" young monk they had met in Cumaná, who had spent years in exile in Esmeralda and who offered them much useful advice for their planned journey up the Orinoco. Upon their return from the Orinoco they had met up with him in New Barcelona and he joined them for the next seven months in Cuba. He, the child he was escorting to Spain, and the collections were all lost off the coast of Africa, a tragedy so haunting that Humboldt mentions it repeatedly. Curiously, the Indians they met or hired had refused to touch the mule carrying the Indian remains, declaring that the beast carrying the bones of their ancestors would soon die. Humboldt never explicitly links their "superstition" to his friend's fate, but he never forgets the coincidence.[42]

Mules don't drive themselves, meals don't cook themselves, passages up river networks or over mountains don't magically open themselves. At every step, Humboldt and Bonpland were both vulnerable to and dependent on the people of the lands through which they traveled. There had to be drivers for as many as twenty mules, or occasionally oxen. On rivers, they needed

to hire paddlers. Carlos del Pino, a Guayaqueria Indian they met even before they set foot in South America, became their companion and guide for
the next sixteen months, and José de la Cruz, a mixed-race "zambo," joined
them in Cumaná and stayed with them for the next five years as a personal
assistant, accompanying them all the way to France. They needed local guides
who knew the roads and paths, interpreters to decode the many languages
of the natives, and agile collectors to scramble up trees to gather fruits and
flowers. Often their caravan camped out in the open, but in the towns and
missions they needed hospitality, someone to put them all up for a night or a
week, feed and water their mules and themselves. Sometimes it was wealthy
plantation owners who threw their doors open to the travelers. More often
they stayed among the monks and Indians in mission villages, some more
and some less prosperous; at the Convent of Caripe, they realized with em-
barrassment that the monks were going hungry to feed their guests. For some
days, until the rains let up and they could move on, they dreaded the tolling
of the refectory bell calling them to meals. In cities like Caracas, Bogota, and
Quito, where they often stayed for many months, they became the center
of social life, attending balls, parties, and salons. Everywhere the loquacious
Humboldt went, he talked, and he listened. Scattered through the pages of
the *Personal Narrative* are overheard comments, fragments of conversations,
and the words and stories of the Creoles and indigenes who worked with
them and whose lands they traveled. Herbert Wilhelmy relates the anecdote
of the mayor of a Mexican town, who, irritated by Humboldt's insatiable desire for information, turned on him and declared: "'Sir, the viceroy told me
you were a clever man: but I don't understand what you can possibly know
for you ask questions about everything.' 'Yes,' answered Humboldt, 'this is
the only way I can know anything at all.'"[43]

Had Humboldt's interests been narrower, his writing would have been
better. He labored with his prose and never felt satisfied; his friend Schiller
had made it clear he was no poet, and so he apologizes for his lack of sty-
listic elegance. But the real problem was that no single genre could contain
all he wanted to say, even as every genre came with expectations he felt he
must meet. His journal, which he downplayed as little more than careless
and fragmentary first impressions, is a highly personal record of awe before
great beauty, of poetic flights, anxieties, sharp disappointments, and social
sarcasms. Most of this he suppressed. In fact, he didn't want to publish his
"personal" narrative at all, but was finally persuaded to do so by public expectation (scientific travelers were supposed to make public the narrative of
their journey) and by the urging of friends, particularly his Paris publisher,
the exiled English radical Helen Maria Williams, who hoped to recoup some

of the fortune she had lost publishing Humboldt's ruinously expensive specialist works. When he finally went public with his *Personal Narrative*, Humboldt, bowing to the demands of world fame, strove for a high-minded scientific decorum that forbade him from violating anyone's privacy, including his own. It would not do to spill his emotions all over the paper, as he did so freely in his journal and in letters to friends.

Besides, the traveling self was, he thought, not the real center of interest: "Amidst the overwhelming majesty of Nature, and the stupendous objects she presents at every step, the traveller is little disposed to record in his journal what relates only to himself." As Anne Godlewska remarks, Humboldt removed himself from stage center to direct his focus not on people but on the relations between them. Thus his "personal" narrative is oddly depersonalized. He mentions Bonpland only when the action demands it, as when Bonpland saves him from drowning, is assaulted on the Cumaná beach, or nearly dies of fever, and he almost never addresses any of the many companions who joined them for months or even years at a time: Carlos del Pino, José de la Cruz, and Juan Gonzales; Father Bernardo Zea and Nicolas Sotto, who went along on the Orinoco; Francisco José de Caldas, the brilliant young botanist, and Carlos Montúfar, son of a revolutionary, who joined them in Quito—these and others are briefly introduced, then disappear from the narrative except for brief though often revealing glimpses. What could account for such reticence? The very details we hunger for are those Humboldt scrupulously removed, for he thinks that the travelers themselves belong offstage: "I have preserved a few, but have suppressed the greater part of those personal incidents, which offer no interesting situation, and which can be rendered amusing only by the perfection of style."[44]

Yet Humboldt himself never disappears from his narrative. His narrative voice creates a persona, a character, who energetically mediates and interprets the journey for the reader. This point is worth emphasizing, for the ideal of scientific objectivity was just then becoming dominant, with its demand that the narrator's self die out of the text to let the pure truth of nature shine through. Humboldt's conscious depersonalization acknowledges this emerging aesthetic standard, but his narrative voice resists it, constantly reminding his readers that the actions, feelings, and thoughts registered in the prose are his and no one else's. This resistance was at least in part a legacy of his Kantian philosophy, for Kant denied the possibility of pure objectivity. Although reason would always strive "for the unrealizable ideal of perspectiveless knowledge," this was an illusion, a fantasy invited by the very fact that, since we always do have a point of view, we can imagine *not* having one. But the conditions of knowledge make this impossible: the world cannot be known apart from the perspective of the knower. So Humboldt frames his reasoning and his experiences from his own point of view: he tells his audience what he thinks and why, what he's doing and why, and how it all makes him feel. As he would later theorize in *Cosmos*, emotions were important elements in the constitution of the universe, and no physical description was adequate unless it took account of the feelings of the spectator and the beings around him.[45]

This uneasy tension between an objective narrative and a subjective narrator is intersected by yet another complication: the density of the associations that every object he sees brings to his mind. His feet may be on the ground, but his thoughts are on the wing. Since everything is connected, everything he sees reminds him of a dozen other things. He cannot wrestle his multidimensional thoughts and experiences into the single linear narrative of the written page. The effect is kinetic, even dizzying: Humboldt was a hyperlinked, hypertext thinker two centuries before computers. Somehow he had to make do with technologies of inscription that his own methods were rendering obsolete. For instance, how, to go back to his favorite topic, should he describe a group of volcanoes? He must situate them on the face of the planet with exact astronomical observations determining latitude and longitude; measure their heights and magnetic fields; identify the plants on their flanks and the rocks that compose them; note the height reached by plant associations and the groupings of rocks; record humidity, temperature, electricity, transparency of the air, topography, and on and on, including the history of eruptions and native legends about them, the stench of sulphur, the sizzle of burning rocks, and the overwhelming vulnerability one feels on the brink of a volcanic crater, looking in. Should these disparate elements be treated separately? That violated the sense of wholeness he sought. But how to stream them together? When he steps to the crater's edge, should he interrupt the story to lecture his readers on geothermal physics? That seemed "cumbrous," yet without knowledge of physics, all else was mere impression. Thus he tried to craft some way to include both observed facts, the tactile thump of boots on the ground, and the spiraling levels of generality telescoped into every physical detail. He fails, of course, and he apologizes for failing: "Notwithstanding the efforts which I have made to avoid ... the errors I had to dread, I feel conscious, that I have not always succeeded."[46]

Sympathetic readers like Williams were more generous: "M. de Humboldt has in this work displayed," she writes, "more than in any other he has yet published, his peculiar manner of contemplating nature in all her overwhelming greatness." His writings are characterized by his ability to raise the mind to "general ideas, without neglecting individual facts; and while he appears only to address himself to our reason, he has the secret of awakening the imagination, and of being understood by the heart."[47] Head, and heart: the Humboldtian imagination could link individual facts with mind-stretching generalizations and in the same gesture, material reality with poetry and feeling. Head and heart were always present; imagination mediated between them, making science into poetry that was true.

Williams claimed more for Humboldt than he dared to claim for himself. This may be one reason they worked so closely together on her English translation of *Personal Narrative*. Astute readers who note the differences between French editions and her English translation, particularly a Romantic sentimentalism not present in the French, might conclude that Williams's translation is merely unreliable. Given the closeness of their collaboration, Humboldt's fluency in English, and his expressed pleasure in her work, this seems unlikely. The problem of translation, as Bruno Latour suggests, is that it binds together two hitherto different interests to form "a single composite goal," which means that both are forced to swerve a bit from their original intentions. If Humboldt is read as a literary writer, it becomes possible to think that both he and Williams, herself a major Romantic poet, swerved a bit to enable Humboldt to voice sentiments in English, and to an English audience, that French constraints of genre, voice, and politics made unacceptable. Furthermore, British Romantic literary critics have only recently rediscovered Williams (together with her female cohorts). Once her work with Humboldt becomes a topic of analysis, her translations of Humboldt may well be seen as classics of British Romantic prose, all the more interesting for being androgynous hybrids of poetry and science, German and English, masculine rationality and feminine sentiment.[48]

His *Personal Narrative* was but one of several attempts Humboldt made to solve the dilemmas posed by his inclusive vision. To avoid the "errors" he saw in his path, he published the results of his American travels in various genres: a picture book with descriptive notes; vivid personal essays followed by exhaustive scientific footnotes; tables of astronomical observations; catalogues of plants; scientific essays on plant geography, ethnobotany, and zoological observations; political essays on geographical regions (New Andalusia [Venezuela], Cuba, Mexico) which explored their economic, civil, and natural histories. To collate the extraordinary numbers of quantitative data points collected with such care by those seven pages of scientific instruments, Humboldt devised novel means of presenting that data visually: elegant maps that recorded elevation via novel techniques of three-dimensional

representation; cross-sections of landforms showing stratigraphy and of continents showing hitherto unsuspected interior plateaus; graceful, arcing iso-
graphs collating average temperatures with longitude.

The figure that summed up the most data possible was the astonishing
Tableau physique des Andes et Pays Voisins (fig. 3), intended to accompany the
Essay on the Geography of Plants. This thumbnail Cosmos offers a new solution to the problem of describing volcanoes: Chimborazo, Cotopaxi, and an
unnamed third peak rise in the center, hand-painted (in the deluxe editions)
in bands of color to show the correlation of vegetation with altitude, just as
Herder had suggested, and altitude with longitude: tropical jungles at the
bottom, temperate forests midway, polar ice at the alpine summits. Ranged
along both sides of this visual tableau, like annotations on an artist's field
sketch, are the columns of supporting data: heights, temperatures, chemical
composition of the air, measurements of light intensity, sky color, gravita-
tional force, distances of visibility under different atmospheric conditions,
the lower limits of perpetual snow, animals characteristic of ascending elevations, geology, plants cultivated by the inhabitants, and on and on. In the cutaway portion, lists of plant species jigsaw their way up the mountain in loose
arcs, listed by altitude, zone by zone. In the accompanying text, fossil remains
introduce the dimension of deep time; the human need for food brings in
moral and political economies; and the developmental ties between humans
and their environment bring in intellectual history and aesthetic culture.

Given all this, the task of the *Personal Narrative* is to array information
diachronically rather than synchronically. Such a narrative must war with
itself, for its two purposes—to record external, public events; to record internal, private impressions—conflict. If a unified composition is to be achieved,
one or the other must be subordinated. Humboldt tacks from one extreme to
the other, from meditative personal impressions through chronological narrative to dry scientific essays. Abridgements, of which there are several, cut the
scientific "digressions," normalizing Humboldt's experimental and hybrid
form into a balanced and intellectualized chronology. But even this breaks
down as Humboldt paddles up the Orinoco into what becomes his own heart
of darkness. Overwhelmed by the experience, he turns self-consciously back
to the raw and visceral journal he elsewhere suppresses, acknowledging that
what he scribbled in the field (whenever the rains and mosquitoes permitted)
"bears a character of truth, I had almost said of individuality, which gives attraction to things the least important." The jungle closes in palpably on the
lone, narrow, and fragile canoe overladen with a botanist, a cosmic philosopher, a Franciscan monk, a colonial administrator, two assistants, numerous
Indian paddlers and guides, various small wild animals, and a very large dog,

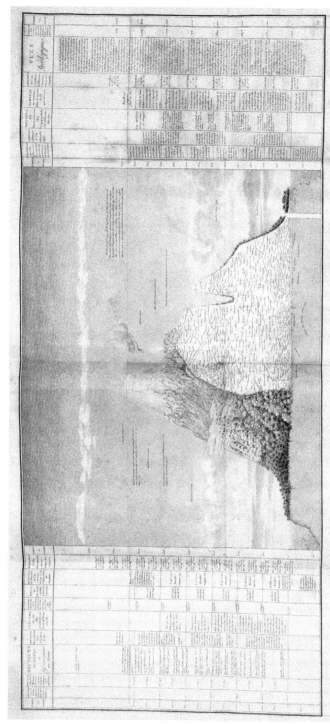

GÉOGRAPHIE DES PLANTES ÉQUINOXIALES.

FIGURE 3 . Alexander von Humboldt, *Tableau physique des Andes et pays voisins, essai sur la géographie des plantes* (Paris, 1807). Rare Books and Special Collections, University of South Carolina Libraries.

while supplies dwindle and even Humboldt's ebullience flags. It makes for gripping reading.[49]

But something went very wrong. After completing three volumes, Humboldt had brought his readers less than two years into his five-year voyage. He seems to have written a fourth volume covering the long trek down the spine of the Andes and into the Amazon, then ordered the manuscript destroyed, at a cost of 9,500 francs. Whether this phantom fourth volume actually existed, and if so why it was destroyed, remains unclear. What is clear is that at the conclusion of volume 3, Humboldt knew what was going to come next: a book that, written twenty-five years after his travels—years that had seen the rise and collapse of Bolívar's revolutions—would skip what others had already said and abstain from "polemics." Instead this fourth volume would "retrace to the imagination the physical representation of the Cordilleras and the plains, the force of an agitated and powerful nature, which fertilizes and destroys alternately, the eternal influence exerted by the configuration of the earth, the course of the rivers that furrow it, the vegetable layer that covers it in the social state, and finally, the institutions and the destinies of nations." This might well have been his masterpiece. Why might he have destroyed it? Could it have been too revealing of his personal feelings for Montúfar? Too sympathetic to the South American revolutionaries? His earlier volumes included blunt discussions of prerevolutionary Venezuelan politics, and the lands at the center of volume 4, Colombia, Ecuador, and Peru, were still hotbeds of unrest. Did the direct intervention of conservative political interests force his hand?[50] In any case, his narrative stops midcourse, and he concludes with a book-length essay on Cuba, leaving his readers hanging. Darwin, reading and rereading the Williams translation on the *Beagle*, wrote home begging to be sent the next volume, never dreaming there wasn't one.

It would have been a difficult volume to write. Humboldt's journals record bitter poverty and starvation among the Andean peoples. As Humboldt and Bonpland pursued their journey, they were increasingly troubled by political unrest and the chaos of war. Once they nearly lost their lives to an armed band of escaped slaves, naked men in chains. Fears of rebellion ran high: memories of Tupac Amaru's 1780 uprising of Peruvian Indians were still fresh, and Toussaint Louverture's successful Haitian slave revolt (which Humboldt celebrated) was on every slave owner's mind. In New Barcelona, they lived for a month with Pedro Laviç, whom they had earlier visited in a Caracas prison where he was held for sheltering the revolutionary conspirator José María España, executed only two years earlier with praise for the United States and a call to independence on his lips. In Quito, Humboldt's residence, the home of Montúfar's father the Marquis de Selva Alegre, was the distribution point for revolutionary literature.[51]

Looking back on all this, in the wake of the terrible and bloody American revolutions he had hoped reform would prevent, Humboldt found that composing his pages meant reliving much pain. The gracious and enlightened Venezuelan plantation owners whose liberal principles so impressed him had lost their lands, and often their lives, fighting in the wars to bring into being new nations—even, if Benedict Anderson is correct, the new concept of nationhood. His two brightest young friends, Caldas and Montúfar, had both been executed. Humboldt mourned the loss: "Already at Quito the most virtuous and enlightened citizens have perished victims of devotion to their country. While I am giving the description of regions, the remembrance of which is so dear to me, I meet at every step with places, which recall to my mind the loss of a friend."[52] His *Personal Narrative* is a memorial to a time, if not of happiness, at least of peace, and the internal divisions he witnessed in South America made him doubt that peace could come there again in many lifetimes. The freedom and joy he and Bonpland had known seemed lost forever. Even as he wrote of their adventures together, his beloved Bonpland had returned to South America, where he unwittingly stumbled into a border war between Argentina and Paraguay. Bonpland was attacked, nearly killed, enchained and imprisoned, and there he remained despite the cries of Humboldt and all of intellectual Europe for his freedom. They never saw each other again.

Yet the mood that pervades his American writings is joyful. Dogged by illness right up to their departure, Humboldt was transformed by the tropics: as he wrote to Willdenow in 1801, "The tropics are my element, and I have never enjoyed such uninterrupted good health as in the last two years." His narrative is full of boundless energy and curiosity, and the occasional moments of depression and despair quickly dissolve into the next episode of gleeful adventure or intellectual discovery. Williams celebrates this hopeful aspect of Humboldt's work: "How often will posterity also [like herself] turn from the terrible page of our history, to repose on the charm of a narrative, which displays the most enlarged views of science and philanthropy!" Humboldt in his struggles "obtains a victory that belongs to mankind." She echoes a theme Humboldt had stated years before, when in Berlin's darkest days, his homeland conquered by Napoleon, he had put together what he called his favorite book, *Ansichten der Natur*, a collection of essays recalling the beauty and pathos of the tropics. In his preface he had stated, "It is to minds oppressed with care that these pages are especially consecrated." Humboldt was evoking the wisdom of Kant, who had written that in an age of oppression, the role of the intellectual was to energize the people with hope for better times, lest they lose heart in the possibility of reform and accept, in Muthu's words, "the injustice of the modern world as an inevitable necessity." Kant's model was Humboldt's old friend and teacher Moses Mendelssohn, who despite every reason for a debilitating pessimism had believed it yet possible to create tolerance and civil rights for Jews.[53] Humboldt's language of nature, of harmony and freedom, provided not escape but hope that change is possible, oppression can be lifted, and injustice made right. His New World became a stage on which a new future might be reimagined for all humanity.

2

Passage to America, 1799–1804

> The tongues are very unequally divided in this picture. . . . The native scarcely dares defend his cause, while the strangers with long beards talk much and loud, as the descendants of a conquering people.
>
> ALEXANDER VON HUMBOLDT

Portals and Passages

It seems fitting that Humboldt launched from Spain, for there is a certain Quixotic quality to his venture: the slim and blue-eyed Humboldt with sturdy Bonpland his cheerful sidekick, tilting at the windmills of Spanish ignorance, prejudice, and colonialism on their endless picaresque travels. For in a sense Humboldt's travels never really did end. Even when he wasn't on the road he was always in motion, in transit, planning the next trip. He abandoned national loyalties to become the paradigmatic cosmopolitan, at home everywhere and nowhere, always passing through, a merchant of knowledge with a bag full of notions. In a time of closed ports, armed borders, gunboats, and pirate ships, he alone passed freely, slipping through with a smile and a story like Marnoo in Melville's *Typee.* Shielded by his royal letters of passport, he talked, and listened: friendly, sociable, charismatic, passing from huts to plantations to palaces, bobbing like a cork on turbulent seas. Not for him the fate of Forster, trapped and consumed between France and Germany, or Montúfar, shot in the Revolution, or Bonpland, ambushed in a border war. Unlike his friends who succumbed to the rising forces of nationalism, Humboldt the nomad stayed afloat in a world of political and natural turbulence, learning to skate well, as Emerson said, on thin ice, to be always in passage.

Humboldt traveled not from country to country but through a planetary field of geological, historical, and environmental forces. His coordinates were not political but bioregional: rivers, mountain passes, coastlines, trails, and roads. It is seldom clear exactly what "country" he's in, as he sifts and compares, moving up, down, and across both spatial and temporal scale levels, continents and eons, on the alert for harmonies and resonances he can test to see if they might justly be called laws. To assume he is another Enlightenment universalizing agent writing the metanarrative for all time is precisely wrong, for there is no "global" in this most planetary of thinkers—only the local at every point, generating the patterns and harmonies that combine to a collective whole. His view is not an eagle's flight to God-vision but hovers down lower, darting like a bee from point to point, where everything can be seen and touched and connected, collecting nectar for the hive and cross-pollinating as he goes. This requires hard work, constant motion, and an astonishing memory. Each passage in Humboldt is a series of crossings, person to person, speech to text, chaotic jungle to labeled specimen, crossing in turn overland to a port and across the ocean to a scientific center to be retranslated by someone on his burgeoning team: botanists, physicists, anatomists, astronomers, engravers, colorists. Every crossing was a transformation, a creative act that reinvented as much as it transmitted. As his texts traveled, passing from reader to reader, readers in turn performed their own acts of reinvention, deploying Humboldt for their own needs, from Darwin on the *Beagle* to Thoreau at Walden Pond to John Bachman in the urbane port of Charleston or Susan Fenimore Cooper in rural New York, and beyond to Europe, Africa, Siberia, India, Australia.[1]

This network of passages is universal only as train tracks may be said to cover the world: while they reach from coast to coast, they are easily interrupted at any point. The king or corporation refuses access; the mule falls, the ship wrecks, the shipment is confiscated, the botanist abandons his work to take up gardening, the publisher goes bankrupt, key books go untranslated creating blind spots otherwise inexplicable.[2] Humboldt generated small islands of order, interlinking them in a widening chain until he had built an archipelago of knowledge that reached across the planet. For Humboldt's Cosmos uses the earth, not God, to orient the self. Passages are made in small craft that leak and large ones that dodge blockades, along ocean currents and in storms, down Inca roads and up rivers guarded by Indians with curare-tipped arrows, across bridges and canals natural, built, and imaginary, ever guided by those instruments of science, compass and chronometer, the signposts of space and time that allow Humboldt, and us, to navigate through the confusion of jungles and mountains and politics, our True North ever before us in the perilous passage from chaos to Cosmos.

Humboldt's passage was nearly interrupted before he got started. First the winds threatened to push the *Pizarro* onto the harbor rocks, then a few days out they sighted an English convoy in the distance. Everyone ashore had predicted they would be captured and jailed in Lisbon (Humboldt learned that the landlocked were always predicting disaster), but the *Pizarro* proved lucky once again, and after a tense start made the crossing without incident. Humboldt went straight to work, delighted to find he was "unaffected by seasickness" and grumbling that the ship's wartime blackout kept him from laboring after dark. Since he was on the ocean, he studied the ocean, taking measurements and mining his sources. Its currents, he realized, ran together to produce one "enormous sea river" that made one great "circuitous sweep" along the equator from Africa to America, up the coast then eastward to Europe and southward back to Africa. It would take, he estimated, a molecule of water two years and ten months to come full circle. This great river of the ocean carried New World tropical fruits all the way to Ireland and fetched up barrels of French wine from West Indian wrecks onto the beaches of Norway. It had floated strange bamboo trunks and two mysterious corpses onto the Azores, inspiring Columbus to imagine land to the west. Columbus had launched himself out onto that same great river, and now the *Pizarro* followed in his wake, the ancient currents washing men, like flotsam, around the world. Shallower crosscurrents ran against each other over the deeps, and Humboldt wrote that "our imagination is struck by those movements which seem to penetrate each other, and by which the ocean is continually agitated." Adrift upon deep currents, agitated by shallower ones, Humboldt patiently practiced his measurements, self-orienting in a turbulent world.[3]

The currents delivered them first to the Canary Islands, along an ocean river so reliable that sailors called it "the road of Santa Cruz." Here Humboldt lingers for over two hundred pages, in a book-length essay that abridged editions usually skip, for these islands just off the northwest coast of Africa are part of the "Old" World, cultivated for centuries. What could be of interest here? Yet this island essay is Humboldt's portal to the continental whole. The Canaries were the threshold not only for travelers leaving Europe, but for European imperialism on its passage out across the oceans. Santa Cruz on the island of Tenerife was "a great caravansary, on the road to America and the Indies."[4] While of the Old World, it pointed to the New, offering Humboldt a laboratory of geological and historical change. His opening essay offers an environmental history of how humans and nature—geology, climate, plant physiology—interrelate and co-create each other, in an island microcosm that he will soon scale up to the macrocosm of America.

For Tenerife, too, had been a "new world" once, back in 1402, when Europeans arrived and encountered the Guanches who had lived there for millennia. At first contact the sovereign of the Guanches had welcomed the Europeans "with the same hospitality that Cortez found in the palace of Montezuma," only to be met with the same "coward treachery." For where, asks Humboldt, are the Guanches today? Their mummies alone survive. The Spanish and Portuguese, using Christianity as a pretext, raided them for slaves. After nearly a hundred years of fierce resistance the remnants surrendered, and soon the few survivors perished of disease. The grim fate of the Guanches prefigures that of the American Indians downstream and, Humboldt fears, of the Tahitians on the other side of the planet. History was repeating itself: before Columbus, it was the Guanches of the "Fortunate Isles" on whom Europeans projected their fantasies of Arcadian innocence, flattering themselves with the idea that in some faraway place, "infant societies enjoy pure and perpetual felicity." Which felicity they then destroy. "Like the Guanches of the Canaries," the Indians of Cuba are long gone, and as the Guanches once "groaned under the yoke of feudal government," so do the Tahitians today, in a state of warfare sanctioned by religion that Humboldt was about to see perpetuated, once again, on the Orinoco.[5]

One will not find tropical paradises in Humboldt. He loved the Canaries and couldn't believe how rich and "untrod" they were for the natural historian, and he would happily have stayed on for months. But the four English ships anchored in port argued otherwise. Four or five days and we must move on, said the captain. That gave Humboldt just time enough, for Forster, trained on the Cook expedition to the South Pacific islands, had taught him exactly what to look for, how to make use of island geography to link the particularities of place with its people—to which Humboldt added his historical sensibility. Richard Grove argues that for the generation preceding Humboldt, "the island acted as a symbolic vehicle" for ideas about nature and environmental change, becoming "the central motif" of an emerging environmentalist critique of European imperialism, born of the marriage between exacting observations of positivist science and the Romantic, utopian cult of the tropical island. Ground zero was Mauritius, home of the dodo, that icon of human-caused extinction, where three naturalists, Commerson, Pierre Poivre, and Bernardin de Saint Pierre, converged to develop groundbreaking new theories. Humboldt knew well the writings of all three. He had shed tears of frustration on Commerson's ship, and he and Bonpland wiled away long hours on the Orinoco by reading aloud to each other from Saint Pierre's novel *Paul and Virginia*, the tragedy of two young people who grow up together on a tropical isle in a utopian state of nature, who are riven apart by the betrayals of French society and die in misery, longing for each other.[6] Both lured and repelled by the paradise island fantasy, Humboldt deployed it to translate island economies to the mainland, thence to the globe itself, imagining the entire planet as an island. His visionary leap propelled environmental discourse from the eighteenth-century margins to the nineteenth-century mainstream.

Thus his brief stop became both a portal to his mainland project and a rehearsal of his mature method. On Tenerife Humboldt practiced using the earth and its indigenous people as a vehicle for the critique of European society, extending Kant's political argument from humans to nature. In 1785, Kant had published a short essay, "To Eternal Peace," in which he built on his ethical argument, the "categorical imperative" that all human beings must be treated as ends, not means, to the world order. Simply put, all people are, and must remain, free and equal, or none will be, for if any are oppressed than freedom no longer exists as a universal principle. Kant's "Cosmopolitan or World Law" of "universal hospitality" enjoins that every traveler has the right not to be treated with hostility, and conversely, no traveler has the right to turn their visit into a conquest. The so-called civilized nations commit terrible injustice, for they "simply identify visiting with *conquest*. America, the lands of the Negroes, the Spice Islands, the Cape of South Africa, etc., were countries that belonged to nobody, for the inhabitants counted for nothing." Or as Kant says in an aside worthy of Melville, "The difference between the European savages and those in America is primarily this, that while some of the latter eat their enemies, the former know how better to employ their defeated foe than to feast on them."[7]

In a global world, where humans meet and mingle across the whole face of the earth, no place is an island, and "a violation of law and right in one place is felt in *all* others." Kant sees this not as a utopian dream but as a "necessary completion" of constitutional and international law, a new cosmopolitan social order that represents the next step in the progress of human history. It is Kant's law that Humboldt applies consistently, even as he sees it violated on all sides. All peoples are and have the right to remain free, Indian tribes no less than German principalities. Nature herself decrees that if any are unfree, than none are truly free. One hears the wistfulness in his voice when he honors the capirote, the Canary Island bird whose "heartsoothing song is unknown in Europe" and whom "no effort has been able to tame, so sacred to his soul is liberty." Humboldt hears the singer, but so elusive is the bird that he can never see it clearly. As ever, when human perfidy and greed is too much with him and freedom seems a dream he can hear but not see, Humboldt sings to keep his courage up, and always in such times he sings of nature, taking the broad view that will put the lapses of human history into the hopeful perspective of deep time.[8]

Here on Tenerife, obtaining the perspective of deep time meant climbing El Piton, the volcano that looms over the island's cultivated fields and forests. Humboldt's emotions are mixed, for just as the beauty of the island is marred by the feudal institutions which destroyed the Guanches and still weigh on its inhabitants, so the beauty of the volcano casts a shadow on the green and fertile country: "The history of the globe instructs us, that volcanoes destroy what they have been a long series of ages in creating." Creation and destruction become the themes of his ascent. He sees with a double vision, present and past: the wheat fields of the high plains remind him that the Guanches once cultivated grains, wheat and barley, and drank goat's milk, an insight that proves them to have been from "the old continent," for grain and dairy products were unknown in the Americas before the arrival of Europeans. Place-names also evoke a violent past: the name of the smiling hamlet of Matanza means "butchery, or carnage," and marks the place where the Guanches won a rare victory over the Spaniards. The victory was brief: the Spaniards struck back, took the Guanches captive, and sold them "as slaves in the markets of Europe." The botanic garden points to the future, for here tropical plants were being naturalized to European climates; and here Humboldt steps into his own future as well when he sees at last the far-famed Dragon Tree of Orotava, of which he had dreamed as a child. One of "the oldest inhabitants of the globe," old even in the days of the Guanches, the mighty tree still bore flowers and fruit in the eternal youth of nature. How on earth, Humboldt wonders, did this East Indian tree find its way here? Could the Guanches have once held commerce with Asia?[9]

From gardens and croplands they climbed through juniper and pine woods to a barren plain of choking pumice dust and obsidian blocks spewed by the volcano. Above this plain they set up camp without tents or cloaks and spend a short and miserable night in freezing temperatures. At 3 a.m. they struck out for the summit, scrambling up the steep lava core on all fours until they stood on the sulphurous crater's edge, then stepped inside. While their hands and faces froze, the hot soil burned their boot soles, and the acid fumes stung their eyes; those who sat down to rest stood to find holes burned in their clothes.

Why climb the mountain? Humboldt offers two reasons. First, for scientific research into its fascinating volcanic phenomena. But also to revel in "the picturesque beauties, which it lays open to those who are feelingly alive to the majesty of nature." Yet such feelings defy language, and repeated exclamations of admiration are boring. Better, then, to describe the mountain's unique, individual features, and so convey the source of those emotions. Here on El Piton, those features include the mountain's height, which is just right: climb too high, and all one sees is blue and vaporous vagueness. But the peak of El Piton is just high enough to reveal "beautiful contrasts of form and colour," from the sea to the coast "decked with villages and hamlets," seen through veils of wind-whipped mist. "From the summit of these solitary regions our eyes hovered over an inhabited world," enjoying the contrast between barren rock and smiling fields, and beholding plants divided by zones from the tropical base to the alpine height.[10]

Hovering just high enough to see clearly, Humboldt's restless mind turns on itself in frustration. He had climbed to find answers, "to form a precise idea" of volcanic geology, but instead of order and pattern, he finds confusions and puzzles, endless variations that lost themselves in obscurity. "Notwithstanding the care with which we interrogate nature, and the number of partial observations which are presented at every step, we return from the summit of a burning volcano less satisfied, than when we were preparing to go thither." Where in these tumbled fragments was the whole? Humboldt's aesthetic sense told him there was one. To his imagination the sweeping view from sea to shore to mountainside made a deeply thrilling unity. But he needed scientific unity too, to see not just one single island volcano but "volcanicity," an intellectual view of "Nature in the universality of her relations" that would be, in its way, just as sublime.[11]

Before leaving Tenerife, Humboldt sketches just such a view, moving from the deep geology of the island's volcanic formation and composition (and the chance the sleeping volcano might awaken), to the cooling of molten matter, "gradually adorned with vegetable clothing," to the distribution of plants and the look of the face or "physiognomy" of vegetation to the human eye, to the history of man "and to those fatal revolutions, which have swept off whole tribes from the face of the earth." Here he interjects a long essay on the Guanches and on their conquerors, now become "a moral, sober, and religious people" of a roving and enterprising nature. Dispersed across the Spanish colonies, both "mountaineers and islanders," the Canarian's true strength is revealed not on their island home but in Caracas, or the Andes, or the Philippines, where they have scope to develop their energy and activity. Soon, Humboldt hopes, the "Fortunate Islands . . . will inspire the muse of some native poet."[12] What Humboldt sees from the mountaintop is more than just a view. Every view reveals a history, from the deep time of geology to the shifting migrations of nations to the personality of the viewer, primed by his time to see beauty, meaning, and hope in the landscape, to read beyond the surface to the dynamic play of crosscurrents surging in the depths.

The Casiquiare Crossing

The problem of the navigator is finding where you are. Humboldt fussed endlessly with his quadrants and sextants and chronometers, halting every few hours to break them out and pinpoint, yet again, precisely where he was on the globe. In some ways this was his single most important task, for otherwise you could ever know where you stood. The maps were all wrong,

sometimes wildly off. This had consequences. The *Pizarro* had been bound
for Cuba, but then a deadly fever struck. Before the epidemic carried them
all off, they must hurry to the coast of South America. But where were they?
And where was the coast? Humboldt had one answer, the captain another.
The sailors smiled when Humboldt predicted landfall the next day. But early
the next morning, there it was. So Humboldt, at least, did know where they
were—but what help was that when the charts were wrong? For days they
sailed through dangerous and unfamiliar waters until they dropped anchor
off a flat coast no one recognized. Nor did its inhabitants recognize them.
Two canoes, each hollowed from the trunk of a single tree and carrying eighteen tall and statuesque Indians naked to the waist, approached warily. The
sailors hailed them in Spanish, and the Indians, relieved the ship was not
English, came straight on board. European warfare kept even the Indians of
this distant land on edge.

It was the Indians who oriented them. The *Pizarro* was, they said, anchored
off Coche, an uninhabited island where they were searching for timber. Only
the English used this channel, which explained why they had taken the *Pizarro*
for an enemy ship. The Spanish normally sailed farther north, into the port
of Cumaná, which was where the master of one of the canoes offered to guide
them. Thus Humboldt met Carlos del Pino, his first New World friend, a
Guayqueria Indian "of an excellent disposition, sagacious in his observations, and led by an unceasing curiosity" who would be their guide for the
next sixteen months. They spent their first evening together on deck, Carlos
regaling the Europeans with tales of the plants and animals of his country.
"What riches to our eyes were contained in the canoes of these poor Indians!"
exclaimed Alexander. Carlos handed them coconuts and colorful fish, an ar-
madillo, plantains, and calabash fruit, all proof they were truly in the tropics.
Humboldt had arrived at last.[13]

Disorientation and reorientation would for many months be Humboldt's
constant theme. On the Canaries, the fact that the indigenous people were
extinct had made them relatively easy to encompass. Here on the mainland
Indians were most certainly not extinct, but present and lively, presenting a
real challenge. Humboldt had been led by the theories of Buffon and others to expect a weak and degenerate race, but the Guayquerias were strong,
smart, knowledgeable, and fully at home in their world. And as he would
soon learn, this was but one tribe among hundreds, each one distinct in appearance, language, and manners, opening a bewildering new universe of human possibility. Nature bewildered him too: the light, the trees, the pelicans
and egrets and flamingoes, millipedes eighteen inches long. Everything was

gigantic, on a continental scale, the vegetation so crowded that it seemed "that the earth, overloaded with plants, does not allow them space enough to unfold themselves." Overwhelmed, Humboldt and Bonpland hardly knew where to begin. As Humboldt wrote to his brother, "Hitherto we have been running about like a couple of fools; for the first three days we could settle to nothing, as we were always leaving one object to lay hold of another. Bonpland declares he should lose his senses if this state of ecstacy were to continue. . . . I feel sure that I shall be very happy here."[14]

Even the stars were strange. They had left the North Star behind, but in its place was the beautiful Southern Cross, marking time as faithfully as its northern cousin marked the compass. Everyone knew at what hour of the night the vertical cross began to incline—"How often have we heard our guides exclaim . . . 'Midnight is past, the Cross begins to bend!'" Their reading in Saint Pierre's island tragedy helped to make the strange familiar: "How often those words reminded us of that affecting scene, where Paul and Virginia . . . conversed together for the last time, and where the old man, at the sight of the southern Cross, warns them it is time to separate."[15] In this alien land it was the rocks that felt like old friends, for geological causes had everywhere, Humboldt theorized, produced the same effects. Nothing made the mining engineer of old feel more at home than laying hands on some gneiss or a granite outcropping, unless it was breaking out his familiar scientific instruments and taking measurements. So out they came, and the novelty of viewing the moon through a telescope or making a frog's legs jump with a jolt of electricity drew around the travelers the curious townspeople, who pestered them with questions. Each side seemed odd to the other, and so the new community precipitated by the visitors set about its work, exchanging information and viewpoints.

The instant they stepped ashore, Bonpland collected his first specimen, a seaside shrub named *avicennia tomentosa*. The captain of the *Pizarro* wanted to conduct them immediately to the governor, but first things first: Carlos led Humboldt and Bonpland to his garden to show them a tree whose trunk had, in only four years (so he claimed) reached nearly two and a half feet in diameter. (Humboldt is politely skeptical.) Then it was the governor's turn. Crossing the Indian suburb, all regular streets and small, neat houses freshly rebuilt since the last earthquake, they made their way into Cumaná's center to pay their respects to Don Vincente Emparan, governor of the province of New Andalusia, who was duly impressed by their royal passports. They were lucky: Emparan, a lover of the sciences himself, was pleased to open his province to the travelers and smooth their way, one reason they stayed and trav-

eled in New Andalusia for a year and a half. Before nightfall they had rented
a spacious house with a good view of the night sky.

Their happiness was troubled by one serious flaw: their house overlooked
the Cumaná slave market. Every morning they were forced to witness young
men polishing their skin with coconut oil for purchasers who pried their mouths
open to examine their teeth "as we do those of horses in a market." Humboldt hoped that such painful scenes were vanishing into the past, since the
dehumanizing practice of slavery, once carried on with "desolatory activity,"
had recently been discouraged by Spanish attempts at reform. But the reforms
were not enough: at one point a priest full of lawsuits and metaphysics insisted
on joining them to the next town, where he was due to minister last rights to a
slave condemned to die for leading a rebellion. The "meagre" and "petulant"
little priest harangued them the whole way on "the necessity of the slave trade,
on the innate wickedness of the blacks, and the benefit they derived from their
state of slavery among the christians!" What hope could such a priest offer a
condemned rebel leader? However mild the Spanish legislation seemed from a
distance, the reality on the ground was barbarous, and justice for the enslaved
Africans unknown.[16]

Like a cat exploring an unfamiliar house, they began by making short and
local excursions, orienting themselves to their new environment. Their first
major expedition took them inland to explore the mountains and the Catholic
missions where the Chayma Indians lived. The contrast startled Humboldt to
attention. The coastal Guayquerias, "a civilized tribe of skilful fishermen,"
were descended from the first peoples to meet Columbus. (In a typical error
of translation, Columbus had named them not for themselves but for their
harpoons, or *quaike*'s.) The original tribe had nearly disappeared, taking with
them their language, and their descendants, who had allied with the Europeans, no longer spoke any language but Spanish. Quite in contrast to these
tall, assimilated, and town-dwelling coastal Indians were the hill tribes of the
Chaymas, a peaceful people who had not been subjugated until the mid-1600s.
Humboldt describes them as short, around five feet in height, thickset and
broad-shouldered, the men beardless, with deep, dark, long eyes and high
cheekbones that set him thinking about Asian migrations. The conquistadors'
wars of enslavement had forced the Indians off their land and "crushed the
seeds of rising civilization," until religion moved in "to console humanity for
a part of the evils committed in its name." The displaced Indians were gathered into small communities called missions, where they were regulated by
European standards of decorum: forced to live monogamous lives in huts arranged in rows, to wear cotton tunics (which, Humboldt noted, shamed them

and which they took off whenever out of sight of the missionaries), forced to learn Spanish, forced to regulate their lives by the clock and to labor growing identical crops for export. Both the law and the church treated Indians as children, teaching them an infantilized catechism and giving them little protection and much prejudice. The results, Humboldt thought, were catastrophic. Their population has grown, "but the sphere of their ideas is not enlarged. They have progressively lost that vigour of character, and that natural vivacity, which in every state of society are the noble fruits of independence. . . . They have been rendered stupid, by the effort to render them obedient." From generation to generation, the missions "enchained" the minds of the Indians and halted their intercourse with other nations. As for the Indians, their "gloomy and reserved looks" show "that they have not sacrificed their liberty to their repose without regret."[17]

In a country without inns, the missions took their place, lodging travelers overnight in humble buildings called, ironically, "the king's house." Beyond the towns the secular government ended and the mission system took over. Humboldt's passports were useless here, and without the letter of recommendation from the governing friar at Cumaná, the travelers could never have sojourned anywhere for more than a night. Some missions, he noted, were quite well run; he was struck by the extreme neatness of the first one they visited, by the "grave and taciturn air of the inhabitants" in contrast with the hilarity of the corpulent monk who welcomed them, laughing at their scientific goals and declaring he'd rather have a good dinner of beef. Indeed, he would not let them leave until they had witnessed the Indians butchering the cow he had just purchased, an operation Humboldt found "disgusting." He was happier with the *conucos* or community gardens which the natives were obliged to work in daily, where they raised exotic crops like coffee and indigo along with maize and other foodstuffs. He was especially pleased with the Capuchin monks, for they allowed the Indians to share in the proceeds thus raised. Months later, after Humboldt had studied scores of missions, he offered the Catholic church some unsolicited advice: train the missionaries for their assignments, ditch the useless rules appropriate only to a European convent, let the Indians enjoy the fruits of their labors, allow them more freedom, then step back and watch their civilization rise naturally. One unwary Father asked Humboldt to write a testimonial affirming the good order of his mission. Humboldt demurred, and instead offered him an earful of the above advice. "The president seemed to listen to me placidly. I believe, however, he would have wished (no doubt from zeal for the sciences), that those who gather plants, and examine rocks, would renounce their indiscreet interest in the copper-coloured race, and in the affairs of human society in general." So, sighed Humboldt, does authority always wish, whenever it thinks itself less than firmly seated.[18]

Far from renouncing his "indiscreet interest" in Indians, Humboldt found it increasing, for it struck him that their manners, languages, and origin were of central importance to philosophy. To enter the jungle was to approach, not recede from, the question of "the history of the human race. As we advance into the interior of the country, these subjects will become more interesting than the phenomena of the physical world." And indeed, his narrative is increasingly dominated by human rather than natural history. Humboldt begins his researches into the peoples of South American with a long and fascinating ethnography of the Chayma Indians, the first he was able to study in any depth. Yet he apologizes for his shallowness. Merely seeing Indians is not enough; one must talk with them. This means everything turns on the question of language, but not knowing their language, and staying with them for only a few days, he can scarcely claim to have "penetrated their character," only to have jotted a few notes from conversations and observations. (To future anthropologists—a field that, of course, did not yet exist—he recommended that the responsible student of South American interior tribes must start by learning at least ten different major languages.) When he turned to the monks who had lived with the Indians for so many generations, he found them, maddeningly, no help at all. They lacked any interest in Indian languages, resulting in the most "whimsical" misunderstandings: among the Chaymas, Humboldt witnessed a monk "violently agitated in proving, that *Infierno*, Hell, and *invierno*, winter, were not the same thing." Since to the Chaymas winter means the season of rains, "the *Hell of the whites* appeared to them a place, where the wicked are exposed to frequent showers." The monk's harangues were pointless.[19]

It struck Humboldt that Indians generally had great difficulty learning Spanish, though they picked up other, even very different, Indian languages easily. This, not any defect in intelligence or memory, was why he found them so difficult to talk with. The problem, he thought, must lie in the radical differences in the deep grammatical structure of Indo-European versus American languages, which perhaps had a hereditary base. The implications went quite deep: for one thing, Humboldt concluded, it had been a tragic mistake to insist on Spanish as the lingua franca of the interior. The monks should have used one of the major Indian languages instead. Indians would have learned it easily and diffused it widely, peacefully uniting isolated and widespread tribes into a single pan-Indian identity: "The Indian, in preserving an American idiom, would retain his individuality, his national physiognomy."

The vision Humboldt outlined was, of course, shocking: that Indians should be encouraged to realize their separate identity *as Indians*, and develop their distinctive "individuality" into a united nation—one that could, presumably, unite politically as well as culturally. The Incas had already done this once, as Humboldt pointed out later when he recurred to this idea. When the conquistadors destroyed the network Incas had created, the damage had been deep, but was it fatal? Could a new, modern Indian empire perhaps have been reborn out of the ashes of the old? Too late, to be sure—but what if Tupac Amaru had won? Imagine a pan-Indian movement rolling out of the Andes and across the Americas! Unimaginable, perhaps—except to Humboldt, riding mission to mission in search of clues to the sweep of history.[20]

It was harder to meet those Indians who lived beyond the missions, but Humboldt understood that many Indians maintained their independence. They were called, of course, "savage," a word he explicitly rejected because the difference it indicated, between so-called "reduced" Indians living in the missions and those living in freedom, "is often belied by facts." Humboldt observed that the supposed "savages" often lived lives that were highly cultivated, and Europeans who thought of them as nothing more than roving hunters were wrong: "Agriculture existed on the continent long before the arrival of the Europeans," and existed still in lands where missionaries had not yet penetrated. Thus to align "*Christians, reduced,* and *civilized*" as synonyms against "*pagans, savages,* and *independent*" was to seriously misrepresent the reality of South American nations. Not only did independent Indians often lead quite civilized lives, mission Indians were often as little Christian as "savages" were idolaters; both secretly preferred the worship of nature. As one missionary reported to Humboldt, it was hard to convey the concept of "temple" to the Indians of the forest: "'Your God,' they said to me, 'keeps himself shut up in a house, as if he were old and infirm; ours is in the forest, in the fields, and on the mountains of Sipapu, whence the rains come.'"[21]

Humboldt resisted describing Indians as "noble savages," too. Like Cook in the South Pacific, he observed that natives ranged from handsome, well-socialized, and friendly to misshapen, poorly organized, and hostile—a confusing array of differences that knocked Rousseau into a cocked hat and put into place a new question of how to account for such variation, given—as Cook, like Humboldt, believed—the family unity of the human species. While both Cook and Humboldt used environment to account for difference, Humboldt the philosopher went farther, to consider the steep inequality between cultivated Europeans and—not "savages," note—but "men in a savage condition." At the turtle camps on the beach of Pararuma, where the travelers came across a motley gathering of Indians, white traders, and monks,

Humboldt's thoughts led him to write a passage often quoted to prove his innate racism. This particular group of Indians struck Humboldt as singularly unimpressive:

> How difficult to recognize in this infancy of society, in this assemblage of dull, silent, inanimate Indians, the primitive character of our species! The savage of the Oroonoko appeared to us to be as hideous as the savage of the Mississippi [as described by Volney]. . . . We are eager to persuade ourselves, that these natives, crouching before the fire, or seated on large shells of turtles, their bodies covered with earth and grease, their eyes stupidly fixed for whole hours on the beverage they are preparing, far from being the primitive type of our species, are a degenerate race, the feeble remains of nations, who, after having been long dispersed in the forests, are replunged into barbarism.[22]

What is seldom remarked is Humboldt's framing irony. He poses this as a problem in perception: how was the philosopher to see these apparently "dull" and "hideous" people? Were they humans arrested at a "primitive" stage? Or had once-advanced peoples been dispersed into the harsh environment of the forests, to "degenerate" accordingly? Humboldt states these alternatives not as truths but as propositions equally hard to believe: given that the former is "difficult to recognize," the Europeans are "eager to persuade [them]selves" of the latter. Humboldt himself does not endorse either answer. What he does endorse is the third alternative, barely suggested above, that he is witnessing an "infant society," a direction he pursues by launching immediately into a long essay on the Indian's use of paint in preference to clothing. Here Humboldt presents an elaborate indigenous culture of the body that proves creative adaptation to local conditions and a native sense of beauty.

Agriculture, as Humboldt observed repeatedly, was common in the jungles, but what was not common were towns. This, he pointed out, created the illusion that the wilderness was empty. Not so, he answered: actually, dispersed across the land were more Indians than anyone suspected, more than six million in both Americas. While those of the temperate zone of North America had every reason to dread the whites, who were actively pushing them off their lands, those in the South American tropics where the land and crops are twenty times more productive needed only a tiny bit of ground to flourish. This encouraged small clusters of Indians to live in isolation from one another, in deep forests that hid their numbers, their cultivated plants blending with the wild ones. Thus in the tropics, even densely populated areas still bear "a savage aspect" and man appears, "not as an absolute master, who changes at his will the surface of the soil, but as a transient guest, who quietly enjoys the gifts of nature." Here Humboldt is torn. The romantic who sighed after *Paul and Virginia* wants to see these isolated families cultivate their quiet and gentle lives in peace, but the Enlightenment physiocrat wants to see them meet and mingle, exchange ideas, spread a network of art and commerce across the Amazon and Orinoco rain forests. But Humboldt did not see how this could happen without the agency of the ever-encroaching missions. White traders would inevitably crowd in behind, pushing the Indians out or costing them, like the Guayquerias, their native language. "Such is the progress of civilization," Humboldt worried, slow but sure. And deadly.[23]

It did puzzle him, though, that the missions, for all their uniformity—imposing the same crops, the same huts arranged in the same order, the same organization of daily chores, the same beliefs—nevertheless failed to produce the same Indians. Instead, most missions were a heterogeneous collection of families from different tribes, of different colors and features, speaking different languages. These chaotic and artificial societies proved difficult to govern. He noted that the monks had better success when they united nearby villages that shared the same language. Then one saw true diversity emerge, as the language was preserved and perpetuated. As his brother Wilhelm was teaching the world, who we are is reflected in the words we think and speak, and in turn, the words we speak in and think with react on our ideas and feelings. Since different languages both reflect and construct different identities, the more the Chaymas, Caribs, and Tamanacs preserved their unique languages, the more they retained their "natural physiognomy." "It is this intimate connection between the languages, the character, and the physical constitution," Humboldt concludes, "which maintains and perpetuates the diversity of nations, that unfailing source of life and motion in the intellectual world." The trick lay in perpetuating difference without creating conflict, encouraging diversity yet maintaining intercommunication. Humboldt was drawn by the legend of the marvelous Casiquiare, an impossible river that flowed two ways, joining the great and independent river systems of the Orinoco and the Amazon, opening a channel for the crossing of watersheds, races, nations. Could one in the same way connect Old World and New, European and Indian, without reducing either to the other? If only he had known that indigenous earthworks had reengineered the Amazon, that the Casiquiare itself may have been opened by Arawak labor. What a field of vision this would have opened to his quest! Could he open a Casiquiare exchange of minds, of cultures, linking the diversity of nations?[24]

It was time to find out. Humboldt and Bonpland spent another month in Cumaná getting everything ready for their Casiquiare voyage. Moments of disorientation returned: There was the night the sky rained stars, in a meteor shower that Humboldt later learned was visible in Brazil, and Europe too, and that had frightened the Eskimos in Greenland. How high they must have been, to be seen across half the planet! Then there was the earthquake. They would live through so many earthquakes in Quito, the heart of volcano country, that a good shaking wouldn't even roust them out of bed, but this first one felt to Humboldt like a betrayal. One instant shattered the fond illusion of a lifetime in a stable and motionless earth, like a painful awakening from a dream: "We feel, that we have been deceived by the apparent calm of nature; we become attentive to the least noise; we mistrust for the first time a soil, on which we had so long placed our feet with confidence." Their confidence in men was shaken too. One night while they were walking on the beach a tall, dark-skinned man attacked them, clubbing Bonpland across the temple hard enough to level him to the sand. In the struggle that followed, their attacker drew a knife, just as some fellow beachwalkers came to the rescue. He turned out to be, apparently, not criminal but crazy, and Humboldt and Bonpland were relieved that he managed to escape a few days later rather than suffer in prison for seven or eight years awaiting trial. Bonpland recovered, but it was months before he could bend over to pick a plant without dizziness. Such a crowd of adventures and impressions made Humboldt feel, when they came to leave, that they had lived there not for five months but for many years, and he wrote that "Cumaná and its dusty soil are still more frequently present to my imagination, than all the wonders of the Cordilleras." It was almost like leaving home.[25]

They sailed west along the coast to Caracas, from which they would ascend the coastal mountains, drop down into the *llanos* or barren plains and travel south until they reached a tributary of the Orinoco river, thence upriver, across a portage to the Rio Negro in the Amazon basin, downriver to the Casiquiare and back to the Orinoco—full circle and back out to the coast, a journey of nearly six months and 1,725 miles. Caracas, their launching point, was a capital city and one of the great ports of the Caribbean Sea, which Humboldt thought of as the American Mediterranean. Opening out to North America and beyond to Europe, it was a cosmopolitan city of great wealth and sophistication, though Humboldt was astonished to find no printing press (the first one was established in 1806), while in the United States every town of a few thousand had its newspaper; and he was disappointed to find not one person interested in natural history.

What he did find was a keen interest in politics. They had encountered this in another coastal town to the east, Cariaco, where for the first time "we heard the names of Franklin and Washington pronounced with enthusiasm," mingled with complaints on the sorry present and bright future of New Andalusia. It was all vague and unfocused, and "foreboded, as yet, nothing hostile and violent, no determinate direction"—precisely the time, Humboldt thought, for the colonial government to engineer reform in order to avert the "coming storm." On the road into Caracas he came across a group of citizens "disputing on the efforts towards independence" then unfolding. José María España had just been publicly executed in Caracas for trumpeting an independent republic of Venezuela, modeled on the United States, which would open free trade, abolish slavery and Indian tributes, and call for harmony among whites, Indians, and blacks. España had been hung, his body mutilated, his limbs displayed on pikes on the highroads as a warning, and his wife still "groaned in a prison" because she had given her husband asylum. Humboldt was disturbed by the ugliness of the passion revealed by the roadside dispute: how much the mulattoes hated both free Negroes and whites; how difficult it was to get slaves to obey; how obscene was the wealth of the monks. Discussion ended when a thick fog sent them all to a nearby inn, where the innkeeper admonished them not to talk politics in public. "These words, uttered in a spot of so wild an aspect, made a lively impression on my mind," recalled Humboldt, an impression often renewed in the high Andes. For in South America, the mountains were not refuges from politics, as they were in Europe; even "thither men carry with them their political dissentions, and their little and hateful passions."[26]

It seemed to Humboldt that Caracas was poised on the brink of change, split between two generations, one living in the past, treasuring their prejudices and privileges, another occupied by the future, by "new habits and ideas." Intersecting this generational split were deep divides between Creoles who had come to America only recently, and the deeper-rooted descendants of the conquistadors, yet Humboldt noted a new racial solidarity emerging as both classes proudly declared themselves "white," uniting in opposition to all peoples of color. Humboldt worried that such deep divisions splintered the hopes for peaceful reform, and indeed, the coming storm was not averted. Years later Humboldt looked back on their departure from Caracas with sorrow, for "our friends have perished in the sanguinary revolutions, which have successively given liberty to those distant regions, and deprived them of it. The house which we inhabited is now a heap of ruins." Even as he wrote, a new city was slowly rising on the grave of the vexed but hopeful city he once knew.[27]

Of course they didn't know this at the time, and Humboldt's band of adventurers departed cheerfully, on 2 February 1800, making their way inland through the beautiful cultivated valleys of Aragua, where their wealthy Caracas friends welcomed them to their fine plantations. Humboldt lingered for a spell on the shores of Lake Valencia to solve the mystery of the falling lake level. The locals hypothesized an underground outlet, but Humboldt, primed by studies just emerging from French and British island colonies, recognized that the real cause was environmental. Within the last half century the forests surrounding the lake had been destroyed, the plains cleared, and indigo planted. Trees attracted a cool and misty atmosphere and forests absorbed rainwater. When the forests were cut, the rains ran off in torrents, eroding the hillsides and bearing down the loosened topsoil in devastating floods. The plains, once stripped of the vegetation that had retained ground moisture, evaporated back into the air what little water remained. *Et voilà*: a barren and eroded landscape surrounding a shrinking lake. Humboldt would explain this sequence a second time in *Political Essay on the Kingdom of New Spain*, for he saw it replicated in Mexico. The lesson was clear, if anyone would listen: "By felling the trees, that cover the tops and sides of mountains, men in every climate prepare at once two calamities for future generations; the want of fuel, and a scarcity of water." His words had a powerful impact in the French and British colonies, particularly in India, though it would take decades for them to reach the United States.[28]

From Lake Valencia Humboldt's party traversed the mountains and dropped down into the radically different landscape of the llanos, a bleak region that Humboldt was glad to see, and even more glad to leave behind. The land stretched out in all directions flat as the ocean, to a "boundless horizon" that became oppressive: "Infinity of space, as poets have said in every language, is reflected in ourselves; it is associated with ideas of a superior order; it elevates minds, that delight in the calm of solitary meditation." It also, in the dry and dusty steppes of the llanos, "dejects the mind by its unchanging monotony." Across the plains roamed herds of cows and tough little horses, many of which drowned when the annual floods turned the metaphoric sea into an actual one. It fascinated Humboldt that there were no native species comparable to these, which meant the Indians had skipped the "pastoral" stage and gone straight from hunting to agriculture, upsetting the standard theories of the development of civilization.[29] Herding these animals were men who rode over the savannahs, naked to the waist, living in the saddle—precursor to the cowboy. Venezuela was wilder than even the Wild West: roaming the plains on horseback were also bands of desperadoes, like Bedouin tribes, notoriously violent robbers who had escaped the coastal prisons where suspects waited nearly a decade for trial. Likely the man who had attacked Bonpland on the Cumaná beach was now among them.

At the far end of the llanos they reached the Capuchin mission of San Fernando de Apure. An old farmer offered to guide them overland to the

Orinoco, but sick of riding horseback, they chose the river route, and rented a large sailing canoe called a *lancha* managed by five Indians, four to row and one to pilot. They built a little cabin on the stern, filled it with a month's provisions—chickens, eggs, plantains, cassava, chocolate, sherry, and oranges, fishing gear and firearms (though the humidity would soon render these useless), and casks of brandy to barter with the Indians. The governor's brother-in-law, Don Nicolas Sotto, happened to be in town. Having just arrived from Spain, he was exploring the country; could he join them? Humboldt and Bonpland welcomed him aboard, and Humboldt noted that "his amiable disposition and gay temper" often helped them forget the miseries of the next two and a half months.[30]

They would soon look back on the confined lancha as the Hilton of canoes, when a few days later they reached the beach of Pararuma. Here were gathered Indians, missionaries, and white traders, for it was the height of the turtle egg harvest. The lancha was far too large to navigate the rough waters upriver, but fortunately one of the monks was willing to sell them a canoe, and Father Bernardo Zea, a missionary who knew the river well, agreed to guide them on his way to the Rio Negro, where he hoped to regain his health.

Crowding everyone aboard took some doing. Their new canoe, a hollowed tree trunk, was only forty feet long and three feet wide, and so unstable that whenever anyone wanted to move, the rowers had to lean to the opposite side. Really it was a "new prison," on which they would be trapped for two months. A bit of latticework created a shed on the stern where four people could lie down under a close roof of leaves, but it didn't extend far enough to cover one's legs, so the sleepers got half-drenched during storms. In front were the Indian rowers, seated two by two, and their pilot, singing sad and monotonous songs as they labored. Trunks and scientific instruments were stowed below the shed, the safest place, but to reach anything at all required beaching the canoe and unpacking everything. All about the canoe were hung cages for various small birds and animals, joined by free-roaming monkeys, parrots, and a toucan—a menagerie which grew steadily. Eventually at least one passenger thought it all too much: Humboldt recorded that "Father Zea whispered some complaints at the daily augmentation of this ambulatory collection." One can understand why: whenever it threatened to rain "the little monkeys . . . went in search of father Zea, to take shelter in the large sleeves of his Franciscan habit."[31] Tucked in amongst the rest were Carlos del Pino, José de la Cruz, and Zerepe the Indian translator, at least twelve men in all, plus one large mastiff dog who went for romps on shore and every night in camp howled at the jaguars.

This little ark would take Humboldt through the wildest country he ever experienced. He noted that along the Cordilleran roads of the Andes there were houses every mile or so, but here, they went for days, even weeks, without seeing a single boat, not even a fisherman's canoe—an absence Humboldt found "melancholy and painful." This was not a human world. As they cruised along the river the wild animals came unafraid down to the water to drink and "pace slowly along the shore" until they turned back into the forest through one of the narrow openings visible here and there. "Now it is the jaguar, the beautiful panther of America . . . and now the hocco with its black plumage and its tufted head." Humboldt felt a "peculiar attraction" to these scenes, for it seemed they were "in a new world, in the midst of untamed and savage nature." It was not just the whites who thought so. "'Esse como en el Paraiso,' said our pilot, an old Indian of the missions." It is just as it was in Paradise. Except, Humboldt added, it wasn't: observing closely, he noticed that the animals feared and avoided each other. "The golden age has ceased; and in this Paradise of the American forests, as well as every where else, sad and long experience has taught all beings, that benignity is seldom found in alliance with strength."[32] Humboldt believed that nature was in some ultimate way harmonious, but years before Darwin was born he looked out onto what a later generation would call a "Darwinian" nature, where the powerful preyed on the weak.

What did this wilderness say about the place of humanity in nature? Human action could upset the balance of nature, as when settlers leveled forests and desiccated whole regions, or nature could destroy human communities, as when earthquakes leveled entire cities, as Cumaná had just been and as Caracas soon would be. Man acted as a force of nature, too, both creative and destructive, but his power was vastly inferior. Standing on the mountain that rose above Caracas, Humboldt reflected that here "man no longer appears as the centre of creation. Far from taming the elements, all his efforts tend to escape from their empire." Volcanic eruptions, earthquakes, mighty floods, and fierce hurricanes destroy in a few hours the work of ages. "It is the conflict of the elements, which characterizes in the New World the aspect of Nature." Deep in the inland rain forest Humboldt found the absence of human traces especially affecting. "We almost accustomed ourselves to regard men as not being essential to the order of nature." Plant life developed freely, without impediment; "the crocodiles and the boas are masters of the river"; the jaguars, peccaries, and monkeys traverse the forest without fear of danger from man. "There they dwell as in an ancient inheritance. This aspect of animated nature, in which man is nothing, has in it something strange and sad." The sense of disorientation continued: "We seem to be transported into a world different from that which gave us birth."[33]

Humboldt found that this deep sense of loss characterized New World colonies, for they all, even the United States, lacked "memorials," traditions and remembrances passed from old country to new and down the generations, binding peoples together through memories of a common past. New World landscapes were too alien to bear for long the names and memories of Europe, which with time are "insensibly effaced," weakening both religious and national loyalties. The colonists might have turned to the inhabitants that preceded them, but by destroying their monuments, customs, and memories, the colonists had undercut the traditions of indigenous peoples and disdained all that related to those they had conquered. Thus colonial peoples were left free, Humboldt concluded, but peculiarly rootless. This made their societies vulnerable, for "in the stormy times of a political regeneration, they find themselves isolated, like a people who, renouncing the study of its annals, should cease to derive lessons of wisdom from the misfortunes of preceding ages."[34]

Humboldt's own writing would help correct this by becoming what Benigno Trigo calls a "temporal bridge," bringing present and past, colony and metropolis, back together. Humboldt's works created a literary monument to a time before the Revolution and educated New World readers to the deep connections they had, perhaps despite themselves, begun to forge with their land. His prequel to the *Personal Narrative*, *Vues des Cordillères*, had sought to illustrate those connections and elevate indigenous monuments from contempt and obscurity to prominence and honor. The New World, says Humboldt throughout his massive American project, will find its monuments in its native peoples and in its distinctive New World nature. It is no accident that American colonials, north and south, used Humboldt as their springboard to invent, then enforce, a national and bioregional sense of identity, adapting indigenous monuments and languages to construct the imagined communities whose absence Humboldt lamented.[35]

What Humboldt pointed to first were the marvels he saw everywhere around him in this strange and beautiful land. In the llanos he halted the expedition to study the bizarre electric eels that infested its waterways and potholes. To capture them, his hosts stampeded horses into a pond to endure the eels' paralyzing jolts. Crazy with fear, the horses roiled the waters—two of them drowned—until the eels, their electric energy spent, could be captured safely. Humboldt, who had made a name for himself in electrophysiology, dissected several and experimented on them, trying to assess their power. No

Leyden jar had ever given him so violent and painful a shock as the eel he imprudently stepped on; his joints throbbed for the rest of the day. It astonished
him to discover that the eels could control and direct their discharges, though
he couldn't see how, and he came away certain that mechanical and organic
electricity were identical. Electricity must be behind all animals' ability to
control muscular action. More research was urgently needed.

At his first mission everyone had wanted him to see the man in a nearby
village with breasts, who had suckled a child with his own milk. Humboldt and
Bonpland met them both, Francisco Lazano and his teenaged son, and found
the father's breasts indeed enlarged and wrinkled like a mother's. Humboldt
took this in stride, noting that similar cases, while rare, were well established,
and noting the interest of such anomalies for the question of "final causes"—
that is, why should men have nipples? It wasn't until Darwin that science
could offer a satisfactory answer. More exciting to Humboldt was the miracu-
lous "cow-tree," which also gave milk, thick, nourishing, and tasty. Unlike a
lactating man, a lactating tree was unknown to science, and of all the wonders
he encountered, few excited Humboldt's imagination more. It recalled "all
the powerfulness and the fecundity of nature," a sense of the "marvellous"
that science too often stripped away. Nature appeared different here, "more
active, more fruitful, we might even say more prodigal of life"—certainly of
amazement. At the convent of Caripe he was taken to the cave of the guacharoes, which the natives called "a mine of fat." Once a year Indians harvested
thousands of the plump young guacharo birds and cooked them up into a
fine, clear butterlike oil. Then there was the turtle egg harvest of the Orinoco,
where natives from far away gathered to "mine" the eggs turtles buried under
the sand. Humboldt estimated no less than 33 million eggs were destroyed to
make five thousand jars of oil, then guessed even that number was too low,
given the collateral destruction. Such numbers were alarming. The Chayma
Indians never went far into the guacharo cave, so for all the carnage they committed, Humboldt saw no danger to future populations. But the turtles were
certainly declining. The Jesuits had protected some of the turtle beaches, but
the Franciscans who replaced them let the entire area be dug up without re-
serve, with predictable consequences.[36]

No less marvelous were the legends—of the wild man, of the Amazons,
of El Dorado. Humboldt gently debunked each one while yet asserting some
grain of truth. The "hairy man of the woods, called salvaje, that carries off
women, constructs huts, and sometimes eats human flesh" might actually be
the manlike capuchin monkey known from Esmeralda, well upriver, or more
likely, a large bear, whose footprints look so human. Do not, Humboldt cautions, treat such traditions with disdain, lest the traces of a discovery in natural

history be lost. Take the Amazons: the legendary tribe of warrior women may have been an old traveler's tale, but Indians across the Amazon and Orinoco all repeated a similar story of women living without men. Humboldt thought it likely that women in various parts of America, "wearied of the state of slavery in which they were held by the men, united themselves together, like the fugitive negroes" and became warriors to preserve their independence. Even if this had happened only once, given the way legends spread, that would have been sufficient. In Humboldt's brand of magic realism, even the oddest of marvels pointed back to some fundamental truth in a land that was a tapestry of miracles. As for El Dorado, Humboldt suspected a clever ruse on the part of Indians eager to manipulate the white man's greed: "The natives, in order to get rid of their troublesome guests, continually described Dorado as easy to be reached, and situate at no considerable distance." Gold always lay just over the next hill.[37]

Most marvelous of all were the people. Upriver, everything to do with the human community seemed darker, deeper, and stranger. The Orinoco missions had pockets of stark cruelty. The first morning after Father Zea joined the party, they were awakened "by the cries of a young man, mercilessly beaten with a whip of manatee skin." Zea was beating Zerepe, their interpreter, "a very intelligent Indian" who was fluent in many languages, but who was refusing to go. "We obtained with difficulty the pardon of this young man," wrote Humboldt, who reported Zea's reasoning: unless the strong and hardworking Indians of the upper Orinoco were shown "severity," they would desert the missions to live with the whites downriver, where their labor was much valued. Humboldt objected. Indians "are not great children," anymore than were the serfs of eastern Europe held down by feudalism, and beating civilization into them wouldn't work anymore than it would work for a child. As rational beings, Indians could be enlightened only by the moral force of reason, never by the physical force of the whip. Zerepe stayed with the expedition, becoming a personal favorite of Humboldt.[38]

This was not the only time that he and Zea had words. Zea came from a culture that still conducted raids on the Indians, "*conquering souls*" with the bayonet and carrying off children to serve as "serfs, or *poitoes*," in the missions. When their canoe came to the "Rock of the Mother," Zea couldn't tell them the story of the name, but some weeks later they found another missionary who could. Just three years before, in 1797, missionaries had raided an Indian hut and taken captive a Guahiba mother with her three children, two of them infants, and carried them far away. Though she fled with her children repeatedly, they always recaptured her, until exasperated, they bound her and took her far upriver, away from her children. She broke her bonds and escaped to the rock named after her, but they recaptured her, stretched her upon the rock, beat her again, and carried her still farther upriver. There she gnawed off her bonds, fled the camp, and somehow, miraculously, found her way back overland to her children, sustained by eating ants. There was no happy ending. The missionaries again carried her to an upriver mission, where she refused food until she died. Normally, concludes Humboldt, he dislikes narrating such individual calamities, but for this one story he would make an exception, hoping some benefit might accrue from its publication.[39]

Where Zea would beat the Indians, Humboldt's approach was to kill them with kindness. In his journal he noted how hard it was to get good Indian guides, but he had a method: choose the best you can find and treat them in an open and friendly way. Of eight or nine, most will desert at the first bad weather, but those that remain are all you will need. They will show themselves good, loyal, careful, "and pull you through in case of danger even if they consider you just as vigorous as they." But Humboldt's friendliness sometimes backfired, leaving him baffled. The first time they met a band of Indians on the trail, Humboldt "pestered them with questions," for he and his party were exhausted and anxious to reach the next mission. The Indians answered "Si Padre, no Padre," seemingly at random. "This made us the more impatient, as their smiles and gestures discovered their intention of satisfying us; and the forest seemed at every step to become thicker." Humboldt found no way to reach them, to penetrate their mask of politeness. Then there was the *alcalde*—an Indian designated by the missionaries as an overseer—who, Humboldt began to suspect, was determined to give him whatever answer he thought Humboldt wanted. One day Humboldt put him to the test by asking, Did he not think the river Caripe, which issued from the cave of the guacharoes, returned to it again by running up the backside of the mountain? "After appearing gravely to reflect on the subject, he answered, by way of supporting my hypothesis, 'How else, if it were not so, would there always be water in the bed of the river at the mouth of the cavern?' "[40]

Humboldt also observed that Indians used language to intimidate, as when their old Indian pilot ran on about the snakes, water serpents, and tigers which lurked in the forest just waiting to attack the Europeans. Humboldt dismissed such warnings, but he understood where they came from: Indians used such devices to "render themselves more necessary, and gain the confidence of the stranger. The rudest inhabitant of the missions understands the deceptions, which every where arise from the relations between men of unequal fortune and civilization." Such artifices were the weapons of the powerless, used to ameliorate their condition.[41] Humboldt understood the power imbalance between himself and his informants, an imbalance no amount of friendliness could correct, even as his need for their knowledge and assistance empowered the Indians, who were, for the moment, more in control of him than he of them.[42]

Often Humboldt's communications with Indians were dependent on interpreters who maddened him with their "indolent indifference" and officious smiles. But he observed that the exasperation was mutual, for he and the Indians cared about different things. To the Indians, who sought trees for timber, what mattered about them was the taste of the wood (by which they could identify them) and the shapes of the leaves. To Humboldt, it was the flowers that mattered, for only thus could he place the tree in the Linnaean classification system. With the flowers high out of reach, or loose on the ground, or out of season, he needed the Indians to tell him which flower grew on which tree. His odd obsession with flowers irritated the Indians, who tried to shut him up by insisting, "'All those great trees bear neither flowers nor fruits,'" denying what they had never needed to observe. Each grew exasperated with the other: "They were tired with our questions, and exhausted our patience in turn."[43] Despite everyone's goodwill, the European scientists and their Indian guides were continually at cross purposes, defeating Humboldt at the very fountainhead of knowledge.

Harry Liebersohn, in his account of scientific travelers, speaks of the nineteenth century's growing awareness of the lack of transparency in translation. Instead of being understood as a clear channel of unmediated knowledge, it became "a creative mediation between cultures" in which transporting information back to Europe inevitably modified it, starting with the travelers' own calculations, emotions, and ideological commitments. Yet in the field, he claims, even sophisticated travelers "were too beholden to their belief in friendship between hosts and guests to ask or discern when their informants were withholding, partially disclosing, disguising, or prettifying their words and behavior." Their conclusions would have been different if they had "thought of their friendships as power relationships from which they were trying to extract a scarce commodity—knowledge—and locals were dispensing it with an eye to maximizing their return."[44] This oblivion to power relationships may have been true of many of Humboldt's contemporaries but it was not true of Humboldt, who knew that to get the knowledge he needed he had to negotiate across power imbalances he could not control. He often felt he had failed. As a man of reason, he tried to analyze the motivations that governed the Indians he worked with and the strategies they deployed while negotiating with him.

Money, he observed, was a poor motivator, for it seldom interested forest peoples living in an environment so rich with natural products. Guides hired for money often appeared lazy, but Humboldt thought that was a misinterpretation—actually, they were bored. The Indians' interests and those of the European did not coincide. But when they were interested, Indians became prodigies of energy: "The same Indian, who complains, when in herbalizing we load him with a box filled with plants, rows his canoe fourteen or fifteen hours together, against the swiftest current, because he wishes to return to his family." Humboldt had argued with Zea that Zerepe, the translator whom Zea had tried to beat into submission, could not be motivated by fear. What did motivate Zerepe came clearer to Humboldt when they returned to his village, where Zerepe found the girl he loved had deserted him, abandoning the mission to go "al monte," or to the woods. Humboldt was amused at how quickly the bereft Zerepe recovered: having been born among the Christians, traveled widely, and learned many languages, "he thought himself superior to the people of his tribe. How then could he fail to forget a girl born in the forest?"[45] Zerepe's wide travels and multiple languages made him worldly; in effect, working with Humboldt had increased his cultural capital among his own people.

In this case, their interests coincided and Humboldt could feel that his negotiations were successful. However, another incident reveals a moment of confusion and painful cross-purposes: early in their voyage, the pilot sailing the lancha, showing off to the Indians on shore, ran too close to the wind and capsized the boat, nearly finishing off the expedition then and there. Bonpland rescued Humboldt (who could not swim), and everyone frantically bailed out the lancha with calabashes and gathered up the floating papers and books. Disaster was narrowly averted, for they lost but one volume, a botany text. Of course the Europeans heaped reproaches on the pilot's head, to which "he opposed his Indian phlegm; and answered coldly, 'that the whites would not want Sun enough on those banks to dry *their papers*.' "[46] The pilot's mockery points back to his hurt feelings, while Humboldt's record of the pilot's contempt gently undercuts his own panic, even as it registers the outrage of the scientist confronting the carelessness of the ignorant.

Once again, the obvious solution to the intractable problem of translation was to learn the Indians' languages. European prejudice against native languages was not confined to untutored monks. Humboldt's colleagues generally disdained indigenous languages that lacked a written literature as, in his words, "but the wild cry of nature," too primitive or childish to be of any interest. For decades Humboldt worked to validate the complexity and sophistication of Indian languages, which he asserted were capable of the most nuanced expression. In a letter to Wilhelm he noted that the language of the Caribs "combines richness, grace, power, and tenderness," with abundant words for abstract ideas and every numerical combination, while the Inca language "is so rich in variety and delicacy of expression, that the young gentlemen, when making themselves agreeable to the ladies, usually adopt it after they have completely exhausted the vocabulary of the Castilian tongue." In his *Personal Narrative*, he repeated the claim of "a respectable man, who has perished in the bloody revolutions of Quito, Don Juan de la Rea," who "had imitated with graceful simplicity some idylls of Theocritus, in the language of the Incas." Humboldt adds that he has been assured "that, excepting treatises of science and philosophy, there is scarcely any work of modern literature, that might not be translated into Peruvian." He regrets that his inability to read Mexican impedes his scholarship on the Aztec calendrical system, which he suspects is related to the calendars of East Asia. If only European scholars would learn the language and give the treatises of the Mexican natives the consideration they deserve, learning would be much advanced.[47]

As for the peoples of the lowland rainforests, "friends of humanity" will surely wish to know all that remains of the native populations of South America: Humboldt lists over two hundred tribes in Venezuela alone, an area "a little larger than France," each tribe distinct and with its own language—a gold mine for the linguist. As he points out, just as the botanist must study all plants, not merely the useful ones, to comprehend the organization of the vegetable kingdom, so will the student of languages want to study every one. Notions of the "barbarism" of Indian languages are erroneous. Each language preserves a distinctive individuality, and together they form a rich and interwoven tapestry that will turn inside out the old and now outdated assumptions of linguistics. Everywhere Humboldt went he collected long lists of vocabulary words, grammatical idioms, and expressive turns of phrase, which he published in his *Personal Narrative* and elsewhere, and turned over to his brother Wilhelm for analysis. Out of these materials Wilhelm advanced influential new theories on the development of language, pioneering works in theoretical linguistics and historical anthropology.[48]

The task was urgent, for native languages had disappeared even in Humboldt's young life. The last Atures Indians of the Great Cataracts of the Orinoco were still alive in 1767, but by the time Humboldt arrived, all that remained to show the inquisitive philosopher was an old parrot whom no one could understand "because it spoke the language of the Atures." On hearing this story, one of Humboldt's Prussian friends composed a poem which Humboldt appended to *Ansichten der Natur*:

> . . . As they lived, free, dauntless ever,
> So the brave Aturians died;

And the green banks of the river
All their mortal relics hide.

Yet the Parrot, ne'er forgetting
Those who loved him, mourns them still,
On the stone his sharp beak whetting,
While the air his wailings fill.

Where are now the youths who bred him,
To pronounce their mother tongue,—
Where the gentle maids who fed him,
And who built his nest when young?[49]

Humboldt never pretended to be a linguist. The goal of his comments was to inspire scholars of language, deepen understanding of the development of the great human family, and awaken respect for the dazzling variety of cultures and customs he witnessed during his months-long journey. It struck him that "the strangest contrasts are found blended in this mixture of nations, some of whom live only upon fish, monkeys, and ants; while others are more or less cultivators of the ground, more or less occupied in fabricating and painting pottery, or weaving hammocks or cotton cloth." His narrative is studded with excursions into the marvelous and the curious: there were, for instance, the tribes who painted themselves red, a custom derived from the Caribs who had once extended their empire deep inland. Was it a practical adaptation, wonders Humboldt? But he sees no obvious utility—one might guess that the pigment repels biting insects, but Humboldt sees no such evidence. He does notice that paint is more comfortable than clothing (did he learn this from personal experience?), and that the natives felt undressed without their coat of paint. Sometimes they playfully imitated European clothes—some Indians at Pararuma sported painted blue jackets with black buttons; others were painted in stripes and studded with "spangles of silvery mica." Why red, he wonders, and not blue? The best red pigments were expensive—missionaries made a bundle selling the more valuable *chica* for prices that required a large man to work a fortnight to paint himself properly. "Thus as we say in temperate climates of a poor man, 'he has not enough to clothe himself,' you hear the Indian of the Oroonoko say, 'that man is so poor, that he has not enough to paint (*s'onoter, se majepayer*) half his body.'" It all showed that the Indians had a finely tuned sense of beauty, and also that beauty is relative: Europeans thought white skin was beautiful, while they, with skins of brownish-red, thought red was beautiful. They embellished themselves just as Europeans did, by heightening the characteristic marks of their race.[50]

Other products caught Humboldt's attention, too. Father Zea was eager to show Humboldt the pottery of the Maypures Indians, leading Humboldt to wonder at the prevalence of pottery shards throughout the Americas and at the continuities he saw in design and pattern, particularly the pleasing rhythms of repeated "arabesques, meanders, and *grecques.*" Upriver he found that the Indians of Javita cut and hollowed the gigantic trees of their region by fire and hatchet to make the canoes that ply the Orinoco from one end to the other. At an Atabapo mission Humboldt was intrigued by the rubber balls used in Indian games of tennis, or to make corks "far preferable to those made of the bark of the cork tree." Here were commercial possibilities: he speculated that enough of this resin, dug from the ground near certain tree roots, might be collected from the Orinoco to supply all of Europe with boots and shoes. Go, he recommends, to the Amaguas of the Amazon for advice, for it is they who make the best rubber. Some forest products could cure, like the bark infusion which instantly relieved them of the irritating little insects that had burrowed under the skin of their hands, or the snakebite antidote that cured an Indian hired to portage their canoe. Others could kill. In the village of Esmeralda, the center of curare manufacture, Humboldt was tutored by an old poison-master, "the chemist of the place," in the mysteries of this deadly poison. Humboldt collected a little in a vial, and almost met his demise when it leaked onto his socks. He felt the glutinous matter as he was about to draw a stocking onto his bleeding foot. Let that be a lesson, he cautions, in "the prudence requisite in the conveyance of poisons."[51]

Once the tutorial was finished, the poison-master conducted Humboldt back to the festival celebrating the harvest of a curious and tasty nut named *juvia,* now marketed as Brazil nuts. They were eating more than just nuts: Humboldt was appalled by the large roasted monkeys, skinned and bent into a seated posture, "ranged in order resting against the wall." They displayed "a hideous resemblance to a child." Finicky Europeans preferred to serve their monkey meat at table with head and hands removed. Time and again he was struck by what Indians ate. Once the travelers dropped in on a group dining on bags of dried ants mixed into a cassava paste. While Zea smacked his lips with pleasure, Humboldt reported that "some remains of European prejudices" prevented him from subscribing to such praises. More puzzling were the Otomac Indians, whom other Indians called barbarous because, in their saying, "there is nothing so disgusting, that an Otomac will not eat it." In flush times the Otomacs lived on fish and turtles, but in lean times they lived on dirt—literally: "a very fine and unctuous clay," unmixed with fat or any organic matter, formed into balls five or six inches in diameter and lightly baked. In the kind of cross-cultural comparison that typified his work, Humboldt inserted a lengthy essay on the worldwide custom of eating dirt, including Javanese women anxious to stay thin and Germans who spread *Steinbutter* or "stone butter" clay on their bread. What astonished Humboldt, however, and what he could not explain, is that in the months the Otomacs subsisted on clay, they did not lose weight. "They are, on the contrary, extremely robust," and their bellies stayed flat. Humboldt concluded that though the Otomacs might have somehow adapted to an earth diet, Europeans would never learn to digest it. But what if, say, chemistry made it possible for humans to live on wood? Such a discovery would "dissolve the bonds of society" and "sap the foundations of industry and civilization." Some things are better left as they are.[52]

Then there was the problem of cannibalism. Eating monkeys that resembled children may, Humboldt speculated, make it easier to eat human beings. But then, cannibals didn't think of themselves as cannibals—that is, they didn't think it was fellow human beings they were eating. He noted that when mission Indians met members of an unknown but related tribe, they would say, "They are no doubt my relations, I understand them when they speak to me." But all who were not of their tribe or family they hunted "as we hunt game," and they celebrated battles by devouring even the women and children of their enemies. Evidently they felt ties of family and relationship, but not of a common humanity. Humboldt was startled when a mild and intelligent young Guaisia Indian they had recently hired, "sufficiently civilized" to help him with his astronomical observations, declared that "the flesh of the marimonde monkeys" tasted to him like human flesh, and that the people of his tribe preferred "the inside of the hands in man, as in bears," an assertion "accompanied with gestures of savage joy." Humboldt followed up with more questions: "We inquired of this young man, so calm and so affectionate in the little services he rendered us, whether he still felt sometimes a desire to eat of a Cheruvichaheva. He answered without discomposure, that, living in the mission, he would only eat what he saw was eaten by *los Padres*." Their horrified reproaches had no effect. "It is as if a Brahmin of the Ganges, travelling in Europe, reproached us with our habit of feeding on the flesh of animals." To the young Guaisia, "the Cheruvichaheva was a being entirely different from himself," no more unjust to kill than a jaguar. Only a sense of propriety kept him, while in the company of whites, from dining on his neighbors. Humboldt was "almost terrified" at this calm ability to dehumanize others.[53] Whether the young man continued in his employ, he does not say.

As they navigated this seemingly endless maze of waterways and lifeways, Humboldt was always ready to examine the contents of yet another Indian hut, jot down a few more vocabulary words, or learn the use of some new plant. One night they camped in the recently abandoned hut of an Indian family, and Humboldt inventoried the contents: fishing instruments and nets, pottery, stores of *mani* used to pitch canoes and fix the bony spines of the ray to their arrows, jars of a vegetable milk used as a glossy white furniture varnish. What a useful product! What if it could be colored; one could paint and varnish carriages in a single step. In the forests above Cumaná the Indians hacked incisions in the tree trunks to show Humboldt "the beautiful red and golden yellow woods, which will one day be sought for by our turners and cabinet makers." Everywhere in the tropics were useful new products just waiting to be discovered and commercialized.[54]

But such ethnobotanical researches and the development of indigenous knowledges into a tropical resource base lay far in the future. Back in the narrative present, tensions in the narrow canoe were growing as conditions worsened. Worst of all, dominating everything else, were the biting insects. Humboldt struggled to convey the horrible and constant torment they all suffered: by day the mosquitoes and venomous flies, by night the *zancudos* or gnats, "dreaded even by the natives." Each river system had its own insect geography, but in all of them, the different varieties relieved each other in shifts, so predictable that "we might guess blindfold the hour of the day or night, by the hum of the insects, and by their stings." Whole regions were rendered uninhabitable. No matter how tough one may be, how hardened to pain, "it is impossible not to be constantly disturbed" by the mosquitoes, flies, and gnats that cover face and hands, "pierce the clothes with their long sucker in the form of a needle, and, getting into the mouth and nostrils, set you coughing and sneezing whenever you attempt to speak in the open air." Often Humboldt could not take his astronomical observations through the thick stratum of insect bodies, which hovered "like a condensed vapor" from ground level up to fifteen or twenty feet. Extraordinary lengths were taken to escape them: Zea had constructed a treehouse refuge; Indians on the coast buried themselves in the sand; at Maypures, Indians slept on rocks in the midst of the cataracts. The resourceful Otomacs made mosquito curtains of palm fibers, and on the upper Orinoco, the natives invited Bonpland, who could not prepare his plants because of "the continual torment," to work in their little chambers or *hornitos* where they flushed out the insects with smoke. Everyone suffered, whites and Indians alike; Humboldt recorded that Indians slapped themselves and their comrades even in their sleep.

Everywhere the state of the insects formed the first topic of conversation, from salutations ("How did you find the zancudos during the night? How are we to day for the moschettoes") to jokes ("as for my flies . . . I can boast,

that with one of mine I would beat three of yours"). Why, oh why, Humboldt
wonders, didn't the Christians locate their mission villages in the bug-free
uplands? The "wild Indians" flee the mosquitoes of the river missions, and
whole tribes go "al monte" into the hills to avoid them. But as one Indian
suggested, even the uplands may not be high enough: "'How comfortable
must people be in the moon!' said a Saliva Indian to father Gumilla; 'she
looks so beautiful and so clear, that she must be free from moschettoes.'" As
Humboldt concludes at the end of this remarkable essay, "Our imagination
is struck only by what is great; but it belongs to the philosophy of nature, to
pause at what is little."[55]

Though addressed—at some considerable distance—with good humor
and the patient curiosity of a natural philosopher, the unremitting torture of
the insects struck the ground note of a beat that began to wear down even the
ever-ebullient Humboldt. The very landscape seemed to defeat them. With
no roads other than the "tortuous and intermingled rivers," no villages except those tucked invisibly amid thick forests, and no mountains or even
hills to provide orientation, only the stars in the sky could tell one's location,
and too often the stars were hidden by damp vapors. To be sure, the Indians
were "excellent geographers" and knew exactly where they were, but they
had good reason to hide their knowledge, and even if they were forthcoming, "the various Indian nations, who frequent this labyrinth of rivers, give
them names entirely different." Humboldt was perplexed as they attempted,
in three or four different languages, to identify place-names through an interpreter. Such conditions generated errors in nomenclature that had been
perpetuated for generations, and their maps were "loaded with names arbitrarily shortened or disfigured." It was possible to draw accurate maps from
the reports of the natives, but doing so required not just physical exploration but "a certain etymological tact." The Casiquiare crossing had become
a maze of rivers, tribes, languages, names. Communication ramified and
thinned out through the bewildering network of portages and branchings,
until isolation became a psychological and social state. Families settled a riv-
erbend in ignorance of who lived hidden in the forest—only half a league
away, but two days' navigation. In open countries, rivers served to connect
and generalize "languages, manners, and political institutions," but here in
this "impenetrable" land they had the opposite effect: to "dismember" great
nations and fragment languages, seeding "hatred and mistrust . . . disunion
and weakness. Men avoid because they do not understand each other; they
mutually hate, because they mutually fear."[56]

The farther they went upriver, the more the oppressive jungle closed in.
At first it seemed a paradise; now it was devolving into a Darwinian night-

mare. Even science has trouble redeeming such a wilderness. Somewhere on the Casiquiare the humidity broke Humboldt's trusted and beloved barometer. Now he could only guess their elevation. Bonpland's plants, dried with such discomfort in the smoky hornitos, were rotting away from the humidity; as they approached the Casiquiare, he discovered that he had already lost the better part of his herbarium. Night after night Humboldt stayed up, fighting sleep, scanning the cloud cover for any sign of a break, the break he absolutely needed to take the astronomical measurements that would establish the Casiquiare's position and finally silence the European academicians. Nothing he said would change their minds; extending the European network of cartography was the only way to settle the controversy and vindicate local knowledge. Night after night the clouds refused to yield.

Worst of all, the river was growing unbearably, unnaturally lonely. Shrinking missions, abandoned villages, and neglected plantings spoke to Humboldt, telling him a grim story of depopulation as Indians withdrew deep into the woods in fear of white incursions. On the whole of the Casiquiare, 180 miles, Humboldt guessed not more than two hundred inhabitants could be found, and not more than five canoes passed through it in a year. In a month "we had not met one living soul" on the river away from the missions. Their hands and faces were swollen from insect bites; they couldn't hunt meat, because their gunpowder was too damp to ignite; they appeased their appetite by washing down dry ground cacao with quantities of river water. They found few places to camp on shore and when they did, the wood was too damp to burn. That meant no fires to cook (they were nearly out of food anyway), but worse, no fires to keep the jaguars away. In this trackless jungle without landmarks, even Humboldt lost heart: as he remarked, "The most discouraging of all physical sufferings are those, which, uniform in their duration, can be combatted only by long patience." Humboldt's great adventure had become an endurance test, the kind that kills without romance or glamour. It was here, in the steaming rain forests of the Casiquiare, that Bonpland "imbibed the seeds" of the illness that nearly ended his life.[57]

As the little band approached the limit of their endurance, drinking chocolate and trading jokes to keep their spirits up, their adventure was punctuated by a small but demoralizing tragedy. Their last night on the Casiquiare, Humboldt, Bonpland, Carlos and José, Father Zea, Sotto, Zerepe, and the other Indians and their entourage of monkeys and birds all camped on the forest's edge, cheered for once by a bright campfire and confident that its light would keep away the jaguars howling in the nearby trees. As ever, their great mastiff dog barked bravely to keep the big cats away, then as they crept near, howled and cowered under their hammocks. Time and again the band had been amused by "these alternations of courage and fear in this young animal, which was gentle and extremely caressing." But in the morning, their gentle and loving dog, "the faithful companion of our travels," was gone. They waited long through the morning, and they searched. Had he wandered off and been taken? Could they have slept through his cries of pain? Three days later they came back, spent another night (listening again to the jaguars' constant howling), but "all our researches were vain." Their companion, "which had accompanied us from Caraccas, and had so often in swimming escaped the pursuit of the crocodiles, had been devoured in the forest."[58] In his anxiety to avoid the personal, Humboldt assures the reader that he relates this telling incident only to demonstrate the dangers of wilderness travel and the perfidy of jaguars. But only two other times, both of them when Bonpland's life was in danger, did he allow himself to reveal such concern for another. His choices were stripping him of those he loved. It was time to go home, back to civilization.

Fortunately, they had almost reached their goal. The Casiquiare had indeed taken them from the Amazon system back to the Orinoco, proving its existence beyond doubt, and Humboldt had finally gotten his measurements. Just upstream was the "little hamlet" of Esmeralda, where they could rest and regroup. Originally Humboldt had hoped to ascend the Orinoco to its source, but Esmeralda gave him pause. It was here that he met the master of curare and flinched from eating babylike roast monkey. This was the end of the earth, the last Christian settlement on the upper Orinoco, home to a few score Indians, zambos, mulattoes, banished soldiers, and exiled monks. Humboldt thought it actually quite picturesque, except for the mosquitoes, and rich with commercial potential—but for now, in May 1800, the news was not good. Two tribes of Indians, the Guaicas and the Guahariboes (now known as the Yanomamö), had reconquered their land taken by the whites, and were holding off all invaders with curare-tipped arrows. Discovering the source of the Orinoco was a job for soldiers, Humboldt concluded, not a tired and poorly armed band of philosopher-adventurers. They turned the canoe back downriver to Angostura, which they reached a little over two weeks later, saying goodbye along the way to Father Zea (whose health had not recovered but who seemed resigned to his fits of fever), and to Don Nicolas Sotto, who returned to his family.

After their two months of pain and privation, the little town of Angostura seemed a marvel of bustling commerce, brilliant conversation, luxurious beds, and—oh the miracle!—wheat bread. At first Humboldt and Bonpland felt merely tired, but soon they both, and José too, were struck with a dangerous fever. José was pronounced dead, but revived, as did Humboldt, but

Bonpland sank into weeks of fever and dysentery, and they all feared for his life. Humboldt reproached himself for his foolhardiness: he of all people should have known the risks to Europeans of tropical rainforest air. "Instead of going up the Oroonoko, we might have sojourned some months in the temperate and salubrious climate of the *Sierra Nevada* de Merida. It was I who had chosen the path of the rivers; and the danger of my fellow-traveller presented itself to my mind as the fatal consequence of this imprudent choice." They moved to a plantation outside town, where Bonpland took quinine and slowly recovered, although his recurring fever would force them to halt their travels at least two more times, for six weeks in Bogotá and two months in Santa Fe, and Humboldt would be stricken again too, in New Barcelona. Yet they kept on traveling. Hardly had they reached the coast, with Bonpland still recuperating, than they began to form new plans.[59]

Before they moved on, Humboldt paused to imagine the future of the vast provinces of what would become Venezuela. The region was clearly at a crossroads. As the civilized nations on the coast pushed inland, "the struggle" for "the emancipation and future prosperity of America" would take place on the llanos and in the deep forests whose inhabitants enjoyed, if not yet "liberty," then "a savage independence." Once the fury of war subsided, peace and prosperity would reign and an immense land, ten times the size of Spain and navigable in every direction, would be united by commerce. The Casiquiare, "as broad as the Rhine" and 180 miles in length, would carry grain to the Amazon basin and rain forest products to the coast. Humboldt was wary of material gain for its own sake. What he envisioned was not merely trade for the sake of wealth for the few, but communication and an advancement of true civilization. As he said, the best remedy against greed and exploitation of the land's riches was "the increase of commerce, in multiplying the connections between nations, by opening an immense sphere to the activity of the mind, by pouring capitals into agriculture, and creating new wants by the refinements of luxury." Ultimately he imagined that some future reader would "behold with exstacy" that "populous cities enriched by commerce, and fertile fields cultivated by the hands of freemen, adorn those very spots, where, at the time of my travels, I found only impenetrable forests, and inundated lands."[60]

Future readers have, indeed, followed in Humboldt's footsteps, but their findings give little credit to his powers of prognostication. As Alfred Crosby reports, "Europeans did not have the gear or concepts equal to the Pleistocene challenge of the rain forest." Despite the scientists and the entrepreneurs who followed Humboldt, heat, humidity, predators, parasites, and pathogens all united to keep his dream from being realized. Lotte Kellner reports the 1958 experience of two botanists who retraced Humboldt's journey: "Many of the settlements described by him have been abandoned and no trace exists of Esmeralda although it is still to be found on every modern map. The Indian tribes are unchanged in their habits, and mosquitos, ants and wasps are the undisputed masters of the river territory." Ten years later Douglas Botting traveled the Casiquiare and found that it was, if anything, "even more desolate" than in Humboldt's time: "The jejunes were as maddening as ever, the jaguars as audacious, the forested banks as claustrophobic. But the missions had gone and so had the Indians, and along the 200 mile course of the Casiquiare there lived no more than two other human beings." A 1993 essay reports the same depopulation. Where Humboldt foresaw traffic and factories, these travelers found heat, distance, and the everlasting insects. "We strive, like Humboldt, to maintain an external cheeriness, but something somber and oppressive is pushing down on us. Humboldt was right: there is all life here, but it has nothing to do with us. That the world is anthropocentric seems laughable. . . . And worse: We have no dreams of progress with which to console ourselves." Today we know that the soil is poor, the trees slow-growing, the rain and flooding untamable. The warrior Yanomamö who stopped Humboldt's ascent are being killed off by soldiers, prospectors, missionaries, and tourists. Nowhere does Humboldt's romance seem so deflated, so outdated, defeated, ironically, by both too much nature and too many humans. It is all just as Humboldt left it, but more so: the Indians are still withdrawing from predatory whites, fortune seekers are still advancing to exploit their lands, nature still looms alien and implacable. Did the world leave Humboldt behind? Or did he leap so far ahead that the world has still not caught up? Humboldt's Casiquiare crossroad to this day remains the road not tried.[61]

High Peaks and Hanging Valleys

Humboldt's journey up the Orinoco, with his tales of electric eels, jaguars, and lactating men, captured the public imagination. His trek through the high peaks and hanging valleys of the Andes would make him more famous still as the man who had climbed Chimborazo, ascending higher than any human being, almost beyond the limits of human endurance. Ironically, Humboldt had had no intention of going to the Andes. The plan was to sail to Cuba for rest and exploration, travel north to the Great Lakes and down the Mississippi to Mexico, ship to the Philippines, and go on around the world, like proper global voyagers. But while they were sojourning in Havana, the local newspapers reported that the Baudin expedition had finally

sailed and was on its way to Peru. This news threw the travelers into conster-
nation. They had both been selected for this expedition, and Humboldt had
promised Baudin he would rejoin it. They must make for Lima immediately.
While Bonpland raced to organize their collections and ship them to Europe,
Humboldt rushed to make new travel and financial arrangements, and by
March 1801 they were sailing back to South America, arriving in Cartagena on
the northwest coast. From there they sailed up the Magdalena River, packed
overland to Bogotá, and walked along ancient Inca roads down the spine
of the Andes to Quito, a journey that took them over eight months. Here
they learned that Baudin had sailed from France around the world via Africa
instead. Their hopes for a voyage around the world were dashed. Down to
Lima they went, detouring into the upper Amazon. From Lima they sailed
to Guayaquil on coastal Ecuador, then on to Acapulco. They would linger in
what is now Mexico for over a year before returning to Europe via the United
States.

It was not the journey Humboldt had planned, and his disappointment
shows when he relates how, instead of joining Baudin, he enjoyed instead a
singularly clear day "in the misty regions of Lower Peru," which enabled him
to observe the transit of Mercury across the sun—a "grain of consolation"
amidst "the serious troubles and disappointments of life." As he sighed in a
letter, "We have been made to feel that man ought not to count upon anything but that which he can procure by his own energy." Later he admitted
that he would have missed Baudin anyway, for he had never dreamed their
journey to Lima would take eighteen months. But there was, all in all, much
to show: accurate maps of the Andes, several thousand new plants, a revolutionary new understanding of volcanic forces. The detour allowed Humboldt
to immerse himself in the supercharged atmosphere of Bogotá, where the
great botanist José Celestino Mutis welcomed the travelers into a community
of Creole scientists who were shaking off European tradition and advancing
new ecological concepts, such as the correlation between elevation and climate zones, that Humboldt eagerly appropriated.[62] Yet if he ever wrote the
volume of the *Personal Narrative* covering this part of his journey, it never saw
publication. What survives, in addition to various scientific treatises, are his
journals, still largely unpublished in English, and the collections he cobbled
together from them: *Views of Nature* and *Vues des Cordillères*.

These shorter works would be governed by a very different aesthetic than
the *Personal Narrative*. The Orinoco had presented a bewildering landscape
without relief or landmarks, strung along a network of rivers flowing in all directions (even one that was said to be geographically impossible, that flowed
both ways at once) through a forest that defeated even Humboldt's attempts

to find in it order or meaning. Where was the pattern in that maze? At the end of the daily drone of packing and unpacking, walking, observing, measuring, league after league, all he could do was wish it away, replace it with a utopian reincarnation of the German low country, a tropical Rhine bustling with barges and freehold farmers and light manufactures. For the Orinoco veiled its history. Forests clothed the rocks and hid the people, even from each other; the only visible monuments of its human past were the mysterious petroglyphs carved high on granite masses, too high to reach, among peoples who no longer knew who had carved them, or how anyone had ever reached so high, or how to work metal capable of chiseling hard granite.[63] How else to organize such a landscape than to thread it step by step onto the linear narrative of motion forward through time?

By contrast, the Andes were riddled with landmarks and thick with monuments, history and nature so tightly interwoven that Humboldt believed neither could be understood without the other. In retrospect what was memorable in this mountain landscape was not the steady ongoingness of the journey but the sudden views that halted all forward motion: waterfalls and chasms, mountains and bridges, *teocalli*, statues, rock formations, buildings, monuments carved from the rock and so blended with it that natural shaded into human, human into natural. The high Andes and the Mexican plateau could be structured around the monuments, ruins, and memories of great civilizations—the Aztecs, the Incas, the Muyscas. So Humboldt created a pictorial, descriptive landscape of crossings and passages, a series of images and views animated, like *Natural Bridges of Icononzo* (frontispiece) by the very chasms they attempted to bridge.

Look, for example, at *Passage of Quindiu, in the Cordillera of the Andes* (fig. 4). Through a wild and rugged mountainscape trudges a file of *cargueros*, human beasts of burden bearing their white masters. One of them, his chair empty, meets our gaze, as if it is our weight he must in a moment take up again. Extremes meet: the tumbled wilderness versus the exploited natives, reduced to the status of mules. Is it not Humboldt himself who has briefly stepped off the *carguero*'s back to stretch his legs and sketch this view—the European aesthete, the colonial master carried by human sweat? Indeed not. This lonely and difficult mountain crossing through a "thick uninhabited forest," with no means of subsistence, took ten or twelve days, but travelers needed to carry a month's provisions lest they be cut off by rains. The pathway was barely a foot wide, sliced by crevices carved by streams cascading down the mountainside and darkened by vegetation through which travelers groped in "thick and muddy clay." Few persons in "easy circumstances"

FIGURE 4. *Passage of Quindiu, in the Cordillera of the Andes*, plate 5, Alexander von Humboldt, *Vues des Cordillères* (Paris, 1810). Rare Books and Special Collections, University of South Carolina Libraries.

would travel this route on foot. The trail was impassable to mules—men, *cargueros*, took their place. The wealthy were said to "andar en carguero" as if on horseback. Such language disgusted Humboldt, who refused to be carried. Since the rugged trail would have torn their expensive Prussian boots to shreds, they were "forced, like all other travelers, who dislike being carried on men's backs, to go barefooted." As he remarked, liberalism has its costs: they arrived with feet cut and bloody, filled with holy outrage at "the indolence of whites" who called their men "horses" and saddled them for service every morning.

Yet, as Humboldt puzzled further, the *cargueros* were not simply exploited natives. To begin with, they were not Indians, but whites or mulattoes who formed their own society, "robust young men" who to Humboldt's surprise eagerly embraced what he saw as a painful, poorly paid, and degrading employment. They, however, preferred its independent and vagabond life "to the sedentary and monotonous labour of cities"—"a life of freedom!"—rather like Humboldt himself, who bypassed the direct and far easier passage of one day to take this roundabout and difficult path. The egalitarian German was amused to hear the *cargueros* "quarrelling in the midst of a forest, because one has refused the other, who pretends to have a whiter skin, the pompous title of *don,* or of *su merced.*" In one region the local government mandated an improvement project that would have upgraded the trail, making it passable to mules. The local *cargueros* united to stop the project. Humboldt could not comprehend why anyone would resist such capital improvements, even as he struggled to understand, and even admire, the culture that such progress would displace.

His own expedition made the crossing in October 1801, passing from Bogotá to Quito "on foot, followed by twelve oxen, which carried our collections and instruments, amidst a deluge of rain"—miserable enough, one supposes, and not the image Humboldt presents. The differing views offered by his text and illustration document a complex network of passages: teams of oxen bearing the instruments and collections of science; files of *cargueros* bearing the masters of empire; barefoot and muddy Europeans bearing the claims of heroism, detouring through bogs and across mountainsides. In the illustration, gazes cross: ours; those of the *cargueros* looking at us, at the trail, at the landscape; the white passenger gazing at his book oblivious to the landscape; all together in transit, about to leave a pastoral plain behind and enter a wilderness.[64]

In Humboldt's illustrations nature and culture constantly inform each other. Those images documenting culture allude to its natural context: planetary and seasonal cycles, the construction and representation of bodies, geological sources of buildings, the carriage of pearls from coastal waters to upland cities. In turn, the images representing nature are saturated with cultural references: cultivated fields, distant villages, pastoral herds of European cattle or Andean llamas, travelers and viewers, native guides and European tourists. Compare *Pyramid of Cholula* with *Volcano of Cotopaxi* (figs. 5, 6). The Spanish destroyed the *teocalli* of the Aztecs—the pyramidal structures that housed their gods—but they did not succeed in destroying the more ancient monuments of the Toltecs, of which "the greatest, most ancient, and most celebrated" is the teocalli of Cholula, "the Mountain made by the hand of Man [*monte hecho a manos*]." At a distance it looks like a natural hill covered with vegetation. In the plate, which represents it "in its present ruined state," Cholula rises out of the plain, towering over a scattering of buildings, its form repeating the forms of the foothills in the middle distance, vegetation reclaiming it for nature. The Indians call it "the Mountain of unbaked bricks," Halchihualtepec (pointing to its method of construction), and recall that on this man-made mountain was once a temple dedicated to Quetzal-

FIGURE 5. *Pyramid of Cholula*, plate 7, Alexander von Humboldt, *Vues des Cordillères* (Paris, 1810). Rare Books and Special Collections, University of South Carolina Libraries.

FIGURE 6. *Volcano of Cotopaxi*, plate 10, Alexander von Humboldt, *Vues des Cordillères* (Paris, 1810). Rare Books and Special Collections, University of South Carolina Libraries.

coatl, now succeeded by a small Christian chapel dedicated to the Virgin de los Remedios. Every day "an ecclesiastic of the Indian race celebrates mass . . . on top of this antique monument."[65]

Cotopaxi, the active, even dangerous, volcano, was most certainly not made by man. Here Humboldt's interest seems exclusively natural: the illustration is pure profile, almost diagrammatic, intended to show the "most beautiful and regular" form of the volcano, "a perfect cone, which . . . shines with dazzling splendor at the setting of the sun, and detaches itself in the most picturesque manner from the azure vault of Heaven." Despite this comment, Humboldt's focus is scientific rather than aesthetic. He wishes to consider "the physiognomy of mountains," the "face" that reveals the depths of construction or character so important to geology. Volcanoes worldwide typically have the same regular conic form, a "resemblance . . . founded on an identity of local causes and circumstances," but they also show local differences as the form is "modified" by the nature of the local rocks—granites, micaceous schists, old sandstones, calcareous formations, each imparting a unique local character to the enormous masses. So it is for Cotopaxi, whose classic regularity is broken by an odd mass of rock, "studded with points," rising from the snow.

To understand this unusual rock formation, Humboldt turns not to geology but to local oral tradition: the natives call it "the head of the Inca." Some relate that the rock mass was once a part of the mountain's top, but was blown out by a massive eruption that occurred shortly before the Inca Tupac Yupanqui invaded Quito; they say it is called "the head of the Inca" because the rock's fall presaged "the death of the conqueror." Others, "more credulous," claim the catastrophe happened at the precise moment when the Inca Atahualpa was strangled by the Spaniards, marking the fall of the Inca civilization. Humboldt defers his own scientific explanation, only hinting that he believes that the volcano's form grows by successive layers of lava, making volcanoes self-creating. Intrinsic to the beauty of a mountain's "face" is its uniqueness, and intrinsic to Cotopaxi's unique beauty is its Indian feature, the "head of the Inca," which recalls native oral traditions relating the rock formation to indigenous history. For some the face of Cotopaxi recalls the destruction, for others the triumph, of conquering invaders.[66]

A more personal imbrication of natural site and local narrative is presented in *Fall of the Tequendama* (fig. 7). Like the *Natural Bridges of Icononzo*, this waterfall "bridges" separate landscapes, but here the link is vertical, not horizontal, as the tremendous falls plunge from the temperate zone of oaks and elms to the tropical zone of palms, bananas, and sugarcane. Or as the locals say, from *tierra fria* to *tierra caliente*, from a cold to a warm country. That

FIGURE 7. *Fall of the Tequendama*, plate 6, Alexander von Humboldt, *Vues des Cordillères* (Paris, 1810). Rare Books and Special Collections, University of South Carolina Libraries.

climates and hence vegetation are ranged in vertically stacked "zones" be-came Humboldt's signature idea, memorialized in the famous *Tableau Phy-sique* of Ecuador's volcanoes and shown here more pictorially in *Chimborazo, Seen from the Plain of Tapia* (fig. 8). The fact that Humboldt himself climbed through each zone nearly to the top of Chimborazo—a mountaineering feat unsurpassed for thirty years—helped vivify the idea of climate zones in the public imagination, although Humboldt himself regretted that the climb had produced only limited results for science. It was so cold at the highest point that he and his companions, Bonpland, Montúfar, and an Indian guide, could barely use their instruments. At the end they were scrambling up a narrow, rocky ridge on all fours, their hands cut, dizzy from lack of oxygen, lips and gums bleeding and eyes bloodshot, hurrying to reach the cone that hovered above them. Then a deep and impassable ravine opened at their feet. Accord-ing to their barometer they had reached the height of 19,286 feet, and they felt "isolated as in a balloon"—no insects, no moss, not even a bit of lichen grew on these rocks. Their only companions were the condors, who sailed at such heights "wholly devoid of fear." *Chimborazo from Tapia* shows the start

FIGURE 8 . *Chimborazo Seen from the Plain of Tapia*, plate 25, Alexander von Humboldt, *Vues des Cordillères* (Paris, 1810). Rare Books and Special Collections, University of South Carolina Libraries.

of the climb, porters disappearing into the distance while Bonpland bends to collect a botanical specimen. Humboldt notes that the plate was based on a drawing he made on 24 June 1802, the day after their ascent, following the heavy snowfall that kept them from reaching the summit.[67]

The Falls of Tequendama in New Granada (now Colombia) proved much more accessible. Humboldt begins with the geological information that the falls drained an ancient lake, which he juxtaposes with the geological intuition of the Muysca Indians, who relate how a channel draining the ancient lake was opened by a compassionate god. He dwells on the falls' beauty, which art cannot capture even though, ironically, the cataract "forms an assemblage of every thing that is sublimely picturesque in beautiful scenery." The difficulty is that the "impression" made by the falls is not merely visual; it depends on the whole sensual experience. Humboldt winds down a dangerous path (carrying his instruments, of course) to the very bottom of the crevice at the foot of the falls, though he is kept by the currents from approaching closer than 140 meters from the water's basin, which is lit only by "a few feeble rays of noon." Here, in dim solitude, overwhelmed by the richness of the vegetation and the "dreadful roar" of the falls, he experiences "one of the wildest scenes, that can be found in the Cordilleras." From various collectivities—scientific knowledge, Indian fable, aesthetic theory—he narrows the field at last to one single astonished subjectivity, his own, drowning in beauty—even as, scientist that he is, he remembers to measure the distance.[68]

Throughout these sketches Humboldt foregrounds his own agency. He frequently positions himself physically, slipping for three hours down a narrow path, admiring Chimborazo from the plain of Tapia the day after he climbed it, or sketching Cotopaxi and the Head of the Inca "to the west of the volcano, at the farm of Sienega, on the terrace of a beautiful country house belonging to our friend, the young Marquis of Maenza, who has lately inherited the title of Grandee, and that of Count of Punnelrostro."[69] Humboldt's is ever the centering and organizing consciousness, working overtime to examine monuments, ascertain local Indian narratives, uncover long-buried hieroglyphic fragments, or argue with armchair scholars an ocean away. Nothing speaks for itself here; all things are spoken, and spoken for, in a contested field of differences that everywhere leaves traces of other voices, other presences.

His eye follows and notes such traces. At Ynga-Chungana (fig. 9), he finds "a great number of small pathways cut on the slope of a rock," which lead him to a carved Inca enclosure surrounding a single seat in stone. From this hilltop seat, Humboldt discovers "the most delightful prospect over the valley of Gulan. . . . [evidence that] the prince, who had chosen this site, was not insensible to the beauties of nature; he belonged to a people, whom we have no right to style barbarous." Crossing the river Chambo on the rope bridge of Penipé (fig. 10), he pauses to admire the technology that was in use "long before the arrival of the Europeans," and to note that such rope bridges "are extremely useful in a mountainous country, where the depth of the crevices, and the impetuosity of the torrents, prevent the construction of piers." An engineer himself, he notes with approval the sturdy rope bridge on the route between Quito and Lima, capable of holding loaded mules—a superior native technology that was adapted by Europeans only "after uselessly expending upwards of forty thousand pounds sterling, to build a stone bridge." It seems to him typical that the Europeans failed to listen to the Indians. One of the hieroglyphs he examines (fig. 11) portrays a lawsuit in which a single Indian argues his case through an interpreter before a court full of Europeans. Aztec hieroglyphs represent speech with tongues floating in front of the mouth: the more tongues, the longer the speech. The Spanish have small clouds of tongues; the lone Indian, exactly one. Humboldt's caustic comment closes his analysis: "The tongues are very unequally divided in this picture. Every thing portrays the state of a vanquished country: the native scarcely dares

FIGURE 9. *Ynga-Chungana, Near Canar*, plate 19, Alexander von Humboldt, *Vues des Cordillères* (Paris, 1810). Rare Books and Special Collections, University of South Carolina Libraries.

defend his cause, while the strangers with long beards talk much and loud, as the descendants of a conquering people."[70]

Humboldt's argument is implicit in the ruins, fragments, and tongues he attempts to recover, explicit in his outrage at the Spanish conquerors who began with wanton destruction, proceeded through malign neglect, and ended with imperial silencing. At the time of writing, he informs us, the lone Indian defendant did not have even one tongue. He would not have been in court at all. His place and voice were taken by white lawyers, and the court's decisions were recorded only in Spanish. Even the lone professorship established at the University of Mexico's founding in 1553 for the study of Aztec antiquities "has been suppressed" and the use of paintings "entirely lost"—and with them the titles, genealogies, codes of laws, and lists of taxes that had once bridged the peoples of New Spain with their Aztec past.[71] For this chasm not even Humboldt can build a bridge. His "views" can only refract a lost world, a human path to a civilization severed without cause, the people who trod that path cut off from contributing to the exchange of cultures and ideas that, for Humboldt, constituted progress. Rendered speechless, Humboldt cannot speak for them; he can only witness, and protest.

And yet this picture is perhaps not so entirely bleak. Through the monuments and traces that survive in the present the past does speak, to those who know how to listen. *The Rock of Inti-Guaicu* (fig. 12) shows another fusion, nature swerved into human purpose. Footpaths cut into the rock lead to a solitary mass of sandstone perhaps fifteen feet high, with a perpendicular white surface on which a formation common to sandstone has inscribed two concentric circles. Into this rock mass the Incas carved a series of steps leading to a seat hollowed in the stone, facing the concentric circles. To the natural inscription the Incas added two eyes and a mouth, turning the circles into the image of the sun, "such as at the commencement of civilization we see it figured among every nation of the earth." Just as the Christians see crosses everywhere in nature, so the Incas, who called themselves People of the Sun, saw images of the sun, and they crafted this rock into a site of meditation or worship. The Spanish, anxious to obliterate from the eyes of the natives all that "was the object of ancient veneration," attempted to chisel away the features of the sun. Their chisel marks are still visible, but the sun's features remain visible too. In Humboldt's plate, a family of Indians is visiting the site: a man, a woman holding an infant, a toddler and two young adults approach the base of the steps in postures of awe and respect.[72]

FIGURE 10. *Bridge of Ropes near Penipé*, plate 33, Alexander von Humboldt, *Vues des Cordillères* (Paris, 1810). Rare Books and Special Collections, University of South Carolina Libraries.

FIGURE 11. "An Azteck Hieroglyphal Manuscript Preserved in the Library of the Vatican," plate 13, Alexander von Humboldt, *Vues des Cordillères* (Paris, 1810) [detail]. Rare Books and Special Collections, University of South Carolina Libraries.

In this image Humboldt suggests layers of a deepening past: today's green and overtaking vegetation, yesterday's Spanish chisels, the ancient Inca inscriptions and stonework, the geology of sandstone, arrayed on a timescale that reaches from the present moment through human history to geological foundation. But what of the future? Will these Indian children grow up in the shadow of a lost past, represented by this defaced monument of their ancestors, or will they forge a new path to civilization? In his writings Humboldt makes clear that America's path to the future lies with the exchange of knowledge and the spread of global commerce, and he identified the Creole and European elites as the next generation of leaders—men like his friends Simón Bolívar, and Montúfar, Caldas, and Rea, who had defended the Inca language, all killed in Bolívar's revolutions. In Humboldt's ideology, carried forward from the American and French Revolutions, it was the enlightened descendants of the conquerors who had the power and the vision to lead the new nations of the Americas to political independence and global commerce. These are the men—intellectuals, scientists, landowners, political leaders—with whom Humboldt travels, among whom he lives, and with whom he identifies, men who would fight to displace the Spanish empire and become America's ruling class.

Could Humboldt imagine political agency for the Indian? Ángela Pérez-Mejía argues not: according to her, Humboldt repressed the political reality of American ethnic groups, who "were fighting for their lands against the Creoles," rebelling against the very people he identified as the liberal leaders of nascent movements for independence. However, far from repressing the reality of Indian and slave resistance, Humboldt refers to it several times. The problem as he sees it is that after centuries of abuse and oppression, Indians have been forced out of the political system and effectively silenced. Once they might have acted peacefully to protect their interests, but now their only option is bloody rebellion, which past experience shows will spin out of control into a race war. The only solution is for those in power—the "wealthy and enlightened"—to step forward and act, the sooner the better, but he warns repeatedly that their revolution will fail if it does not draw with it the descendants of the conquered and the enslaved. As he says at the end of

FIGURE 12. *Rock of Inti-Guaicu*, plate 18, Alexander von Humboldt, *Vues des Cordillères* (Paris, 1810). Rare Books and Special Collections, University of South Carolina Libraries.

his *Political Essay*, "The prosperity of the whites is intimately connected with that of the copper coloured race, and . . . there can be no durable prosperity for the two Americas till this unfortunate race, humiliated but not degraded by long oppression, shall participate in all the advantages resulting from the progress of civilization and the improvement of social order!"[73]

In Humboldt's Enlightenment paternalism, the children brought to the shrine of their Inca forefathers might not grow up to lead the revolution, but that revolution would fulfill its promise only if their white fathers allowed them a place, and a voice, in the new social order. Near the end of *Views of Nature*, in "The Plateau of Caxamarca," Humboldt lingers at Caxamarca, the ancient capital city of Atahualpa, the last of the Inca emperors. The Spanish had executed Atahualpa in the public square, and Humboldt examines the stone where drops of his blood were said still to be visible. While the emperor may be long gone, his descendants remain, and Humboldt spends time with them "amidst the dreary architectural ruins of departed splendour." He is especially impressed with the young son of the Inca leader Astorpilca, who leads him through the ruins of the palace of his ancestor. Humboldt offers a biographical vignette of the young Indian: "Though living in the utmost poverty, his imagination was filled with images of the subterranean splendour and the golden treasures which, he assured us, lay hidden beneath the heaps of rubbish over which we were treading." Humboldt asks him whether he does not sometimes wish to dig up the treasure, to relieve his family's poverty.

> The young Peruvian's answer was so simple and so expressive of the quiet resignation peculiar to the aboriginal inhabitants of the country, that I noted it down in Spanish in my Journal. "Such a desire (*tal antojo*)," said he, "never comes to us. My father says that it would be sinful (*que fuese pecado*). If we had the golden branches, with all their golden fruits, our white neighbours would hate us and injure us. We have a little field and good wheat (*buen trigo*)."

Thus he recalls "the words of young Astorpilca and his golden dreams." Humboldt learns that the natives generally believe that to dig up the treasure of the Incas was a criminal act that "would bring misfortune upon the whole Peruvian race." It was best to wait patiently, cultivating their fields: someday the Inca dynasty would be restored. The descendants of the sons of Atahualpa were biding their time, living on a buried El Dorado, waiting for the moment to be right.[74] History had passed into legend and myth. The future had to lie elsewhere.

3

Manifest Destinies

I could not resist the moral obligation to see the United States and enjoy the consoling aspects of a people who understand the precious gift of Liberty.

ALEXANDER VON HUMBOLDT TO THOMAS JEFFERSON

Humboldt's Visit to the United States, 1804

In five years, Humboldt had woven his name into the history of Latin America, and in just over five weeks he would do the same in the United States. Beyond the "moral obligation" he felt to see the world's lone functioning republic, there was every reason to avoid the detour. Humboldt was desperate to get himself, his friends, and his collections safely home to Paris. He was done exploring. His scientific instruments were wearing out, and new ones were not to be had this side of the Atlantic. In five years, he feared, science had progressed so far as to leave him in the dust. And in all that time, he had barely heard from his family, and for all he knew, they hadn't heard from him: the fate of his letters home, or worse, of the manuscripts and collections which they had entrusted to the tender mercies of war, piracy, and shipwreck, were unknown. Heading north risked losing everything they had with them to the British blockades of U.S. ports—assuming his ship was spared by the notorious Atlantic storms.

But what was all this compared with the chance to meet Thomas Jefferson and glimpse the free republic he had helped build? Back in Havana, their last stop, the American consul Vincent Gray (who knew exactly what Humboldt had to offer the United States) convinced him that America's welcome was worth the risk. Humboldt decided to make the pilgrimage. While Gray posted letters of introduction to his boss, secretary of state James Madison, Humboldt, his friends, and their massive collections all shipped to Philadelphia on the Spanish frigate *Concepcion*. They managed to evade the British, but not the hurricanes. On the terrible morning of 9 May, Humboldt thought his worst fears were coming true: "I felt very much stirred up. To see myself perish on the eve of so many joys, to watch all the fruits of my labors going to pieces, to cause the death of my two companions, to perish during a voyage to Philadelphia which seemed by no means necessary . . ."[1] The storm raged for a week, but finally the seas calmed and soon they were sailing up the Delaware River.

Beyond the riverbanks stretched a vast forest interrupted by marshy lowlands, giving way on the west bank to, in Herman Friis's words, "numberless neat farm-houses, with villages and towns . . . in some places cultivated down to the very edge of the water." As the ship rounded a point of land, Philadelphia came into view, a handsome city of seventy-five thousand built of red brick with paved streets lined with rows of poplars, heavy with the fresh green leaves of late spring's heat. The ship sat offshore in quarantine until Humboldt, growing impatient, wrote to Zaccheus Collins, botanist, geologist, and customs inspector, begging they be allowed ashore. Collins was out of town, and another three days passed before he opened Humboldt's letter and hastened, in a shower of apologies, to arrange for Humboldt, Bonpland, Montúfar, and de la Cruz to come up to the city. The four travelers landed the next day, 24 May, and the American Philosophical Society still preserves the customs form attesting that onto the bustling wharf on lower Market Street they unloaded two trunks of "wearing apparell," their beds, and seventy-seven boxes of "Plants and Collections for Natural History." The local newspapers announced the arrival of the *Concepcion* bearing "Molasses, sugar, logwood," plus one "Baron de Humbott" and his party, fresh from the Spanish colonies.[2]

As soon as he stepped ashore Humboldt posted a letter to Thomas Jefferson, then flush with popularity in his third year as president of the United States. Humboldt knew Jefferson's work from his studies at Hamburg, and the family library at Tegel included a copy of *Notes on the State of Virginia*, so Humboldt's first words to the president weren't entirely flattery: "Your writings, your actions, and the liberalism of your ideas . . . have inspired me from my earliest youth." His letter was carefully calculated to be irresistible. As Ingo Schwarz notes, his motto was "Läuter gehört zum literarischen Handwerk," "An author must ring bells in order to get attention." So he rang every bell he had: his arrival not from Cuba, but from Mexico, the country most on Jefferson's mind after the Louisiana Purchase; the presence of his friend "Citoyen Bonpland," representing the French Revolution that Jefferson had famously supported; the Casiquiare's paradoxical existence, the Orinoco's mysterious petroglyphs, the travelers' astonishing ascent of Chimborazo "higher than any other human being before us." As if that weren't enough, he added, "I would love to talk to you about a subject that you have treated so ingeniously in your work on Virginia, the teeth of mammoth which we discovered in the Andes." Humboldt knew that finding mammoth teeth near the equator would inflame the curiosity of the man who had written that mammoth remains had never been found farther south than Tennessee. As a final touch he signed himself "Baron Humboldt, Member of the Berlin Academy of Science"—a title he virtually never used, and an affiliation which technically he could not yet claim.[3]

Jefferson took the bait, and wrote back immediately promising a warm welcome in the nation's infant capital. While Humboldt waited for Jefferson's reply, the elite of Philadelphia—itself the U.S. capital until four years before, and still center of American science and intellectual culture and of a nation seen as the world model of democracy—swept him up and showed him the town. The members of the American Philosophical Society, the premier learned society of the United States (founded by Benjamin Franklin in 1743), adopted Humboldt as one of their own, voting him to full membership at their next meeting. Humboldt, Bonpland, and Montúfar toured John and William Bartram's botanical gardens, the largest and best in the country, and Charles Willson Peale's natural history museum, also America's largest and best. It was probably in the fine new building of the Library Company of Philadelphia (founded by Franklin in 1731) that Humboldt shouted for joy when he read the announcement that his irreplaceable manuscripts had arrived safely home. Dr. Caspar Wistar of the College of Pennsylvania Medical School made Humboldt the guest of honor at one of his posh Saturday night "Wistar Parties," and Peale honored Humboldt with a dinner at his museum that gathered all the naturalists he could scrape together on short notice, including John Bachman, then a self-conscious sixteen-year-old embarrassed to find himself in the company of the great ornithologist and artist Alexander Wilson, his engraver Alexander Lawson, his friend the naturalist George Ord, and the two Bartrams. In old age Bachman, by then one of the nation's leading naturalists, would reminisce fondly about that evening, the first time he was permitted "to look upon the countenance, to press the hand and listen to the interesting words of this great philosopher," whom he saw every day while Humboldt remained in Philadelphia, whom he visited in Europe, and with whom he corresponded for over fifty years.[4]

Peale took charge of the trip to Washington, making the arrangements and serving as guide along the way. He hoped—successfully, as it turned out—that in the enthusiasm of the moment, Humboldt would extract a pledge from Jefferson that the federal government would purchase Peale's Museum and move it to Washington. Peale, Humboldt, Bonpland, and Montúfar were joined by two of the APS's more prominent members: Dr. Nicholas Collin, pastor of the Old Swedes Lutheran Church, and Dr. Anthony Fothergill, a retired English physician. Peale kept a careful diary of the trip, in which he recorded that they set off at 7:50 a.m. on 29 May by the mail stage, at a cost of eight dollars apiece. They rode through the day and on through the night, Humboldt entertaining them nonstop in his polyglot speech, reaching Baltimore in time to breakfast with Peale's daughter Angelica. Peale's son Rembrandt showed them around and plied Humboldt with questions about mammoths in the Andes. On the first of June they set out for Washington (Peale complained bitterly of the rotten roads), which they reached in time for a late dinner. The next morning, Peale, Humboldt, and company waited on the president in the unfinished White House. Since Jefferson's reply had not yet reached Humboldt, they must have been relieved to find him home, and pleased to invite them all to a state dinner the next day. In 1804 America's capital city still existed mostly in the eyes of its boosters. Humboldt no doubt endeared himself to his hosts when he was shown the view from Capitol Hill and instead of complaining about the unbuilt buildings in the "vast Serbonian bog" before him, declared "that never had he beheld a more beautiful panorama."[5]

The weather had turned unbearably hot and muggy. Of the lot, probably Humboldt alone, so used to the tropics that ever after he liked to heat his rooms to nearly eighty degrees, was comfortable. Over the next several days they paid their respects to the British and French ambassadors (the Spanish ambassador had decamped to Philadelphia to protest the Louisiana Purchase); they toured the city and the Capitol, meeting with Dr. William Thornton, its architect and, of course, another APS member; at the Navy Yard they toured a warship, which Peale thought an "unpleasant memento of death and destruction." At Gilbert Stuart's studio they admired his unfinished portrait of the scientist and revolutionary Joseph Priestley; at the Patent Office they admired the skill and ingenuity of American inventors. (One of them was Peale himself, who delivered on this trip the polygraph he had designed for Jefferson, by which the statesman generated copies of all his outgoing correspondence.) Often their hosts were Samuel Harrison Smith and his wife Margaret Bayard Smith, energetic young Philadelphians who at Jefferson's request had relocated to Washington so Samuel could run the left-liberal, pro-Jefferson *National Intelligencer*. Margaret made their home the center of Washington society and became the memoirist of Washington's early decades. The travelers also dined several times with Dr. and Mrs. Thornton: the first time, Anna Maria ordered fish, strawberries, a half gallon of Madeira and a half dozen bottles of claret for the occasion, at which she played on her piano and sang while William showed off his botanical drawings, one of which—appropriately, the fever-reducing cinchona that had saved Bonpland's life—he sent to Humboldt as a parting gift. One day the party set off for Mount Vernon, where the visitors

drank in the view while Peale mourned for the good old days when he sat and
drank with George Washington. Peale introduced Humboldt to Billy Lee, the
last of Washington's slaves, who had been granted freedom and an annuity in
his will. On their return to the city the two coachmen, potted on grog, got into
a drunken race, Peale and Dr. Collin begging them to slow down while a tipsy
Dr. Fothergill "very imprudently" egged them on.[6]

Humboldt spent most of his time in Washington visiting with Jefferson
in the White House, where the two engaged in what Kent Mathewson aptly
calls "an open-ended seminar on a wide range of questions and topics." (The
legend persists that they met at Monticello, but there wasn't nearly enough
time for so long an excursion over such bad roads.) The guest list for the
first presidential dinner included Madison and Thornton and it was, Peale
recorded, a "very elegant" affair. Teetotaler that he was, Peale was pleased to
note that no toasts were drunk and that the table talk avoided politics, turning instead to such "agreeable" topics as natural history, "improvements of
the convenience of Life," and the manners of various nations. The next day
the discussion grew serious, as Humboldt spread his maps and papers before President Jefferson, Madison the secretary of state, and Albert Gallatin,
the secretary of the treasury—the three men who were, in Henry Adams's
words, "the true government" of the United States. Gallatin wrote his wife of
the "exquisite intellectual treat" he received that day from Humboldt, whose
breadth of knowledge was "astonishing" and whom "we all consider . . . as a
very extraordinary man." Gallatin and Humboldt later became close friends,
but at first Gallatin was a little put off: Humboldt "was not particularly prepossessing to my taste, for he speaks more than Lucas, Finley, and myself put
together, and twice as fast as anybody I know, German, French, Spanish, and
English, all together. But I was really delighted, and swallowed more information of various kinds in less than two hours than I had for two years past in
all I had read or heard." Humboldt spoke, Gallatin continued, "surrounded
with maps, statements, &c., all new to me, and several of which he has liberally permitted us to transcribe." Although Humboldt had been traveling in
an embargoed kingdom under the exclusive permission of King Carlos IV, he
freely shared his findings with New Spain's ambitious northern neighbor, including a copy of a lengthy statistical summary on Mexico which was a trove
of information, and the most extensive, accurate, and detailed map yet made
of the region. Humboldt lent the map to Gallatin for copying, and General
James Wilkinson—a double agent for the Spanish who was just then plotting
with Aaron Burr to lead a revolt, invade Mexico, and establish an independent kingdom on the Mississippi—seems to have made an illicit copy for his

own use. Much to Humboldt's annoyance, plagiarized versions of the map appeared in print in the United States well before he published it in 1811.[7]

The dinner guests may have avoided politics, but Humboldt had in fact arrived at a terrifically charged moment. Just the year before, Napoleon had been planning to move French forces up the Mississippi, until the wholly un-foreseen success of Toussaint Louverture's Haitian Revolution destroyed his army. Without warning Napoleon offered the entire territory of Louisiana, which Spain had just three years before retroceded to France, to the United States, for the bargain-basement sum of fifteen million dollars. Spain was furious; England was alarmed, especially when it learned the sale was financ-ing Napoleon's war on them; overnight the size of the United States doubled. Even as Humboldt and Jefferson were meeting, Jefferson's hand-picked team of Meriwether Lewis and William Clark were heading up the Missouri River to explore the new territory, and the United States and Spain hovered on the brink of an undeclared war over the exact location of the boundary between them.[8]

As Humboldt spread out his maps of Mexico, the keenest question on the minds of the three American heads of state was the location of that disputed border and the nature of the lands it did, or did not, include. Three days after their first meeting Jefferson wrote Humboldt urgently that "the question is this": Spain claims the land from the Mississippi west along the Red River, we claim the land from the Rio Grand north. In between lies all of Texas. "Can the Baron inform me what population may be between those lines of white, red, or black people? And whether any & what mines are within them? The information will be thankfully recieved [sic]." Indeed the baron could. Texas was barren, depopulated, and lacked mines or ports. It had little of politi-cal or economic value to offer either country, except a buffer zone between them. The timing of Humboldt's visit seemed providential. To begin with, he brought incalculably valuable knowledge that bore directly and immedi-ately on the political and economic future of the nation. What was more, his methods and instructions to future exploring expeditions would guarantee that never again would the U.S. government fail, as it did with the Lewis and Clark expedition, to leverage from them staggering amounts of scientific in-formation. In his future letters to Humboldt, Jefferson would marvel at his most fortunate timing, to have made countries known to the world "in the moment they were about to become" actors on its stage.[9]

There were also some lighter moments. Jefferson's daughter Martha and her children were visiting the White House, and one evening Humboldt, free to come and go as he pleased, walked unannounced into the drawing room to find Jefferson romping on the floor with a half dozen grandchildren. "When his presence was discovered Mr. Jefferson rose up and shaking hands with [Humboldt] said 'You have found me playing the fool Baron, but I am sure to you I need make no apology.'" Letters and diaries show glimpses of Humboldt in various other drawing rooms of Washington: after a dinner with James and Dolley Madison (at which James's famous wine cellar was much in evidence), Dolley wrote her sister, "We have lately had a great treat in the company of a charming Prussian Baron von Humboldt. All the ladies say they are in love with him. . . . He is the most polite, modest, well-informed and interesting traveller we have ever met, and is much pleased with America." Margaret Bayard Smith wrote her sister-in-law that Humboldt was "a charming man. . . . An enlightened mind has already made him an American, and we are not without hopes, that . . . he will spend the remainder of his days in the United States." Into her commonplace book she pasted a souvenir silhouette of Humboldt given her by Peale. Jefferson's private secretary, William A. Burwell, recorded another revealing moment when Jefferson walked in on Humboldt, who was breakfasting at the White House, waving a newspaper "filled with the greatest personal abuse of himself; which he handed to Humboldt requesting he deposit [it] in a museum in Europe so that all could see 'how little mischief flowed from the freedom of the Press,' given that despite the stream of such abuse Jefferson's administration had never been more popular."[10]

The delegation from the APS returned to Philadelphia on 9 June, leaving Humboldt and friends to spend a few days in Washington on their own. Jefferson invited them all, with his cabinet members, to a farewell dinner, and Humboldt spent his last day in Washington making farewell visits before taking the coach to Lancaster, Pennsylvania, then the largest inland settlement in the United States. Here he spent several days with Andrew Ellicott, who had surveyed the boundary of Washington and trained Meriwether Lewis in mapmaking, and the botanist Gotthilf E. Mühlenberg.[11] Humboldt's route took him through some of the United States' most pleasant and fruitful country, settled largely by German immigrants. It must have felt like home.

By 18 June Humboldt and company were back in Philadelphia, where the first order of business was to secure passage to Europe. Humboldt learned that the *Favorite* would soon be leaving for Bordeaux, and he rushed to tie up loose ends and say his farewells. Over several visits with the famous physician Benjamin Rush—the obstinate and passionate reformer who had ridden to the First Constitutional Convention with John Adams and signed the Declaration of Independence next to Benjamin Franklin—Humboldt shared his speculations over the moral influence of New World gold and silver. They must have discussed at length the environmental causes of such diseases as yellow fever and malaria, for this pet theory of Rush's dominates Humboldt's writings on disease. There was just time to sit for a portrait by Peale, who was gratified by the chance to prove that he could still paint. Still on display at the College of Physicians in Philadelphia, it shows a boyish and rosy-cheeked Humboldt with tousled hair and lively eyes, fresh from his travels before returning to European fame, adulation, and increasing cares. When his friends learned of Humboldt's own skill as an artist, the APS membership was invited to Philosophical Hall for a special showing of his landscape sketches.[12]

Then there was the matter of their passport, a sticky issue during wartime. Humboldt feared that the British, who were searching American vessels for French property, would take "Citoyen" Bonpland prisoner. After being stonewalled by the British consul, Humboldt appealed directly to Madison, who wrote up a passport requiring all U.S. ships to let the Europeans pass "without hindrance," and all "friendly powers" to respect them as promoters of useful science. Gallatin hastened to return Humboldt's precious maps and to put in his hands a packet of statistical information on the United States, which Humboldt promised to publish to "tell the world how altogether admirable and benevolent your financial administration is." (For decades Gallatin continued to send Humboldt updates, which Humboldt featured prominently in his works, fulfilling his promise to spread the word of America's success across Europe.) Finally, as the ship stood at anchor at the mouth of the Delaware, Humboldt scribbled out the first detailed and comprehensive sketch of his American travels as a favor to John Vaughan, secretary of the APS, who must have been frustrated by Humboldt's rapid and multilingual speech and begged him to put it all down in writing. Vaughan translated Humboldt's hurried French into (rather stilted) English and published the account later that year. The resulting essay bristles with themes and ideas that Humboldt would spend the next twenty years developing.[13]

Humboldt left the United States certain he would return. He must have raised hopes that he would stay, for Margaret Bayard Smith is not the only one to venture that after his travels were over he would adopt the United States as his home; her friend William Thornton remarked in a letter to Vaughan, "But I am sorry the interesting Baron has pocketed all South America. I wish he could have rested his Limbs a while & published his works here.—The Treasures of knowledge he has amassed are worth more than the richest gold mine." But as Humboldt was writing to Madison, no matter how much he loved "this beautiful land . . . my duties call me back to Europe and I dare not linger any longer." In a few years, once the route from the Missouri to the Pacific was open, he would, he promised, return to venture through the upper Midwest to the coast as far north as Alaska. Soon he repeated his promise to Vaughan: "Whenever

I think of seeing you again, I get a deep longing for roaming over the western territories, a plan for which Mr. Jefferson . . . would be just the right man to aid me." But, he sighed, his dream of "large projects" in the far West must wait for a couple of years until he had published his current materials (a job that delayed him not two years but thirty). He had other reasons for wanting to return to the United States, which his friends, including Schiller and Goethe, were proclaiming was the future of the arts and sciences. In his farewell letter to Jefferson, Humboldt spoke more broadly, not just of plans for exploration, but of the moral imperative embodied by the United States, which had offered him "the consoling experience" of witnessing true social progress, "whereas Europe presents an immoral and melancholy spectacle." Not that he was unaware of the problems shadowing America. In his farewell letter to Thornton, Humboldt reminded his Washington host of the "abominable law" permitting the importation of Negro slaves into the Carolinas. The laws of humanity dictated that the United States abolish slavery, an act that would cost, he estimated, little more than a dip in cotton exports. Humboldt was good at making such mercantile calculations, but they always enraged him: "But alas! How I detest this Politics that measures and evaluates the public good simply according to the value of Exports! A Nation's wealth is just like an individual's—only the accessory to our happiness. Before being free, we must be just, and without justice there can be no lasting prosperity."[14]

In short, there was much work to do if the promising young republic was to realize Humboldt's hopes, and he expected to return to contribute to its progress. Yet the delays kept growing: to Jefferson in 1811 he wrote regretfully that he could not come quite yet. Two years later he wrote wistfully to Madison (now in his second term as president) that amidst the calamities of European politics "I like to think of that pleasant part of the world which I regard as my second home, where you uphold the light by the power and wisdom of lawful government."[15] Even in Europe Humboldt continued to think of himself as "half an American," though once the *Favorite* sailed for France he would never again set foot on American soil. As the years passed and his youthful hopes faded, he worked to inspire a rising generation of young Germans and Americans to take his place, to help America realize the promise he could only imagine, and encourage, from abroad.

The Humboldt Network

Humboldt stepped ashore in Bordeaux, France, on 3 August 1804, to instant celebrity. He had left France a republic; he returned to find it an empire, ruled by an emperor who despised him. Napoleon harassed Humboldt constantly, but could not let him go, for he, too, was "a French conquest to be retained if possible as a permanent and valued possession." Humboldt did persuade Napoleon to grant a lifelong pension to Bonpland, who became gardener to the Empress Josephine (who did, in fact, collect plants). After dithering on the South American plant collection so long that Humboldt had to farm it out to others, Bonpland eventually returned to South America, where on the Paraguay frontier he was captured, held, and finally released to live out a long life in peaceful poverty. In 1810 Montúfar left Paris for Ecuador, where he enlisted in his father's rebel army only to be captured and executed by the Spanish. José de la Cruz, their steadfast servant, returned to Cumaná in 1805. As for Humboldt, he soon was off to Rome to visit his brother Wilhelm (then the Prussian ambassador to the Vatican), returning at last to Berlin late in 1805, where the king appointed him a royal chamberlain with a pension and no expectation of service.[16] Then came the terrible winter of 1806–7, when Napoleon conquered Prussia and occupied Berlin. Humboldt responded by writing *Ansichten der Natur* to rally his compatriots, and agreed to serve as a mediator with the French. In this role he traveled to Paris with Prince Wilhelm in 1808, and once their business was complete he was given special permission to remain in Paris as long as he wished. The world capital of science would be his home until 1827.

Meanwhile, amidst the travels and turmoil, he plunged into scientific work. With Gay-Lussac he analyzed the chemistry of the atmosphere; with Biot, the phenomenon of geomagnetism. Of all his masses of material, he chose to publish his *Essay on the Geography of Plants* first, for he thought that by showing an imaginative and aesthetic union of scientific phenomena of every description, it would be the best introduction to his life-long project. "The world likes to *see*," he wrote, "and I there exhibit a microcosm in a leaf." The essay, with its spectacular graphic cutaway of Chimborazo showing plant zonation with altitude, laid, in one stroke, the foundations for modern plant ecology. Humboldt's circle of friends and associates widened as he placed his materials in the hands of specialists: plants (after Bonpland lost interest) to his old friend Willdenow and to Willdenow's student Karl Sigismund Kunth; astronomical observations to Jabbo Oltmanns, zoology to Pierre-André Latreille, biology to Georges Cuvier, Indian languages to his brother Wilhelm, and so on. In this way, Humboldt became the de facto center of an entire research front, the central and inescapable figure of French science. His closest friends included the brilliant young radical physicist François Arago, the evolutionary theorists Jean-Baptiste Lamarck and Geoffroy Saint-Hilaire, the great and innovative botanists Auguste-Pyrame Decandolle and

Antoine-Laurent Jussieu, the geologist René Just Haüy, Guizot the politician, and François Gérard the artist, with whom Humboldt studied for some years (culminating in a beautiful and sensitive self-portrait, destroyed in World War II). Humboldt loved to identify and promote young talent. Adolphe Quetelet, for instance, built on Humboldt's and Arago's work to found the science of statistics; Justus Liebig founded organic chemistry and claimed he owed his career to Humboldt; Charles Lyell worked with Humboldt in 1823 and likely derived from him the concept of dating rocks from fossils; Louis Agassiz was about to give up on his scientific career when Humboldt took him under his wing.[17]

Through all his Paris years Humboldt was publishing. It was a ruinously expensive program, and when the final volume came out in 1834 he had long since spent himself into poverty. There were thirty volumes in all, with 1,425 maps and plates, published under the umbrella title *Voyage to the Equinoctial Regions of the New Continent, from the years 1799–1804, by A. de Humboldt and A. Bonpland*: ten volumes of historical materials, including the *Personal Narrative*; zoology in two volumes; *Political Essay on the Kingdom of New Spain*; two volumes of astronomical, barometric, and trigonometric observations; *Essay on the Geography of Plants*; and finally, all the botanical studies. The cost was staggering. Agassiz estimated production costs alone at $250,000, with complete sets selling for $2,000. Only Napoleon's *Description de l'Égypte* cost more. Helmut de Terra calls Humboldt "a St. Michael in mortal combat with a dragon that spewed bills and debts," and in the end Humboldt could not afford a complete set of his own works.[18]

Thus in 1826, when King Frederick Wilhelm III demanded that Humboldt return to the fatherland, he could not refuse. Publication costs had put him deeply in the king's debt and the king's pension was his only income. However distasteful the move, Humboldt, the apostle of freedom, felt obligated to comply. Early in 1827 he moved to Berlin, where he was expected to wait daily on the king's pleasure, entertain the royal family in the evenings, and commute to the royal residence in Potsdam so often that Humboldt said he felt like a swinging pendulum. Humboldt became the king's own Voltaire, just like his father's; the king even had Humboldt installed in Voltaire's garden palace at Sans Souci, where just as in the old days, Humboldt drew around himself writers, artists, and scholars. Yet, loved and trusted as he was by the king and crown prince, Humboldt was hated by the court and, for all the royal favors he could command, he was completely without political power. His circle of friends shrank: among them were the steadfast Mendelssohn family and Karl August Varnhagen von Ense (now married to Humboldt's boyhood friend, the salon hostess Rahel Levin), a fellow liberal to whom he could confide his disgust with court life. (When Humboldt's secret letters to Varnhagen were published, their bitter sarcasms kindled glee among radicals and shocked conservatives; some said Humboldt's reputation for measured diplomacy was ruined forever.) As ever, work was his release. Upset by the ignorance of science in Berlin, he offered a course of university lectures intended to jump-start German science that was so popular he offered a second course at no charge, before a grandstand audience of thousands. He resisted calls to publish them (for the most part he had extemporized from notes), but the idea took hold: out of these lectures was born his last great work, *Kosmos*. Meanwhile, though the East India Company had banned him from India, he did manage a second expedition, to central Russia in 1829. It was everything his South American voyage was not, a state-sponsored and tightly controlled "travelling circus," but it produced important new insights into climate, geology, geomagnetism, and (typically) Chinese literature.[19]

Humboldt's efforts to coordinate international scientific research started to pay off: in 1828 he helped reorganize Germany's struggling scientific society, the German Association of Naturalists and Physicians, by holding a gala scientific conference in Berlin that attracted several hundred guests from across Germany and Scandinavia. Speaking as its newly elected president, he urged that future meetings abandon the restricted membership of old and foster a free exchange of ideas in open conversation, balanced by meetings of specialized sections at which papers would be read and discussed. Humboldt's redesigned association became the model for the British Association for the Advancement of Science, founded three years later, and in 1848 its American equivalent, the American Association for the Advancement of Science. In 1836 Humboldt launched the world's first international scientific collaboration when he persuaded the British to join the United States and Russia in setting up a series of permanent scientific stations where observers would record magnetic and meteorological observations, creating the first global image of the earth's geophysics and climate. The twenty-first-century understanding of climate change and global warming is a direct descendent of Humboldt's visionary work and his speculations that climate changed over time.[20]

After returning to Berlin, Alexander had grown increasingly close to his brother Wilhelm and his wife Caroline. It was a hard blow to them both when Caroline died, like their mother, of breast cancer, in 1829. Moved by their fate, Humboldt wrote a friend that cancer was "the most hopeless and fearful disease by which poor humanity can possibly be afflicted." His greatest grief, though, was when Wilhelm died in his arms in April 1835, probably, given the uncontrollable shaking of Wilhelm's arms and legs, of Parkinson's. "I

never thought my old eyes had so many tears!" exclaimed Alexander to Varnhagen. On he labored, completing *Examen Critique*, his history of the discovery of the New World, and adding more material to *Kosmos*. Embittered
by decades of reactionary politics, trapped in the Berlin court, Humboldt
watched the Revolutions of 1830 and 1848 largely from the sidelines, trying to
help from behind the scenes. Occasionally he stepped forward to offer symbolic leadership: during the uprising of 18 March 1848, as he wrote Arago, "I
was attacked four times by armed men, who did not know me and who had
not read *Cosmos*." Four days later he led, alone, stooped, his white hair bared
to the March winds, the funeral procession for the 183 martyred heroes of the
revolution.[21]

By year's end the king had repudiated the new constitution and conditions were still worse. As his king slid toward insanity, Humboldt slid toward
despair: "I am sad to say that at the age of eighty I am reduced to the banal
hope that the noble and ardent desire for free institutions is maintained in
the people and that, though from time to time it may appear to sleep, it is as
eternal as the electromagnetic storm which sparkles in the sun." The old man
of 1789 felt lost in time. In 1849 he wrote to a friend, "Like the bird perched
above the foaming cataract, of which you have so sweetly sung—the last of
the Atures—so am I now left the sole survivor of an extinct race."[22] He felt he
had returned from the banks of unknown rivers, the wild forests of the Orinoco and the snowcapped burning mountains of the Andes only to find the
civilized universe slipping backward into despotism and oppression. Like the
parrot of the Atures, he was the last to speak the language of an extinct race.

In the United States, though, there was yet hope. From 1804 on, Humboldt's
widening correspondence—by old age he was receiving over three thousand
letters a year, and replying to most of them (think of the postage!)—included
scores of U.S. citizens, and it was said that while he would turn away princes,
he would never turn away an American. Just how many Americans were part
of the Humboldt network will never be known. Many of his letters were lost
crossing the Atlantic, and he destroyed nearly all the letters he received: as
he jotted on one of the few he saved, a letter from Jefferson, he had "the bad
habit . . . of tearing everything up in order to be able to carry everything with
me, no matter what life's conditions are." Given that in both Paris and Berlin
his mail was routinely opened and his apartments periodically searched, he
also had good reason to obsess about protecting his privacy. Much of what
survived Humboldt's housekeeping has also been lost: John Bachman's library and papers were burned during the Civil War, including, Bachman

lamented, his letters from Humboldt; John Frémont's letters, many from Humboldt, were also lost to fire; after his death Humboldt's own library and personal papers were largely destroyed in a warehouse fire; the Humboldt family library and archives were gutted when Tegel was burned and looted in the last days of World War II. Out of what must have been many thousands of letters only a few hundred survive, yet these remnants open a keyhole view onto a ramifying economy of favors requested and enacted, introductions and thanks, information and sentiment and flattery. In this economy Humboldt emerges as a potent force behind the scenes of power and privilege, advancing his projects across a wide front, one gift at a time: a book, a favor, a kind word. He attached his name not just to a science, but to an entire research front that incorporated the labors of hundreds: Jorge Cañizares-Esguerra writes that Humboldt's thirty volumes "should be read not only as the product of a genius working in isolation but also as a summary of the Spanish American Enlightenment." The cumulative effect is quite literally incalculable.[23]

Of his original U.S. American hosts, Humboldt stayed in the closest touch with Jefferson, Madison, and Gallatin. Jefferson was, of course, his first American correspondent, and the two exchanged letters until Jefferson's death in 1826. (Fortunately, thanks to Peale's polygraph, we have copies of both sides of the correspondence.) Humboldt led off by sending Jefferson several letters and a gift copy of his *Essay on the Geography of Plants* that never made it out of war-torn Europe. Finally in May 1808, a package got through: the first volume of his *Political Essay* on Mexico, hot off the press, with some astronomical work from the Andes and a note expressing gratitude that, in the midst of his country's "misfortunes," Humboldt still had the memory of Jefferson to inspire hope. President Jefferson waited until the following year, the eve of his retirement, to reply, then assured Humboldt that none of his previous letters had been received and indeed, he'd had no idea where Humboldt was. Given Humboldt's equivocal political situation, he promised, with some heavy sarcasm, not to "implicate" him by passing along any of the "horrible heresies" of American republicanism.[24]

Humboldt replied immediately and joyfully, and having established each other's location and mutual goodwill, the two entered into a sporadic but continuous exchange of lengthy, thoughtful letters that continued the conversation begun in the White House. Humboldt sent Jefferson successive volumes of the *Political Essay* and begged Jefferson to reciprocate with an autographed copy of *Notes on the State of Virginia*. Jefferson complied, and Humboldt thanked him for the treasure—which, alas, has since vanished, either looted or destroyed. They shared their hopes and fears as events unfolded that they each helped shape. Often they worried about the outcome of the bloody and violent revolutions in Mexico and South America. "What kind of government will they establish?" wondered Jefferson; will they know liberty without "intoxication"? Will they do justice to their Indians? Humboldt feared that the very violence of the conflict would force people to retreat to "the established pattern" of the old social order. Yes, replied Jefferson: the "mutual hatreds and jealousies" in South America would likely result in military despotism and mutual enslavement of warring casts characterized by "profound ignorance & bigotry." Jefferson was not optimistic: "History, I believe furnishes no example of a priest ridden people maintaining a free-civil government." Then his gloom turns to defiance, as he stakes out the Monroe Doctrine ten years ahead of Monroe: "In whatever governments they end, they will be *American* governments," no longer embroiled in Europe's endless wars. "America has a hemisphere to itself: it must have it's [*sic*] separate system of interests, which must not be subordinated to those of Europe." For instance, "we" had been pursuing a "benevolent plan" for the Indians, teaching them peaceful arts and industries that would soon have amalgamated them with us. But British policy "defeated all our labors for the salvation of these unfortunate people" by seducing the tribes "to take up the hatchet against us." Now we are forced "to pursue them to extermination" or drive them "to new seats beyond our reach." As Jefferson writes, his anger rises: such "confirmed brutalization, if not the extermination," of the Indians will "form an additional chapter in the English history of the same colored man in Asia, and of the brethren of their own colour in Ireland and wherever else Anglo-mercantile cupidity can find a two-penny interest in deluging the earth with human blood."[25]

Jefferson's candor with Humboldt is breathtaking—Humboldt too was impressed, for he loaned this particular letter to Madame de Staël to share among her friends—but for his own part, Humboldt, ever the diplomat, was more careful. His letters were briefer and often business-oriented, delivering thanks or requesting patronage for a favorite. In one letter he upbraided the explorer Zebulon Pike for plagiarizing his Mexican map. Jefferson soothed him with a gentle reminder that Pike died a hero and in any case meant only to enlarge knowledge, not gain "filthy shillings and pence." In another letter Humboldt confided that "I have not really felt happy since I left your wonderful country," and romanticized the great statesman's exemplary retreat to Monticello. Whereas they freely shared their worries over the future of U.S. America's Indians, only once did Humboldt broach the one subject that came between them: slavery. In an early letter he apologized to Jefferson for denouncing U.S. American slavery in the opening pages of his *Political Essay*, promising to retract his criticism now that Congress was moving toward "total abolition." As he added, "I was carried away by my devotion for the cause of the blacks and this need not cause me to blush." But the move toward abolition stalled, and Humboldt let his criticism stand. Thereafter in his letters to Jefferson he avoided the topic altogether. After all, he knew and idolized Jefferson's principles, and he of all people knew only too well how principles could be compromised by circumstance. Humboldt would develop his campaign against slavery on other fronts.[26]

No such warmth developed between Humboldt and Madison. Humboldt's earliest letters were effusive as ever, and Madison cooperated in securing his passport, but when Humboldt again took up the correspondence, Madison had ascended to the presidency and Humboldt was approaching him on a delicate matter, the defense of one David Bailie Warden, an Irish scientist who immigrated to the United States in 1799 after being arrested for participating in the rebellion against England. Warden went to Paris in 1804 as secretary to the American minister to France, and though he is forgotten today, for four decades he was an important intellectual and cultural mediator between Europe and the United States. Humboldt befriended him and wrote repeatedly to Jefferson, then Madison, defending Warden's character against his political enemies (certain aristocrats seem to have found Warden insubordinate). Madison's replies, if any, have not been recovered, but clearly Humboldt put the president on the spot: as Humboldt wrote in 1813, on his third attempt, "It would please me greatly if the opinion of a simple explorer of the Orinoco had some weight with Your Excellency." The sarcasm may have been unfortunate. Warden's promotion to minister to France was denied, and Humboldt's letters apparently stopped, except for two quite formal letters of introduction. In 1822 Madison wrote Humboldt a rather chilly note of acknowledgment for sending a book, then suddenly, in 1833, a much warmer letter reminiscing on the connection forged nearly thirty years before, assuring Humboldt of his welcome should he return to the United States—and requesting Humboldt's kind reception of one Professor Hoffman.[27] This was a correspondence of utility, of favors granted and, once, a favor denied.

Where Humboldt's friendship with Madison shut down, with Gallatin it opened up, with far-reaching consequences. After 1804, they, too, fell into several years of silence, but their friendship was renewed in 1814 when Gallatin was sent to London to negotiate peace with England. In Europe all the sympathy was on England's side, and it was generally agreed that the upstart Americans needed to be taught a lesson. Fortunately Humboldt was part of the Prussian delegation, and in any contest between old Europe and new America, he would always stand with America. In Paris he worked out a plan with his friends Madame de Staël and the Marquis de Lafayette to aid the American cause, which he carried into effect with some significant success. The Treaty of Ghent was signed on Christmas Eve 1814, and early in 1815 Humboldt was able to congratulate Gallatin on a job well done. With the negotiations over, Gallatin moved to Paris to serve as the American envoy to France. The two became intimate friends within the Paris circle of Lafayette, de Staël (who was Gallatin's cousin), Chateaubriand, and their associates. Gallatin's son James recalled that Humboldt was virtually a member of the family, and in his biography Henry Adams remarked that "Mr. Gallatin never was so happy and never so thoroughly in his proper social sphere as when he lived in Paris and talked of Indian antiquities with Humboldt."[28] Soon they did more than just talk. When Gallatin retired in 1823 to return to the United States, it was Humboldt who suggested he take up his studies of American Indians and turn them to account. Gallatin did exactly that. For nearly three decades he worked on the Indian tribes of North America, completing several foundational monographs, and in 1842 he founded the American Ethnological Society (still going strong today), naming Humboldt an honorary member. Thus it was under Humboldt's inspiration and encouragement that Gallatin founded the science of American ethnology, a role for which he is still honored.

A new chapter opened in the growing Humboldt network when, after the last aftershocks of the Napoleonic wars subsided, a wave of ambitious young Americans traveled to Europe to study. In the vanguard were George Ticknor and Edward Everett, friends who sailed together in 1815 to study at Göttingen University, where old Professor Blumenbach was still holding forth. When Harvard offered Ticknor the Smith professorship of French and Spanish, he took off to tour Italy, Spain, England, and France. While in Paris he called on Humboldt, initiating a friendship that lasted until Humboldt's death. In his journal Ticknor sketched his first impressions: Humboldt was "one of the most interesting men in the world, and the idol of French society." He was impressed with Humboldt's firm step and the "decision and force" of his every movement, and with Humboldt's extraordinary schedule: he worked through the night, slept whenever weary, ate only when hungry, and joined multiple dinner parties every evening not for food but for "pleasure and amusement." Only a strong constitution and the reputation of a king could take such a punishing life, and Humboldt, Ticknor concluded, had both. The two must have met often. In May Ticknor recorded a dinner with Benjamin Constant (Ticknor also met de Staël, Constant's estranged wife), where the company had assembled to hear Humboldt read from his travel writings. The next morning Ticknor recalled Humboldt's "genius and modesty" and his "magical" descriptions of the Orinoco, the holy solitudes of nature, and the missionaries. In June he noted that Humboldt had helped him out of some unnamed "difficulty" with the Paris police, occasioned, Humboldt suspected, by an English spy.[29]

Ticknor returned to the States to become one of the era's most important cultural mediators. He taught at Harvard from 1819 to 1835, where he made foreign languages "the live literary center of the college" for such students as Emerson and Thoreau. He twice more spent time with Humboldt, the first time when the Ticknor family went on a long tour of Europe and Humboldt showed them the sights of Berlin, including Berlin University (founded by, and now named for, his brother Wilhelm). Ticknor was impressed by the "great deference" shown everywhere to Humboldt, and joked that his own family's stock had risen considerably since they had been seen in his company. Back in the United States, he set to work on his *History of Spanish Literature,* and, extending the network, wrote William Hickling Prescott a letter of introduction to Humboldt, who in turn assisted Prescott with his epic histories of the conquest of Mexico and of Peru, and publicly praised the books of both Ticknor and Prescott as good examples of cultural progress in the New World; to John Lloyd Stephens, Humboldt commended Prescott as a historian without equal in England or Germany.[30]

Their last visit was in 1856, when Ticknor was in Europe buying books for the new Boston Public Library and Humboldt again smoothed the way for his old friend. After a sentimental day spent at Tegel, Ticknor wrote to Humboldt that "since 1817, when I first saw you in Paris, you have been foremost among my *memorabilia.*" To Prescott, he reported that their good friend (now eighty-seven years old) was much aged but as active as ever. When Humboldt knew he had not long to live, it was to Ticknor that he wrote a loving farewell letter, requesting he translate and publish it as an answer to all his many friends in the United States, "colored as well as white," whose letters he could not answer personally. Ticknor hastened to do so, and assured Humboldt that no fewer than half a million copies had been distributed across the country.[31]

As Ingo Schwarz has pointed out, both Ticknor and Humboldt were reform-minded scholars who maintained a wide circle of cross-Atlantic friendships and shared a deep interest in science, literature, and culture; both were militant mediators between Europe and the United States "in a day when nationalism was as fashionable in literature as in politics." Both were major forces in what I have elsewhere called "the Culture of Truth," a cosmopolitan, high-minded, reform-centered clerisy that sought to rise above divisive sectional and national interests to create a worldwide network of progressive intellectual leaders, authors, and teachers. Ticknor demonstrates how this American clerisy aggressively sought ties with European leaders, and his friendship with Humboldt illustrates how each used the other's influence and connections to further their ideals.[32]

The same could be said of Ticknor's friend Edward Everett, who stayed behind at Göttingen to become in 1817 the first American ever to be awarded a PhD. He, too, found his way to Humboldt, visiting him the following year in Paris. Humboldt gave Everett a letter of introduction to his brother Wilhelm, and after Wilhelm's death Everett wrote Alexander a heartfelt letter thanking him for this kindness. For Wilhelm and Everett—the former, one of Europe's most famous intellectuals and diplomats, the latter "a youthful & obscure stranger"—had become life-long friends. Although personally closer to Wilhelm, Everett nevertheless played a key role in promoting Alexander von Humboldt in the United States. Soon after taking up his own Harvard professorship in 1819, he told a callow but promising young student to read Humboldt's works. That student, Ralph Waldo Emerson, embarked on a career-long reading of Humboldt and in turn introduced his works to Henry David Thoreau. By 1823 Everett was editor of the august *North American Review*, and in this role, having already introduced Goethe to America in a lengthy and erudite review, he undertook to do the same for Humboldt. In the first important notice of Humboldt by a U.S. American, Everett urges *everyone* to read his books: "Works so important to America as those of M. de Humboldt, deserve, if any where, to be known and prized in this country." In an essay full of humor and pleasantries (and a few barbs at the slow and erratic publication of Humboldt's massive works, which Everett realizes is the fault of the dauntingly expensive printing process), Everett calls Humboldt "a philosopher in the truest sense of the word," and after providing a remarkably complete and well-informed survey of Humboldt's writings, he closes with the pages from *Personal Narrative*, chosen to be "instructive as well as amusing," detailing the miseries of Humboldt and company with the mosquitoes of the Orinoco.[33]

As Everett's career veered from letters to politics (he served five terms in Congress, a term as governor of Massachusetts, and six terms in the Senate, with a spell as president of Harvard for good measure), the Humboldt network continued to ramify: his brother Alexander Everett had just been appointed minister to Spain by John Quincy Adams, and soon Alexander invited America's most famous writer, Washington Irving, to Madrid, in order to write a biography of Columbus. Irving, who had been knocking around Paris since 1823 (where he got to know Humboldt) happened to be looking for work, since he had lost a fortune speculating in South American mines (an investment bubble widely blamed, unfairly, on Humboldt). A biography of Columbus promised to secure him permanent literary fame and at least a modest fortune. As Irving labored on the first Columbus, he mined Humboldt, the so-called second Columbus, for source material, and briefly suspended his labors under the mistaken impression that Humboldt was writing a rival work. In fact, Humboldt was collecting materials for *Examen Critique*, and the two forged a friendship, working together to determine the route of Columbus and his first landing. Irving helped Humboldt solve an important puzzle—why was the New World named "America" rather than "Columbia"?—by locating the 1507 map containing the first appearance of the name "America," and in turn Humboldt traced the map to the German mapmaker whose use of the name "America" had propagated Vespucci's name rather than Columbus's throughout the world. Humboldt deprecated his own book, which even his closest friends called a picture without a frame (it has never been translated into English). Some critics condemned Irving's biography of Columbus in similar terms, but Humboldt himself praised it: while his own treatment was weighed down by petty details of nautical astronomy and natural description, Irving's "has demonstrated that the power of imagination in the higher intellect does not necessarily preclude the ability to devote oneself with success to the more serious study of the historian." Thus both Irving and Humboldt profited from each other's works to make original contributions to the historiography of America—although Humboldt's contribution remains virtually unknown in the United States.[34]

Shall we take one more pass through the byways of the Humboldt network? In May 1821, George Bancroft, yet another young American on the make who took off to study at Göttingen, wrote Harvard's President Kirkland of his delight in meeting Humboldt: "The more I see of Mr. de Humboldt, the more I admire him; he does understand the art of talking to perfection." Humboldt makes the driest subjects interesting, can manage a political discussion with the masters, yet he "talks to the ladies with as much ease as if he had passed years in frequenting salons and drawing rooms, instead of climbing Chimborazo and exploring Mexico." It happened that just before writing this letter, Bancroft had been visiting with Wilhelm as well, who mentioned that he wanted to open a correspondence with the American philologist John Pickering (whose *North American Review* article on Native American languages and customs had impressed him), to share materials on North American Indian languages. Bancroft the matchmaker, who knew Pickering, put the two in touch with each other, and soon Pickering and Wilhelm von Humboldt were exchanging closely written and detailed letters comparing Indian grammars and vocabularies. Pickering deeply influenced U.S. American linguistics, to which he brought a strong orientation to German thought, and he in turn influenced Wilhelm von Humboldt's thinking on language; they shared a radical new idea, a comparative study of the world's languages.[35]

Wilhelm von Humboldt seems to have been a regular reader of the *North American Review*, for in the same issue as Edward Everett's 1823 review of his brother's works, he found a lengthy article on the work of one Jedidiah Morse, like Pickering a scholar of Indian languages. Morse's report detailed the economic and educational progress of the Cherokee, Creek, Choctaw, Chickasaw, and Seminole tribes, and as Daniel Walker Howe states, "advised that they be left in peace to continue it." Such sentiments resonated deeply with Wilhelm's ideas, and soon after the article appeared he asked Pickering to forward his compliments to Morse, who, amusingly, mixed up his Humboldts and replied not to Wilhelm but to Alexander, taking advantage of the situation to introduce his sons Richard and Sidney. Soon Sidney improved the opportunity by following up with a letter of his own to Alexander von Humboldt, together with a copy of his geography textbook, which Humboldt received graciously. However, it was neither Richard nor Sidney whom Humboldt actually befriended, but their older brother, the artist and inventor Samuel F. B. Morse, who teamed up with Humboldt to develop the telegraph and lay the transatlantic cable that would make possible instantaneous communication between Europe and the United States.[36]

We're not quite done. While Wilhelm von Humboldt and Pickering were swapping information on Native American languages, so were Alexander von Humboldt and his old friend Gallatin, and Alexander was funneling all Gallatin's material to Wilhelm for his magnum opus on North American Indian languages—never, alas, completed. As with the Morses, the Pickerings' Humboldt connection carried on to the next generation: John Pickering's nephew Charles would ship out as the zoologist on the Wilkes expedition of 1838–42, the United States' first large-scale, government-funded Humboldtian exploring expedition. Charles Pickering combined linguistics, ethnology, botany, and zoology to build on the work of his uncle John, of Gallatin, and of both the Humboldt brothers to analyze the migrations and dispersal patterns of the world's indigenous peoples by tracking the migration and evolution of calendars, astronomical observations, languages, and food plants. This interdisciplinary biogeographical methodology was first proposed by Alexander von Humboldt in his *Essay on the Geography of Plants*, which Edward Everett had singled out for particular praise in his 1823 article: "M. de Humboldt has . . . taught us, that if, in one series of observations, the history of our race is written in the heavens above us, it may be traced in another on the surface beneath our feet."[37] In later years, Pickering's and Gallatin's Humboldtian approaches would draw them, with John Bachman, onto the losing side of the American debate over polygenesis: while scientific and popular opinion swung to the emerging dogma that only whites were fully human, these three, alone, held steadfast to Humboldt's insistence on the essential unity and common descent of all human beings of any race and color—a view that Humboldt's greatest follower, Darwin, would soon dramatically confirm.

As these dizzying passages through the Humboldt network suggest, the links visible today are barely the tip of the iceberg. Chance connections ramified in the most improbable of ways, tying together nations, fields, and generations along Humboldt-inspired trajectories. The borrowings and appropriations were mutually opportunistic. Americans leveraged their prestigious Humboldt connections to acquire international status, exchange favors, and circulate their works abroad; on his end, Humboldt used his correspondence with Americans to advance his own projects and to advocate for causes that particularly interested him: the economic and moral effect of the global circulation of American gold and silver; the abolition of slavery; the best means and location for laying the transatlantic cable that would finally bond Europe and the United States in space and time; the dream of connecting the Atlantic and Pacific via a canal across the Isthmus of Panama, a pet project Humboldt promoted tirelessly starting in 1803. This was not a correspondence of courtesy, a polite exchange of sentiments and mere goodwill. These were letters that made things happen, that marshaled talented young men, and women too, into the causes Humboldt believed in. His letters became passports that smoothed the way for his protégés everywhere on the globe, and marching orders for two generations of scholars, historians, scientists, politicians, and artists, who saw in Humboldt's global transdisciplinary program enormous, apparently infinite, opportunity. Entire careers, entire fields, opened up in these decades under Humboldt's diplomatic but persistent prodding.[38] It was, truly, the age of Humboldt, as his contemporaries recognized. It would take some six decades to play out the meaning and consequences of the ideas he first shared in those whirlwind weeks on the shores of America in the spring of 1804.

The Many Faces of Humboldtian Science

More than careers and fields opened up in the age of Humboldt: so did half a continent. In 1803, the United States was a coastal nation whose citizens mostly lived where they could smell the sea. The French occupied the center of the continent from Canada to Louisiana; the Spanish held Florida and the Southwest clear to Oregon country (which the British and Russians also claimed), while the Russians held Alaska. Had that map crystalized into permanence, North America would have been an uneasy federation of at least five nations, acting variously in concert, or not, with hundreds of Indian tribes. By Humboldt's death in 1859, the French were long gone, the British were confined to Canada, the Spanish to a much-diminished Mexico, the Russians were soon to sell Alaska to the United States, and the Indians were fighting rearguard battles in the far West. The coast-to-coast "manifest destiny" of U.S. America had been inscribed on the map and on the land. Humboldt's tools and methods made this possible, and when non-Americans got in the way, his republican ideology was used to rationalize sweeping them off the stage. His followers ran the gamut from lone romantic explorers and sensitive poet-artist-scientists, to keen and calculating bureaucrats who propelled the United States onto the stage of international science, to cowboy and gunboat diplomats of manifest destiny. Some, like John Charles Frémont, managed to be all three at once.

Common to all was some version of what has become known recently as "Humboldtian science." Yet its first and in some ways most important model does not, today, look much like "science" at all—more like applied economics with a moral, historical, and environmental twist. For after completing the brief *Geography of Plants*, intended as a microcosm for all his work, Humboldt turned next to the macrocosm of Mexico, publishing the massive *Political Essay on the Kingdom of New Spain* from 1808 to 1810. The vast, closed kingdom of New Spain, which stretched from Panama's northern border north and west through Texas and Colorado to Oregon, was about to revolt and step onto the world stage, and Humboldt's book reached U.S. readers in 1811, just as insurrectionary movements for independence were spreading across the region that now, thanks to the Louisiana Purchase, formed America's southwestern border. Contrary to the assumption of most current scholarship, it was this book, not his *Personal Narrative*, that made Humboldt's name "a household word in the educated circles of the Western world." This means, as Nicolaas Rupke observes, that Humboldt rose to international fame not as the romantic explorer of tropical forests and snowcapped mountains, nor as the "cosmopolitan universalist" of *Cosmos*, but "as a colonial surveyor of Spanish Central America, appealing to Eurocentric political and economic interests."[39]

Yet Rupke's analysis doesn't quite capture the distinctive nature of Humboldt's remarkable survey, which is based on his premise that, in Humboldt's words, "the progress of population and welfare of the inhabitants" cannot be understood without knowing "the physiognomy of a country, grouping of mountains, extent of plains, elevation which determines its temperature; in short, whatever constitutes the construction of the globe."[40] As he would assert graphically in *Vues des Cordillères*, the face of nature is simultaneously natural and human. *Political Essay* is a full-scale environmental history, a holistic survey that threads human purpose into the matrix of nature. It develops a proto-ecological view of a major region based on ecology's originating discipline, economics, which then meant not simply financial administration, as in today's much more restricted usage, but something closer to the original Greek: *oikonomia*, "household management."

Humboldt began his great work on Mexico with the problem of how to represent the landscape through the medium of cartography. Existing maps were wildly inaccurate, and though truly accurate maps were, in the present state of ignorance, impossible, one had to try. Repeating the lesson he had learned on the Orinoco, where he found the Indians were "the best geographers of their country," Humboldt recommended that mapmakers embrace local knowledge as well as scientific surveys. Muleteers, for example, may not know absolute distances, but "they learn from long habit if one distance be the third or fourth or the double of another." There was also the problem of depth: how could a flat map show three dimensions? Elevation may have been of little interest in Europe, but it was critical in America, whose high interior plains determined climate and thus the possibilities of cultivation. So spatial depth was important. But so was time depth: both the statesman and the scientist had to understand geology. In fact, fairly representing the New World required so much information that one map wasn't enough: the delicate hatched lines that defined the slope and undulation of the ground obscured boundary lines and rendered place-names illegible. Ideally each region needed two maps, one physical, one geographical. Yet every map, Humboldt warned, even the most detailed, was a selection and a compromise.[41]

In volume 1, Humboldt is interested in the shape of the land and how it came to be named (he is particularly interested in the indigenous origin of the word "Mexico"), its various climates, peoples, and cities, its transportation networks that, by water or land, connect or divide regions, its trans-Atlantic and trans-Pacific global connections with Europe, Africa, Japan, China, India. In successive volumes he details each of New Spain's "intendancies" or political units in turn—Mexico, Puebla, Guanajuato, Durango, Old and New California, and so on—then turns to economic categories: agriculture, mines, manufactures and commerce, diseases, military defenses. Everywhere he emphasizes six basic themes: first, the social construction of race, particularly of "whiteness" in a land of racial mixture, and the influence of race on emergent nationalism amidst bitter conflict between ethnic groups, particularly Indian,

Creole, European, and the enslaved African. Second, the history of Spanish voyages of exploration, particularly those that plied the Pacific coast well into Alaska, and Spanish attempts at colonization in those regions. Third, the environmental consequences of abusive colonial land practices, particularly deforestation, which as in Venezuela's Lake Valencia had caused the desiccation of entire regions. Fourth, the history and productivity of Mexico's mines, including their labor practices. Fifth, an indictment of the false wealth produced by the mines, for gold and silver are the corrupted fool's gold of feudalism: the true wealth of Mexico lies in its agriculture, "the only capital of which the value increases with time," especially if crops are thoughtfully cultivated according to conditions of climate and planted to promote the needs of nourishment rather than global commerce. Finally, a portrait of New Spain's commerce and manufactures, together with an outraged indictment of sweatshop labor practices.[42]

Humboldt's message throughout is that once the immense, wealthy, and highly cultivated country of Mexico is "freed from the fetters of an odious monopoly" enforced by Spain, it will become an economic powerhouse with boundless potential, commanding economic traffic across every hemisphere. His *Political Essay* is, in short, an instruction manual for the economic development of an independent, postcolonial Mexico, and it ends on the eloquent warning that the prosperity of both Americas depends on sharing all the advantages of civilization and social progress with the "copper coloured race." He clearly foresaw that New Spain was about to become an independent nation, and for his role in creating it, Mexico made him and Bonpland honorary citizens in 1827 and continues to this day to honor them as popular heroes. That Mexican independence would precipitate a war with the United States Humboldt did not foresee. In his closing pages he wrote hopefully that there was no need to discuss the interior defenses of Mexico, "for the principles of wisdom and moderation by which the government of the United States is animated, lead us to hope that a friendly arrangement will soon fix the limits between two nations, who both possess more ground than they can possibly cultivate." Those words reflected his adulation of Washington, Jefferson, and Madison. He did not anticipate that his beloved United States would soon elect such presidents as Jackson, Polk, and Buchanan. As he watched from a Europe steeped in imperial design, he saw his statistics and maps, his models and even his own disciples, hatch something new: an imperial America.[43]

For it was Humboldt's fortune to arrive in the United States just in time to be swept into what William Goetzmann, the canonical historian of western exploration, calls the "era of imperial rivalry," a period of international competition for control of the West which began with Lewis and Clark and ended when the United States annexed Oregon and northern Mexico. Goetzmann's many volumes show that the United States won that rivalry by making the best and most effective use of Humboldt's science. Was Humboldt to blame, then, for American imperialism? True, in every contest with Europe, he leapt to defend the United States; but in the contest between his two homelands of the heart—the United States and Mexico—he was torn. As always when his friends fought each other, he fell silent lest he alienate himself from either of their affections. At the height of the Mexican War in December 1847, his old friend George Bancroft, then minister to England, put him on the spot: Just whose side was he on? Humboldt equivocates: The Mexicans gained independence before they were ready for it. They lacked a history of liberal institutions; they lacked union. "For us to come down and take all Mexico he deemed impossible or rather an unwise design, but all the north to latitude 35 he thought we ought certainly to have. Such opinions so strongly expressed he could not publish; for he holds a situation at the Prussian court and is, moreover, a Mexican." At that very moment General Winfield Scott, who had taken the same route as Cortés, was headquartered in the very palace Humboldt had memorialized in his engraving of the Great Square of Mexico. (The marines who captured the palace called it the Hall of Montezuma, an ironic gesture to the annihilated Aztecs repeated every time the "Marines' Hymn" is sung.) Four weeks later, days before the Treaty of Guadalupe Hidalgo was signed, Bancroft wrote excitedly to President Polk that Humboldt was pleased with Polk's decision: "The amount of territory you demand, he deemed to be legitimately due to us." That Bancroft cared so much for Humboldt's opinion is revealing, but by then whatever Humboldt thought was moot. The new border was a done deal that would precipitate both nations into civil wars. This could hardly have been the "friendly arrangement" Humboldt had hoped for nearly forty years before, but he rationalized that a United States administration would in the long run bring more liberal institutions to the territories it claimed.[44]

Humboldt was not unaware that in sharing his Mexican information he was overstepping the bounds of his royal passport. As he wrote to Jefferson, the *Political Essay* "was dedicated to King Charles IV so as to pacify the attitude of the Madrid government toward certain individuals in Mexico who furnished me with more information than the court would have regarded proper." The information he handed over first to the American government, then four years later to the world, gave the United States two tools whose value was beyond calculation. First, for all its shortcomings, his map of Mexico was still the best ever drawn, opening to American eyes the road to the gold and silver mines of Mexico and beyond to the Pacific. For forty years it guided a succession of exploring expeditions into what would become the American Southwest. Humboldt cautioned that immigrants lured into the barren lands of Texas by the mystique of Mexican mines would find themselves stranded and the mines still distant, but he knew no words of his would stop the movement, and of course it didn't: as soon as Mexican independence removed the fear of Spain, Americans started filtering into Mexico and the erosion of Spain's northern empire had begun.[45]

Second, not only did Humboldt give the United States information, he gave it methods. Lewis and Clark's journey to the mouth of the Columbia may have been a triumph in the eyes of its bicentennial celebrants, but in terms of useful science, it fizzled. The explorers lacked the training or instruments to draw accurate maps, to collect and prepare scientific specimens or gather detailed ethnographic or linguistic information, to take geological and geographical measurements, or to publish their results to the world. Jefferson found it necessary, seven years after their return, to apologize to Humboldt: "You will find it inconceivable that Lewis's journey to the Pacific has not yet appeared, nor is it in my power to tell you the reason." Jefferson certainly tried. He had sent Meriwether Lewis, his personal secretary, to Philadelphia to be crammed full of natural science, and his instructions to Lewis were in the best eighteenth-century tradition: they should explore the route to the Pacific, observe latitude and longitude at key points "with great pains and accuracy," study every aspect of the Indians—names and numbers and extent of Indian nations, "language, traditions, monuments," occupations, food, clothing, diseases, laws, and the state of their morality and religion—also "the soil and face of the country," its plants, animals, minerals, fossils, waters, volcanoes, and climate. According to Goetzmann, Jefferson's instructions keynoted "a more flexible and economically mobile American approach to the West" that allowed American explorers to beat out competing nations, focused as they were on specialized interests such as advancing the fur trade. Thus despite its scientific weakness, Lewis and Clark's wilderness adventure succeeded admirably, for they "injected the United States into the struggle for a national empire."[46]

However, once the United States had been injected into that struggle, victory would demand something more, and this is what Humboldt's methods supplied. Jefferson's old-fashioned instructions redacted his own approach in *Notes on the State of Virginia*, as if the goal were a gentlemanly survey of a land and its notable peoples and productions with the intent of recruiting settlers; the Virginia model dates back at least to Robert Beverley's *History and Present State of Virginia* in 1705, and arguably to John Smith in 1607. Humboldt had certainly been influenced by Jefferson's book, but he used his training in Enlightenment scientific exploration (exemplified by Captains Cook and Baudin) to transform the old gentlemen's survey into something new, a disciplined and rigorous collaboration by members of a team, each trained to use the latest and most accurate instruments and to pursue a specific specialty through to publication. This collaborative approach evolved out of Humboldt's and Bonpland's shoestring expedition, in which Bonpland focused on plants and insects while Humboldt emphasized astronomy, geology, zoology, and ethnography. To publish his results, Humboldt had been forced to recruit a team of specialists in Paris. By the time truly "Humboldtian" expeditions were underway, starting in the late 1820s, the scientific team was not back in Paris but out in the field, its various specialists getting wet, slapping mosquitoes, sleeping on the ground, and tinkering with instruments just as Humboldt and Bonpland had done in the Orinoco and Andes.

Historians have given this transformative new paradigm the name "Humboldtian science." Strictly speaking this paradigm did not originate with Humboldt, whose sources included Herder, Kant, Georg Forster, Willdenow, and Jefferson, as well as Creole scientists from South America and Mexico. Nor did it completely characterize his own practice, for Humboldt's emphasis on the role of aesthetics and perception was seldom taken up by his scientist followers. Yet it does capture the exacting precision of Humboldt's science and the way it was propagated throughout the world. The notion of "Humboldtian science" was first introduced in 1959 by Goetzmann, who stressed that Humboldtians "always looked for underlying patterns, unities, and laws which linked all parts of the globe and cosmos, practically, philosophically, aesthetically, and spiritually," making " 'Humboldtian science' synonymous with 'romantic science.' " The term entered the mainstream of the history of science through Susan Faye Cannon, who credited Goetzmann for originating it but criticized him for not asking *why* Humboldt insisted on such far-flung collections and such fanatically accurate measurements. In her view, Humboldtian science is "the accurate, measured study of widespread but interconnected real phenomena in order to find a definite law and a dynamic cause." Humboldt was revolutionary not for inventing all the pieces, which he derived from hundreds of sources, but for "elevating the whole complex into the major concern of professional science for some forty years or so," roughly the first half of the nineteenth century. For young scientists in those decades, jumping aboard Humboldt's program meant riding "the latest wave of international scientific activity: they are being cosmopolitan."[47]

Historians of science have been arguing ever since about the exact nature and dimensions of Humboldtian science. Here it will suffice to propose that the complete Humboldtian follows four commandments: Explore, Collect,

Measure, Connect.[48] First, explore: the Humboldtian believes that nature is an ensemble of complex forces and phenomena that cannot be fully understood by chopping it up and parceling it out into a laboratory. Nor can parlor theories say anything useful about outdoor nature. The scientist must be immersed in the sensual particulars of nature and open to the challenge of the unpredicted, the surprising, even the overwhelming. Nature speaks: the scientist must go out, and listen. Second, collect: patterns are difficult to detect in the rich chaos of outdoor nature. One must collect samples—plants, rocks, insects, fossils, fishhooks, and pottery—and bring them back to a central location for description and comparison. Nor does the collector pile up heaps of just everything, magpie fashion. As Darwin several times observed, if data is to be useful it must be gathered for or against an idea. Collecting is not a mindless grabfest but an effort of intelligence that creates, in effect, new organs of perception. The same can be said of the third commandment: one measures not everything, but for or against a theory. As collecting takes up concrete objects and relays them to a distant space for reassembly into a new order, so does measurement render the abstract qualities of nature into tangible and portable quantities which can, like specimens, be assembled elsewhere for comparison and identification of patterns. To take one instance, Humboldt was obsessive about barometric measurements because only thus could he determine elevation, a key component of climate. All his measurements—of the blueness of the sky, transparency of the atmosphere, strength, depth, temperature, and direction of ocean currents, and on and on—were directed toward identifying subtle interactions of hitherto unsuspected natural forces that could have major consequences. For example, he documented that soil temperatures rose when plant cover was removed, increasing the rate of evaporation, which led to the desiccation of once-fertile croplands. Only exact and painstaking measurements of temperature, rates of evaporation, humidity of the air, rainfall, and cloud cover could move this notion from unfounded speculation to an established causal relationship that could, and soon did, govern land use policies limiting deforestation.[49]

Finally, connect. The goal of all the myriads of single specimens and minute observations and careful measurements was to detect patterns that would point to underlying laws. The smallest of things could, collectively, lead to the largest of ideas. Humboldt's method led to extraordinary conceptual leaps. One of his followers, Adolphe Quetelet, invented the science of statistics by applying the theory of astronomical measurement to human societies. Teams of observers stationed across Russia, the United States, and the British Empire measured subtle variations in the earth's magnetic field over space and time; those measurements, reduced by Edward Sabine in London, revealed the existence of magnetic storms which correlated with the eleven-year cycle of sunspots. As Sabine wrote, "We find ourselves landed in a system of cosmical relations, in which both the sun and earth, and probably the whole planetary system, are implicated." (It was Sabine's wife, Elizabeth Sabine, who translated *Cosmos* into English.) Charles Darwin, following Humboldt to South America, brought back collections of birds that, compared together in a London laboratory, revealed a startling pattern that no existing theory could explain. Darwin's search for connections led him to formulate his theory of evolution by natural selection, and to work out his theory in detail, he spent many years collecting every species of barnacle in the world and analyzing their anatomical connections.

The most powerful way to find pattern in a mass of data is to cast it into visual form. Often what puzzles the mind can be seen by the eye. As Anne Godlewska observes, Humboldt's science was not just a way of thinking, but a way of seeing that combined "both conceptual depth and rigor with holistic vision," creating "systematic rather than geographic time/space" out of immense volumes of data. Humboldt pioneered and developed innovative new ways of presenting data visually, in thematic maps, charts, and graphs: cutaway cross-sections of continents to show systems of land forms and elevations; maps of errors in locating cities and coastlines; graphs of the flow of precious metals from the New World; tables correlating population and territorial size, or elevation of plant species and a multitude of environmental factors. Most famous of all, Humboldt developed the use of isolines, such as isotherms connecting points on the map of the same temperature. Thus every weather map in the morning paper or nightly newscast traces back to Humboldt. The implications could be huge: his map of global climates showed an "isothermal zodiacal belt" across the northern temperate zone that correlated exactly with the course of westward expansion. This added fuel to the fire of Manifest Destiny, since square in the middle lay the continental United States. Geography became geopolitics.[50]

Humboldt's program was far beyond the resources of any single individual. Even he, the independently wealthy and tireless polymath who slept four hours a night and worked around the clock, quickly spent himself into poverty. To realize it meant committing the resources of whole nations and legions, armies, of workers. This is precisely why Humboldt put his stamp on the age: both the United States and the British Empire made Humboldtian science the government-sponsored research front of its day, the equivalent of the moon shot or the human genome project. Goetzmann estimates that at its height in midcentury, the U.S. government subsidy of Humboldtian science "must, at times, have represented perhaps one-quarter to one-third of the federal budget."[51] Glamour and funding drew in the best available talent, turning science from the avocation of wealthy gentlemen amateurs to a paying middle-class profession pursued, increasingly, through university training, which produced not Humboldtian generalists but specialists dedicated to pushing the limits of knowledge in one limited and highly disciplined field. Too much for one man, Humboldtian science demanded, and got, teamwork and collaborations. The old-fashioned lone naturalist was displaced by a new cadre of ambitious young specialists who knew how to network. How, in fact, to link up with the Humboldt network: from its beginnings in Jefferson, Gallatin, Pickering, Ticknor, and Everett, the network exploded until in fifty years it encompassed virtually all of professional science in Anglo-America and much of it beyond.

What's more, this process of professionalization split Humboldtian science in two. It created a need for the ambitious specialist, but it also appealed deeply to amateurs and enthusiasts with a natural history bent, or even just a vivid appreciation for the poetry and beauty of nature. Thus in this period the professionals did not work in a vacuum but in an environment of widespread public interest and sympathy. How else could elected representatives have been willing to allocate such staggering sums to the cause of science? The exciting new advances in science were the talk of the streets and the parlors, classrooms and lecture halls. In an age of ever-cheaper newspapers, magazines, and books, and of growing middle-class leisure to read them, to be literate and educated was to be involved in Humboldt's new science.

By Land and by Sea

One can watch the growth and splintering of Humboldtian science across the procession of voyages and explorations it inspired. American exploration did not revolutionize itself overnight, as one can see from the expedition that followed on the heels of Lewis and Clark. In 1806, Jefferson commissioned Zebulon Pike to explore the Spanish borderlands under the authority of General Wilkinson, the man who had swiped Humboldt's map, which he passed along to Pike. Pike probably didn't know about Wilkinson's conspiracy with Aaron Burr (then fresh from killing Alexander Hamilton in a duel) to seize and rule an inland empire carved out of the Louisiana Purchase. The plot collapsed when Wilkinson, the double agent, tipped off both Jefferson and Spanish authorities. Burr was tried for treason, acquitted on a technicality, and allowed to return to his law practice in New York. Pike made it as far

as the Rockies (where he tried and failed to climb Pike's Peak), and when
he was finally captured by the Spanish, they marched him to Mexico City,
ironically enabling him to reconnoiter all the country in between before the
Spanish released him to return to the United States. He published his plagiarized map in 1810, causing Humboldt's huff of protest; his accompanying
report included the significant information that, just as Humboldt had said,
the western plains were arid as Africa, a "Great American Desert" useless for
settlement. Pike was killed in the War of 1812 while invading Toronto under
the command of Wilkinson, who had been cleared of all charges.[52]

The next major federal expedition was, by contrast, a fully Humboldtian
affair, with the goal of reconnoitering the borderlands above New Spain. It
consisted of a team of scientists and artists led by Major Stephen H. Long,
member of the newly appointed Army Corps of Topographical Engineers.
With him were the zoologist Dr. Thomas Say (the first American to use fossils to date geological strata), and the naturalist Titian Peale (just a toddler
when Humboldt had visited his father). Goetzmann reports that this was "the
first time that an official corps of discovery which included trained scientific
specialists had ever been sent into the West." As they steamed up the Missouri in 1819 at least one observer was impressed: "Botanists, mineralogists,
chemists, artisans, cultivators, scholars, soldiers; the love of peace, the capacity for war; philosophical apparatus and military supplies; telescopes and
cannon, garden seeds and gunpowder; the arts of civil life and the force to
defend them—all are seen aboard." They ascended to the Rockies, climbed—
and yes, measured—Pike's Peak, and returned, their welcome home marred
by the fact that instead of descending the Red River, as instructed, they had
mistakenly followed the Canadian, leaving the putative border with Spain,
which had just been negotiated by John Quincy Adams, still unknown. Their
scientific report, published in 1822, was compromised by the loss of their scientific notebooks, but it did contain much new information and an updated
map. Furthermore, they confirmed Pike's observation of the arid nature of
the high plains. James Fenimore Cooper mined the details of the Long report
for his bleak descriptions of America's inland empire in his Leatherstocking
novel *The Prairie* (1827).[53]

At the same time, there was a very different kind of Humboldtian voyager
on the loose whose goals were less national or imperial than personal. Inspired by Humboldt's example, and often by his personal assistance, a series
of Germans came to see and record the American West for themselves and
explain the meaning of the new lands and their indigenous peoples to the
world, often developing a critique of the American imperialism that they observed as outsiders. These "children of Humboldt," in Goetzmann's phrase,

would have a tremendous impact on European and American perceptions of the West.

The first seems to have been Paul Wilhelm, Duke of Württemberg, who had grown up longing to follow Humboldt to Mexico and the Andes. Duke Paul knew Humboldt's writings well, and strove to emulate his hero: as he wrote, "The extraordinary harvest to science which [his] journey brought consequently urged other natural scientists and geographers to follow in the paths blazed" by Humboldt and Bonpland. The revolutionary violence in Humboldt's old haunts turned the duke northward, and he pursued his scientific studies up the Mississippi to the Ohio and Missouri Rivers as far as South Dakota. He paused in St. Louis to drop in on General William Clark, now the superintendent of Indian affairs, where he witnessed a meeting between General Clark and the leaders of several tribes. Duke Paul, who believed the Indians were the descendents of powerful and civilized states, watched as the Potawatomi chief Junaw-sche Wome gave a speech fighting for the very life of his people. He was struck by the Indians' outward composure, yet he observed their faces closely enough to notice their inner turmoil, concluding that their alleged stoicism was actually an artful strategy of self-control during difficult negotiations necessary "to serve the best interest of their nation. Junaw-sche Wome seemed most deeply moved. Several times I noticed tears in his eyes, especially when his son arrived." His wide-ranging account, rich with such thoughtful and sympathetic observations, was published in 1835 as *Travels in North America, 1822–24*. Duke Paul went on to make six more excursions through the Americas, with jaunts to Brazil and Australia, China, and Egypt. When he died suddenly of pneumonia in his German castle in 1860, he left behind rooms full of unpublished notes and undescribed specimens, said to have been the largest collection ever assembled by a private individual. Most of it was destroyed in World War II.[54]

In 1851, just outside St. Paul, Duke Paul ran into a young Prussian, Balduin Möllhausen, whose romantic ideals had inspired him to explore the American West's exotic diversity of peoples and cultures. This chance meeting led Möllhausen straight to Humboldt's Berlin parlor, though by a rather winding road: Duke Paul hired him as an assistant, and the two shared many hair-raising adventures, the most serious of them when they were separated in a blizzard. The duke was rescued by a mail coach, but Möllhausen, alone, snow-blind, and starving, wandered for weeks on the Kansas prairie until rescued by a band of Oto Indians. He spent the winter with the Omahas, recovering his strength (and if his later novels are to be believed, falling in love), and later he met up again with the duke in New Orleans for a spell of botanizing. In 1853 Möllhausen returned to Berlin escorting a shipment of wild animals for the Berlin

Zoo. The zoo director, impressed, introduced him to Humboldt, who soon became, in Ben Huseman's words, his "most important role model, mentor, and patron," opening the doors to his future career as an explorer, artist, and collector—and husband and father: Möllhausen married Humboldt's foster daughter, Caroline Alexander Siefert, in 1855.[55]

Under Humboldt's sponsorship, Möllhausen—to whom Humboldt was writing with deep and fatherly affection—returned twice more to the United States, first in 1853–54 to join Lt. Amiel Weeks Whipple on his Army Exploring Expedition in search of a railroad route along the 35th parallel to the Pacific, and again in 1857 to join Lt. Joseph Christmas Ives's expedition to the Grand Canyon. In 1858 Möllhausen published an American edition of his account of the Whipple expedition, including a preface by Humboldt, who expresses his pleasure that California, once under the rule of the monks, is now "in the hands of a restlessly active, enterprising and intelligent population" destined to join in commerce with China, Japan, and Siberia. Möllhausen lived out the rest of his long life in Berlin, where he wrote forty-five novels and eighty novellas based on his experiences in the American West, beginning with *Der Halbindianer* (The Half blood), which exposes the prejudices of the U.S. Americans against "every darker colored skin and the consequences arising therefrom." Though never translated into English, Möllhausen's fiction was immensely popular in Germany, where he was given the title "the German Fenimore Cooper"—though unlike Cooper, Möllhausen was describing landscapes he had experienced for himself.[56]

The extent of Möllhausen's influence on German ideas of America, the Western frontier, and American Indians seems to have been profound, but never assessed; what makes this more remarkable is that his success helped inspire Karl May, who (though he never actually went there) turned the American West into a popular culture industry, writing seventy-six volumes which have been translated into thirty-three languages and have sold, to date, over a hundred million copies. Though May's West of the imagination made him "the best-selling German author of all time," he is only marginally better known to American academics than his literary predecessor Möllhausen. The powerful identification of Germans with the victimized American Indian evidenced in this literature needs further exploration; it seems possible that Humboldt was a powerful carrier of this current, perhaps even a catalyst.[57]

This desire to portray Indians sensitively and sympathetically, even to identify with them on some deeper level, is evident as well in a similar Humboldtian duo, Maximilian, Prince of Wied, and the artist Karl Bodmer. Prince Maximilian grew up fascinated with natural history, studying, like Humboldt, with Blumenbach at Göttingen. After striking up a friendship with

Humboldt in Paris (the two met in 1804), he honored his mentor by leading a very similar scientific expedition into Portugal's closed colony of Brazil, where like Humboldt he served as his own expedition artist. As he admitted, his artistic skill was not up to such demands, and Humboldt advised him not to go to North America without an artist who could record the landscapes and peoples with the meticulous accuracy demanded by Humboldtian science. Maximilian hired the young and unknown Swiss artist Karl Bodmer, and from 1832 to 1834 the two traveled from Boston to the Mississippi, which they followed all the way to Fort McKenzie at the foot of the Rocky Mountains. Along the way they documented every aspect of the native cultures they encountered, spending five months with the Mandan tribe, soon utterly destroyed by the smallpox epidemic of 1837.

Back in Europe, Maximilian personally funded the publication of their work in his *Atlas*, published in 1839–41, which Goetzmann calls "perhaps the finest work ever executed in the Humboldtian tradition." So extravagant were these volumes that almost no one could afford them; Humboldt tried to pump up sales with an endorsement, but competition with Catlin's work and the Panic of 1837 were not to be overcome. However, Bodmer's sensitive and beautiful images, which turned ethnographic portraiture into high art, took on a life of their own. Unprotected by copyright laws, they began to circulate in U.S. American periodicals, to be copied and recopied so widely that they became iconic of the American Indian. Bodmer himself withdrew into the Barbizon to paint animals and forest scenes, highly regarded in Europe but unknown in the United States; his original Indian portraits were rediscovered in the twentieth century, and they continue to be widely reproduced and to influence fundamental perceptions of the Indians of the American West. One curious result of the scientific, literary, and artistic work of these German followers of Humboldt was that for a time Europeans were more knowledgeable about the trans-Mississippi country, particularly of the peoples who inhabited it, than were the Americans who were so busy conquering and subduing it.[58]

Such interests were not confined entirely to Europeans. Bodmer and Prince Maximilian just missed meeting the Pennsylvania painter George Catlin, who had steamed up the Missouri himself the year before. Catlin had been living dutifully as a lawyer until he threw it over in 1823 to move to Philadelphia and study painting. There he encountered a delegation of Indians on their way to Washington, and Catlin was galvanized into action. For the rest of his life, he dedicated himself to traveling among Indian peoples and documenting them through painstaking and detailed portraits. In the years immediately following the passage of the Indian Removal Act in 1830, Catlin traveled the American

West, painting as he went, his mission aided by an obliging General Clark, who let him set up a studio in his St. Louis headquarters and took him to meet various upriver tribes. Catlin wasn't just documenting a vanishing race—he was fighting for their right to exist. In 1834, the government issued an arrest warrant for Catlin for protesting the removal of the Cherokees from Georgia; he took his case all the way to President Jackson, who threw him out of his office. In 1836, he set about showing his work to the world to raise public support to restore the Indians to justice. His traveling show visited New York, Washington, Philadelphia, and Boston; then in 1839—when the Jacksonian-controlled Congress rebuffed his requests for support—he took it to England and on to Paris in 1845. There he was befriended by Humboldt, who as Catlin wrote, "took a great deal of interest in the Indians, having seen and dealt with so many in the course of his travels." A charming frontispiece to his account of his years in Europe shows Humboldt doffing his top hat to embrace the hand of a resplendent Indian chief. Audiences loved the show, but mounting debts forced Catlin in 1853 to sell his collection. By then his wife and son had both died. Stranded and alone, Catlin took the advice of Humboldt, who had suggested he travel to South America to paint the Indians there. Catlin's reborn career took him from the Tierra del Fuego at the southern tip of the continent to Kamchatka in the Siberian north, and by his death in 1872 he had documented 128 tribes in North and South America, earning sufficient respect from the U.S. government to be given a studio in the Smithsonian. His last years were shadowed by charges of fraud: in 1856 Humboldt wrote Catlin, then in Uruguay, warning that Henry Rowe Schoolcraft, under the auspices of the Bureau of Indian Affairs, had declared that Catlin's paintings of Mandan religious ceremonies were sheer inventions. Humboldt assured Catlin that he had talked it over with his old friend Prince Maximilian, and the Prince entirely corroborated Catlin's account. From then on Catlin used Humboldt's and Prince Maximilian's letters of testimony to defend his reputation against Schoolcraft's charges.[59]

There were others: Frederick Wislizenus, forced to flee Germany in 1833 because of his liberal views, who joined the German community of St. Louis, where he became friends with the best botanist in the West, George Engelmann, another of Humboldt's associates, who had also fled the repressive government of 1830s Prussia and who was busy building a Humboldtian network of knowledge in the West. Both these scientists befriended the Swiss mathematician and astronomer Joseph Nicollet, who had studied with Humboldt in Paris (where he discussed Humboldt's theories with Quetelet), then sailed in 1831 to the United States, determined to practice true, field-based Humboldtian science in a way no longer preferred in a France increasingly spellbound by the triumph of laboratory science.[60] And there was Albert Bierstadt, the German artist who immigrated to the United States to become the competitor of Frederic Church, both of them inspired by Humboldt to paint gigantic canvasses of sublime nature in a genre that soon included Thomas Moran, who joined Ferdinand Vandiveer Hayden on his famous 1871 expedition to Yellowstone. In short, the aesthetic and visual vocabulary of the American West would be shaped by the ideas of Humboldt, particularly by the Germans, who found its indigenous peoples and vast landscapes as exotic, and even more thrilling and emotionally moving, as anything in the Andes or the Amazon. If Humboldt had pocketed "all of South America," as Thornton complained, he had left the North a wide-open field, ripe for fresh and independent applications of his methods and ideals, in an environment where the sublimity of nature and the fascination of its peoples gained urgency from the speed with which they were being swept away by America's imperial drive to the West. As Prince Maximilian snorted when a search of Philadelphia bookstores turned up almost nothing on America's indigenous peoples, it was "incredible how much the American race is hated by its foreign usurpers."[61]

How, then, does one connect such idealistic European portrayals to the more canonical American perpetrators of Manifest Destiny? Joseph Nicollet is a key transitional figure whose career reveals much about the development of Humboldtian science in the United States. He was both the quintessential Humboldtian romantic adventurer, thrilled by the sublime landscape of the prairies and acutely sensitive to its native peoples, and also the quintessential Humboldtian professional scientist, whose expertise, unlike the independently wealthy Duke Paul or Prince Maximilian, found its support in service to the state. There would always be room for the lone romantic adventurer—this part of the tradition would be reborn in Thoreau, John Muir, and countless others—but the more disciplined version of the Humboldtian professional scientist came to dominate for the simple yet compelling reason that it was funded on a massive scale by the federal government. Nicollet is poised right on the crux, pointing in both directions.

Nicollet had moved from Paris to London, where he was working with John Herschel and Edward Sabine on measuring the shape of the earth (the science known as geodesy), when financial ruin brought on by the Revolution of 1830 precipitated his abrupt departure for greener fields in the United States. His stated goal was to contribute "to the progressive increase of knowledge in the physical geography of North America," following in the footsteps of Humboldt, "whose procedures in barometric determination of heights and astronomic determination of locations were the heart of his own work." When he found that no state support was forthcoming, Nicollet pursued his goals on his own, traveling upriver from the mouth of the Mississippi, in hopes of refining the crude measurements of Lewis and Clark and Major Long. The map he eventually produced so impressed secretary of war Joel R. Poinsett that in 1838 he took the extraordinary step of appointing Nicollet—a civilian and a foreigner—to lead the first exploring party sent out under the direction of the newly reorganized Army Corps of Topographical Engineers. (It helped that West Point, where Army Corps members were trained, was steeped in French science and the nationalistic, practical, militarized instructional techniques of Napoleonic France.) An attempt to discourage Nicollet by depicting the ravages of smallpox among the Missouri River tribes like the Mandans only strengthened his resolve: he must hurry and vaccinate all the Indians he could find. His journals portray a morning spent vaccinating three families of Indians camped near Pipestone Quarry, now in Minnesota: "Fathers, mothers, and children all present themselves with eagerness. . . . Sixteen in all, happy morning! An act of justice and an act of kindness."[62]

Nicollet preferred to work alone, and he refused to lead a military expedition. He did, however, request an assistant, and Poinsett gave him the best, a young and ambitious lieutenant in the Army Corps and a native of Charleston, South Carolina: John Charles Frémont. This was young Frémont's first venture west. It was thus Nicollet who trained the man who would become America's most famous, and infamous, explorer, passing along to him Humboldt's interdisciplinary approach to science. As Nicollet wrote to Poinsett at the close of the expedition, he had given Frémont not only science but "the large view of politics, commerce, farming, and so on, which the regions we have explored may offer in the future to the government and the people of the United States. At the least opportunity, I made him a part of my long studies and the results of my experience." Frémont drank it all in: in his own letter to Poinsett, he wrote that he could scarcely express his delight in having been assigned to Nicollet. "Every day—almost every hour I feel myself sensibly advancing in professional knowledge & the confused ideas of Science & Philosophy [with] wh[ich] my mind has been occupied are momently arranging themselves into order & clearness." It was Nicollet who turned Frémont into a disciplined scientific explorer.[63]

They were joined by two of Nicollet's friends, the German botanist Karl Andreas Geyer (yet another member of the St. Louis German community) and the Viscount de Montmort, plus Eugene Flandin, a son of the French consul. This compact international scientific team steamed up the Missouri to Fort Snelling (near St. Paul), and thence across southern Minnesota into South Dakota. Frémont recorded a typical workday: as they entered a village, Nicollet would be surrounded by Indians "and with the aid of the interpreter getting them to lay out the form of the lake and the course of the streams entering the river nearby, and, after repeated pronunciation, entering their names in his notebook; Geyer, followed by some Indians, curiously watching him while digging up plants; and I, more numerously attended, pouring out the quicksilver for the artificial horizon, each in his way busy at work." At Pipestone Quarry they all carved their names onto a rock (still visible today), and across the prairie Nicollet carved the Indians' place-names onto his map.[64]

Nicollet repeatedly expressed his respect and admiration for the Indians and his attempts to still his own voice to listen to what they were saying. As he wrote of tribal elders, "It is enough only to take the trouble to listen to them—to know how to listen to them and to control the conversation without offending their feelings and the order and twists and turns of their ideas; they become mute as soon as we try to make them speak and think as we do." His were the instincts of the ethnographer, and many of his Indian materials found their way into Gallatin's hands. Nicollet was baffled by the inability of whites to get along with Indians: "It takes so little to make them your friends that I cannot conceive why so many whites blunder in their dealings with them. All this comes without doubt from failure to understand or realize the ways of the respective parties. Kindness and language are the two avenues for reaching the heart of the Indian. A little tobacco and a few words will do what an army cannot do." Though many of Nicollet's extensive manuscripts on the Chippewa and Dakota nations remain unpublished, his editors find them "remarkable in their detail and sophistication" and judge that "Nicollet's contribution to scholarly interpretation and understanding of Indian life and language cannot today be overestimated."[65]

Nicollet led a second expedition in 1839, again with Frémont, that tested and culminated his system of taking barometric measurements and resulted in his definitive map of the upper Mississippi watershed. It also produced a moving meditation on the landscape with which his name would be forever associated: "May I not be permitted," he asked in his government report, "in this place, to introduce a few reflections on the magical influence of the prairies?" What follows is pure Humboldt, a passage that could have been lifted from the pages of *Views of Nature*. No matter how many hours or days or weeks one spends moving from "wave to wave over alternate swells and depressions, . . . one never tires; pleasurable and exhilarating sensations are all the time felt."

> The security one feels in knowing that there are no concealed dangers, so vast is the extent which the eye takes in; no difficulties of road; a far spread-

ing verdure, relieved by a profusion of variously colored flowers; the azure of the sky above, or the tempest that can be seen from its beginning to its end; the beautiful modifications of the changing clouds; the curious looming of objects between earth and sky, taxing the ingenuity every moment to rectify;—all, everything, is calculated to excite the perceptions, and keep alive the imagination. In the summer season, especially, everything upon the prairies is cheerful, graceful, and animated. The Indians, with herds of deer, antelope, and buffalo, give life and motion to them. It is then they should be visited; and I pity the man whose soul could remain unmoved under such a scene of excitement.

In his private notebook, Nicollet added that "Indians, with their private morals, the laws of their families, their customs, and the moving language they speak which is so little known, are the heroes here." He thought that the characteristics of the land and the nature of its people reflected each other—on the prairie, open, gentle, smiling, and gracious; in the mountains, "somber, sad, severe, and awesome." There among so many difficulties, man, instead of believing himself "free and the master," feels himself "dominated. . . . He defends himself, struggles, is victorious, but he is not in command. He is not the master. . . . In a word, he is only a little reed in the forest and a little stone in the valley." In which environment is man happier? The question is useless, Nicollet concludes: each man is happiest in the land "*where he passes his infancy and youth.*"[66] We are shaped, in our shared culture and our most intimate soul, by the land we grew up in. Nature is not outside us; it presses against us, "impresses" us, penetrates us, and alters our very soul. Nicollet shows himself moving in the landscape, immersed in the prairie, surprised by it, changed by it. His meditation is Humboldtian to the core, in poetic language that Humboldt would have envied.

So successful was Nicollet's 1839 expedition that the powerful senator from Missouri, Thomas Hart Benton, with an eye to the acquisition of Oregon, urged he lead further expeditions across the mountains to the Pacific. But Nicollet was dying, his health broken by the malaria he had contracted years before on the lower Missouri. His old friend, Ferdinand Rudolph Hassler, head of the Coast Survey, coddled the weakening explorer while Nicollet finished his great map, fighting to retain in the published version the delicate feathery hatched lines that, in the best Humboldtian style, would represent the elevation of the sites he had measured. His assistant, William H. Emory, went on to lead his own expeditions across the American Southwest, employing Nicollet's Humboldtian techniques to construct the much-needed map that would, at long last, render Humboldt's obsolete. Emory also learned from Nicollet the importance of watersheds, advising in his 1848 government report that the settlement of the arid lands of the Southwest could be accomplished only with tightly enforced and centralized control of water rights, a conclusion that anticipated John Wesley Powell's mandate three decades later. And it was also Emory who, inspired by Humboldt's followers Prescott and John Lloyd Stephens, began the serious study of Southwest archaeology, with the odd result that for the next decade, the army would be the United States' leading agent of archaeological research.[67]

Nicollet, then, would not lead the expeditions to the Pacific. That role would fall to his student and assistant, John C. Frémont, whose ascendency was capped when he married Senator Benton's daughter Jessie in 1841. While Nicollet is too little known, there is, of course, an immense literature on Frémont. Suffice it to say that he led a succession of five expeditions, first to the Rockies, then beyond to California, in an attempt to establish the best routes to the far West. In 1845, on the third of these (perhaps under secret orders), he finessed the Bear Flag Rebellion by American settlers in California into an American claim on the entire region, which after the gold strike of 1848 was admitted to statehood in 1850. Frémont's daring got him court-martialed, and though he was found guilty on all counts, President Polk commuted the most serious charges. Frémont resigned in protest, funded his last two expeditions himself (both of them disastrous attempts to force a crossing of the Rockies in winter), became California's first senator, and ran for president in 1856, losing to Buchanan. His exploits won him public adulation and his court-martial brought him sympathetic outrage. Whereas Nicollet had preferred to travel light and stay independent, repudiating any military appearance, Frémont seized the warrior role. Where Nicollet seemed, to Frémont's astonished eyes, impervious to the "stir of frontier life," Frémont reveled in its romance. Under his leadership the Humboldtian exploring expedition matured into both a fully functioning scientific enterprise and a disciplined and focused wing of the military. Civilization, and the guns to protect it: Frémont, the heroic, masculine conquering hero, turned exploration into empire.[68]

By land and by sea: while Nicollet, Frémont, Emory, and others were pushing overland to the Pacific, another expedition, the most costly and extravagant of all, was sailing from Norfolk, Virginia, around the tip of South America to Antarctica and beyond to Australia and the South Sea Islands, Hawaii, the northwest coast of North America to East Asia, and on round the world. The long and curious story of the Wilkes expedition has been ably told by William Stanton, and its Humboldt connections have recently been established by Aaron Sachs. The story begins with one John Cleeves Symmes, Jr., who around 1820 began to wonder where martins went in wintertime. The answer came to him as a revelation: nowhere *on* the earth, but *in* it. The earth must be hollow, and open at the poles. The theory would be easy to prove: go and look! In the 1820s Symmes traveled the country to promote a voyage to the Arctic. He made one key convert, J. N. Reynolds, who seized on Symmes's theory to push for a polar expedition, allowing the scaffolding of "Symmes' Hole" to drop away to reveal the sheer beauty of an American polar expedition. Reynolds leaned hard on the theories of Humboldt, whose assertion that the equator was not the zone of maximum heat led to speculation that the poles might not be the zone of maximum cold. Perhaps there were open polar seas, just waiting to be explored. In 1828 Reynolds won the support of president John Quincy Adams (who had hobnobbed with Humboldt in Paris), the first president to call for permanent, federally funded scientific institutions, and though defeated for reelection that year by Andrew Jackson, Adams continued to support Reynolds from his seat in the House of Representatives. Mounting the expedition took ten more years, during which time Reynolds took a shakedown cruise to the southern polar seas. By the time the expedition actually sailed, political infighting had removed Reynolds from the manifest and installed the martinet Lieutenant Charles Wilkes at its head. The squadron of six ships weighed anchor on 18 August 1838, from Norfolk, Virginia, just as Nicollet was crossing the Minnesota River.[69]

Reynolds had guaranteed the Humboldtian nature of the expedition by insisting it include not a handful of scientists but a whole team, sufficient to do full justice to the exponential expansion of scientific knowledge. His 1837 address to Congress called for "an enlightened body of naval officers, joining harmoniously with a corps of scientific men," who would, "like stars in the milky-way, shed a lustre on each other, and all on their country!" Instead of harmony and light, a contest erupted: the navy initially refused to admit civilians on board, though it quickly became obvious that the best, even the only, men for the job had never joined the military. Reynolds successfully recruited Asa Gray, America's hot young botanist, only to see Gray quit in disgust over the protracted delays in sailing. James Dwight Dana, who became the United States' most prominent geologist, stayed with the mission, as did John Pickering's talented nephew Charles, the linguist Horatio Hale (who would in old age mentor Franz Boas in North America), the naturalist Titian Peale, and various others. They lacked, however, an astronomer. The obvious person was Nicollet, but as Stanton writes, nothing came of that proposal, probably because of the navy's tacit understanding that their expedition was closed to foreigners.[70]

As Stanton shows, the story of the Wilkes expedition did not end with their return in 1842. The explorers had achieved genuine success: they established the existence of a southern continent, Antarctica; surveyed various

South Sea islands with an accuracy unsurpassed until the twentieth century; proved that America must wrest Puget Sound from England if the nation was to have a secure port in the North Pacific. (The preferred site had been the mouth of the Columbia River, until their flagship, the *Peacock*, ran aground and broke up in its treacherous waters). In 1843, Frémont's second expedition connected with Wilkes's survey of the Columbia, closing the circle of America's continental inventory. And all the scientists aboard succeeded in collecting: thousands of specimens, hundreds of boxes of them, tons, whole shiploads, so many that some were still being unpacked and described well into the twentieth century. It was a staggering achievement, but as the Lewis and Clark expedition showed, unless the results were cast into print and circulated around the world, it would all, beyond a puff of American pride, go for naught. Publication demanded money. The congressional appropriations rolled in, and "the scientifics" got to work.

What happened next changed the face of American science. The journey had siphoned off much of the best young scientific talent in the United States and given these young men the best training in the world. Back home, the Wilkes corps, joined by the legions needed to help prepare the public reports, created in one fell swoop a new class of being: an entire cadre of professional, elite scientists trained, experienced, expert, and, perhaps most important, globalized from their bones to their fingertips. The work they performed would be up to the best European standards. Humboldt made sure everyone in Europe knew this by reporting their results in his books and praising them endlessly in salons and letters. Such praise had real impact in a democracy where progress in the sciences was dependent on public funding and support.

The sheer bulk of the collections presented another problem, which in itself forced the reorganization of American scientific institutions. Where to house them? There was quite literally no room, anywhere. Samples were put on display in Washington's fine new Patent Office building, where they attracted crowds, including an impressed Ralph Waldo Emerson, who checked them out in 1843 and pronounced them "the best sight after the Capitol." In the end the crisis was solved by turning the newly built Smithsonian Institution into the United States' first national museum, finally realizing the vision Peale and Humboldt had impressed on Jefferson. In 1858 the Wilkes expedition's global collections joined the Smithsonian's North American collections to create one of the world's great museums. Meanwhile the living plants found their way into the National Botanic Garden, and the dried plants constituted the National Herbarium, where they became the training ground for generations of botanists.[71]

The question posed by Humboldtian science in the United States had never been asked before: what was the role of science in a democracy? Was it a morally uplifting pursuit that should be available to all the people? Or a keen-edged, dynamic profession for the few who could afford the university training it increasingly demanded? Women were, of course, almost completely excluded from professional science, although many played important roles in its larger support structure. Frémont's government-sponsored reports became national bestsellers largely through the assistance of his coauthor, his wife Jessie Benton Frémont, who perforce becomes one of the era's more significant authors. Humboldt's works reached his English-speaking audience almost entirely through the agency of women, who in untangling and translating the thickets of his prose demonstrated a deep understanding of cutting-edge science: Helen Maria Williams, Elise Otté, Edward Sabine's wife Elizabeth. Some women, like Mary Somerville and Maria Mitchell, against all odds rose to prominence as authors and scientists in their own right; Humboldt counted them both as friends and encouraged them both. Women with artistic talent served as colorists for the many thousands of scientific illustrations generated by the new wave of science. But until higher education was opened to women, their direct involvement with science would be largely limited to such amateur, popular pursuits as botanizing and shell collecting. The female imagination was guided not into the rigorous halls of science nor its rugged western expeditions but into some of natural history's gentler paths. Humboldt, with his keen aesthetic and poetic sensibility, was waiting for them there, too.[72]

For Humboldt found himself straddling both sides of this question. To him, science was both moral uplift for the many, *and* the arcane pursuit of the few. He tried to keep both sides in communication, even as they battled each other over positions and funding. The most spectacular instance was the rivalry of Alexander Dallas Bache and Matthew Maury, each self-conscious, even militant, Humboldtians. Bache, great-grandson of Benjamin Franklin, came to power in 1843 as the dynamic new leader of the languishing Coast Survey, which became under him the largest and most powerful of the federal bureaus. He also became one of the most aggressive defenders of science as a profession, staging (with Louis Agassiz, Joseph Henry, and Benjamin Peirce) a coup that took over the American Association for the Advancement of Science in 1851, and struggling to wrestle control of science away from the military. This put him in direct conflict with Lt. Matthew Maury, USN, director of the Naval Observatory. Maury, who lacked scientific training, had adopted Humboldt's methods and extended them to the oceans, creating a new science that he named, on Humboldt's suggestion, "physical geography of the sea"—now known as oceanography. Maury's books were hugely popular, but even more popular were his charts of winds and currents, which he constructed on the Humboldtian model, collating the reports of thousands of common shipmasters and sailors. But when he sought to extend his meteorological research to land by recruiting farmers, Bache intervened, and Maury appealed to his hero for aid. Normally Humboldt was happy to give Maury unstinting praise, but on this occasion he fell silent, once again refusing to take sides. How could he? Maury represented one side of his ideal, a practical and elevating science generated by and for the common people. Bache represented the other, a rigorous and professional science fully integrated with the latest European advances. This was not a choice Humboldt could make.[73]

If one of Bache's goals was to wrestle science away from the military, one of Maury's goals was to use the military to spread American power beyond its borders—more particularly, to provide a haven for Southern slave owners. The son of poor Virginia farmers, Maury fully believed in the rightness and justice of Southern slavery. When the Compromise of 1850 jeopardized Southern expansion into the new western territories, his hopes turned to South America, and with the full support of the navy, Maury recruited his son-in-law, William Lewis Herndon, to lead an expedition to the Amazon. This was hardly a scientific expedition. Herndon's route, from Lima over the Andes and down the Amazon, was well known and well traveled. All down the river he encountered the cast-off horses from an American circus that had preceded him a few months before. As his navy instructions made clear, science was to take a back seat to the true purpose of the expedition, which was to investigate the commercial possibilities of the Amazon for Americans. The instant Herndon crossed the continental divide he was thrilled by the prospect: "I thought of Maury, with his researches concerning the currents of the sea." Recollecting how the waters of the Amazon mingle with those of the Mississippi, Herndon plucks a bit of green moss, tosses it into the water, and follows it in imagination "down through the luxurious climes, the beautiful skies, and enchanting scenery of the tropics" to the Atlantic, hence whirled by currents up to the Gulf of Mexico, there perhaps to meet "with silent little messengers cast by the hand of sympathizing friends and countrymen" high in the Mississippi headwaters "or away in the 'Far West,' upon the distant fountains of the Missouri." For Herndon can attest to the truth of Maury's assertion that "the Valley of the Amazon and the Valley of the Mississippi are commercial complements of each other," each supplying to commerce what the other lacks. "They are sisters which should not be separated."[74]

As for Indians, the lieutenant disliked them. They were little more than "drunk and lazy" vehicles for his contempt, and he was happy to report that they melted like snow before the rising sun of white man's improvements. "This seems to be destiny. Civilization must advance, though it tread on the neck of the savage, or even trample him out of existence." Yet this created a problem, for without Indians, whence would come the labor force that would transform the tropics? Herndon quotes Humboldt's vision, from the introduction to his *Personal Narrative*, of a peaceful and prosperous South America, its waterways plied by barges of grain and produce. But how was the prophecy of the great and profound Humboldt to be realized? By slave labor, of course. "The Indians will not work." They would rather die than exert themselves. But "the negro slave seems very happy in Brazil," and if Brazil would only throw off its "puerile" fear of Americans and invite settlers to the Amazon, Humboldt's vision would finally come to pass. Southern planters, who looked with "apprehension" at the future of slavery in the United States, "would, under sufficient guarantees, remove their slaves to that country, cultivate its lands, draw out its resources, and prodigiously augment the power and wealth of Brazil." Herndon's report, published in 1854, popularized the notion that Maury's current charts proved the Amazon was a natural extension of the United States. As a *New York Daily Times* article put it, "The winds and currents of the sea are such as to place the Atlantic ports of the United States on the wayside of nations either going or coming from the mouth of the Amazon." Mexico had fallen, at least much of it. Cuba would be next; why not Brazil?[75]

Thoreau sneered that Herndon had got the wrong handle on civilization: what was wanted were not, as Herndon had said, tobacco and slaves, "artificial wants to draw out the great resources of the country," but "a high and earnest purpose in its inhabitants." That alone would draw out nature's true resources. Do we want culture or potatoes, illumination or sugarplums? Humboldt, who had written his American friends that there was no prosperity without justice, tried to see the better, not the worse, side of Herndon and company; he wrote Herndon's father-in-law Maury that he was "warmly interested" in Herndon's report, together with a whole stack of others—over the years Emory, Frémont, the Wilkes expedition, Whipple's railroad survey, and a host of others all come in for his effusive praise. As he added approvingly, "What Government of Europe can boast in modern times, of having originated so many solid works for the Geography of the interior of the continent!" But if he had read Herndon closely, he surely had wept to find his own name dragged into such nationalistic jingoism. Yet there it was, and there it would remain, in the mind of the American public and of generations of scholars thereafter. How shocking to read, for instance, that the "Humboldt code" militated one to "kill and lay waste everything pertaining to the tribes, whenever found—no trials, but at arms; no prisoners; no red tape."[76]

This is an obscene perversion of Humboldt, but his explanations and theories were all too easily turned toward exploitation. His speculations about the possibility of an open polar sea launched expeditions in search of trade routes and new lands to colonize; his analyses of Mexican and South American mines led to a financial bubble that ruined many, Washington Irving included; his observation that Western civilization had consistently moved westward along an isothermal temperate belt became justification for the Manifest Destinarians' push to the Pacific and beyond to China and Japan. Remarkably, his notion that desiccation follows deforestation led his followers to conclude that *re*forestation would bring the rains. Those arid western plains could be reclaimed. As Nicollet wrote, the prairie grasslands were "the work of the Indians who destroy by fire the rich vegetation to assure themselves of animal food." Leave the prairies alone, and with time "the forests will come to restore the flow of the meteorological agents and modify the harshness of the climate. . . . Providence makes us feel the effects of the barbarity of man in depriving us of water, a natural consequence of the destruction of the forests." It was not Humboldt but his followers who made this leap in logic, beginning with his German immigrant friend Wislizenus, continued by dean of American science Joseph Henry, and culminating in 1864 with George Perkins Marsh's full-fledged environmental argument pointing to the dangers of deforestation and the potential for human restoration of ecological damage. The U.S. government began a campaign to reforest the barren plains, which evolved into the pseudoscientific motto that "rain follows the plow." Under its spell a generation of sodbusters broke the high plains and prairie grasslands, but the rains never came. The government-planted trees all died of drought. Humboldt must have been wrong. Perhaps a similar inversion of Humboldtian logic resulted in the Humboldt code: if human communities were indeed so deeply integrated into their natural environment, then laying waste to that environment would wipe out the human beings. It was effective logic, but it has nothing to do with Humboldt.[77]

Whether he intended them to or not, the tools Humboldt had crafted opened America to empire. Henry Adams located the loss of American innocence not to the Civil War but to the years between 1800 and 1815, by the end of which "the Rights of Man occupied public thoughts less, and the price of cotton more." In 1804 Humboldt had hoped against hope that America had its priorities right, for recognizing the Rights of Man would cost it little more than a dip in cotton exports. In 1809 he had apologized to Jefferson for telling the world that the U.S. Congress, and Jefferson himself, lacked the "power" to abolish slavery. But all too soon he discovered that his first guess was the right one. Given the choice between cotton and humanity, the power of the people had chosen cotton. Here lay the weakness of democracy. Humboldt's own politics leaned toward the radical wing of liberalism (the one that would lead some of his followers into socialism and beyond), which sought the intervention of a progressive state to protect the freedoms of the oppressed. His ideas were born at a moment when the New World was teaching Europe to throw off feudal institutions and embrace the modern nation-state, and Humboldt envisioned the Creole leaders of these new republics cutting across the old lines of class and race to create an egalitarian culture based in a shared geographic and historical environment. He did not condone what actually happened: class lines were not dissolved but reified into polarized racial divisions, elevating "whites" and quite literally dehumanizing all other races. Humboldt's precarious model republic decided that in a world contested by European imperialism and domestic Indian nations, aggression was its best defense: protecting America meant absorbing the weak and warring with the strong.[78]

Humboldt's ideas and ideals thus played into a very different history than the one he had foreseen. The split in his scientific followers was reflected on the wider stage of American history: those who celebrated his name—Gallatin, Everett, Ticknor, Prescott, Webster, Adams—tended to be Northern Whigs who favored economic diversification, strong government, and social reform, protested the Mexican War, and defined themselves against the expansionist, proslavery and anti-Indian policies of Jacksonian Democrats. Yet Jacksonians too seized on Humboldt's theories and methods, interpreting, like Bancroft, Maury, and Herndon, his global and egalitarian cosmopolitanism through a nationalistic and racist lens.

As America expanded, Americans all across this spectrum used Humboldt's name for the communities they were founding. German immigrants named new towns after Humboldt from Pennsylvania to California, Texas to Saskatchewan. Free State immigrants to Kansas trumpeted their antislavery politics by naming their town Humboldt. Frémont, who ran for president on an abolitionist platform and had used Humboldtian science to settle and claim the far West, repaid his distant hero by giving his name to the Great Basin watershed. Had the final vote at the Constitutional Convention of 1864 gone a bit differently, the center of American gambling culture would have been not Nevada but the great state of Humboldt. Such acts of naming were part of the cult of Humboldt that peaked in the United States in the 1850s: in 1858, the secretary of war, John B. Floyd, presented Humboldt with an elaborate album of nine maps showing the nation's Humboldt place-names. "Never," he wrote, "can we forget the services you have rendered not only to us but to all the world. The name of Humboldt is not only a household word throughout our immense country, from the shores of the Atlantic to the waters of the Pacific, but we have honored ourselves by its use in many parts of our territory, so that posterity will find it everywhere linked with the names of Washington, Jefferson, and Franklin." The American ambassador who walked in on Humboldt while he was still holding the just-opened letter reported that he took the news in stride: "I wish you to know," joked Humboldt, "that I am a river about 350 miles long; I have not many tributaries, nor much timber, but I am full of fish."[79]

Despite his self-deflating humor, the point remains: since the French Revolution, Humboldt had dreamed of a future in which republicanism would sweep the globe and bring liberty and equality to all its peoples. While he never wavered from his support for America, the word that named this ideal, the United States worried him deeply. In public he honored it and was venerated in return; in private to his closest friends he was brutally honest:

> In the United States there has, it is true, arisen a great love for me, but the whole there presents to my mind the sad spectacle of liberty reduced to a mere mechanism in the element of utility, exercising little ennobling or elevating influence upon mind and soul, which, after all, should be the aim of political liberty. Hence indifference on the subject of slavery. But the United States are a Cartesian vortex, carrying everything with them, grading everything to the level of monotony.[80]

The inspiration of his youth had been the principles of Jefferson and the heroism of Washington, but in the bitterness of age he saw himself, too, dragged down into the vortex. In the hands of too many of his followers, his words of peace had become swords; his words of liberty, chains; his words of justice, a dollar sign. What then would be the nature of his legacy?

Finally Shall Come the Poet:
On the Interface of the Two Cultures

> After the noble inventors, after the scientists, the chemist,
>   the geologist, ethnologist,
> Finally shall come the poet worthy that name,
> The true son of God shall come singing his songs.
> WALT WHITMAN

For generations, scholars have hailed the antebellum era as the moment when American writers turned away from Europe to develop their own, national, literature. Had not Emerson declared, back in 1837, that "our day of dependence, our long apprenticeship to the learning of other lands, draws to a close," and proclaimed the emergence of that new being, "the *American Scholar*"? Yet recent scholarship has shown that no one was more "planetary" than Emerson, who ranged through world literature from the Italians, French, and Germans to the Persians, Hindus, and ancient Greeks, and who, like his teachers Ticknor and Everett, was deeply informed by his European travels. Many scholars now focus on the ways American literature interweaves national territory with intercontinental space. From this perspective, the sublime national narrative of "manifest destiny" no longer seems so inevitable, so destined. It appears instead as a strategic policy pursued by a particular subset of American leaders, at a time when the very boundaries of the nation were in flux, subject to contingencies of global competition for territory. When "American" explorers packed up Humboldtian tools and methods and turned west, they left the United States behind to enter foreign and contested lands. There was no automatic assurance that the frontiers they entered might consolidate under a U.S. flag: expanding those national boundaries meant jockeying for power and position with France, England, Spain, Russia, and Mexico while keeping an eye overseas on Africa, China, Japan, and India. From its first inception as a series of colonies, through consolidation via a war of independence against England with France at its side, to the Louisiana Purchase and the War of 1812, to the contests over Texas and California and Oregon and beyond, America was a dynamic and evolving product of contending global forces.[1]

No one understood this better than Humboldt, the old partisan of the French Revolution who had in 1790 hurried from Paris to bone up on Jefferson in cosmopolitan Hamburg, and whose writings would position every region he traversed in relationship with the contending forces of politics, economics, and moral and physical nature. Humboldt became iconic in the Americas by putting them both literally and metaphorically at the center of the world map. To writers in the United States, the crucial question of "Who we are" meant addressing who we once *were*—variously English, French, Spanish, Dutch, Scottish, Amerindian, African; who we *are* in the dynamic global circle of nations; and who we were *becoming*, once we had joined and completed that international circle, linking East with West, making the globe into one. Just as Herndon had dropped a bit of green moss into the upper Amazon to imagine it swirling away to join the waters of the Mississippi, so did Thoreau imagine the ice of Walden Pond melting into the sacred Ganges: his pure Walden ice is wafted by favoring winds "past the site of the fabulous islands of Atlantis and the Hesperides . . . and the mouth of the Persian Gulf, melts in the tropic gales of the Indian seas, and is landed in ports of which Alexander only heard the names."[2] Had any country, Humboldt wondered, done more to explore its interior? Was any other national literature framed so fully by the paradoxes and problematics of Humboldt's science?

This is not the place for a traditional literary close reading of individual authors and works. I would like to step back and offer instead a "broad reading" that sketches patterns and connections, a Humboldtian "physiognomy" of the American literature that took shape while Humboldt was at the height of his power and influence. From this conceptual hilltop—one of middling height, as Humboldt recommended—one can see American writers working to bring into being an America that would mark a difference between "Old World" and "New." The sign of that difference would be "freedom," and that prophecy of freedom would transform the world. In *Common Sense*, observes Eric Foner, Tom Paine rallied colonists to war by invoking the "millennial hope for the coming of a new world, the vision of a perfect society," a secular Utopia. "We have it in our power to begin the world over again," wrote Paine; "the birthday of a new world is at hand." While freedom might initiate American difference, soon that difference would propagate to all nations and all peoples in a spreading revolution that would revitalize the globe. Jefferson too wanted to begin not a new nation but a new era, "to legislate as though eternal peace were at hand" even though the world was torn by war. Thoreau caught the spirit when, after a long immersion in Humboldt's works, he wrote in his journal, "How novel and original must be each new mans view of the universe—for though the world is so old—& so many books have been written—each object appears wholly undescribed to our experience—each field of thought wholly unexplored—The whole world is an America—a *New World.*"[3]

Washington Irving also caught the spirit in his biography of Columbus, whose "troubled life" was "the link which connects the history of the old world with that of the new." Irving's narrative of Columbus became, metonymically, the narrative of America: noble ideals and magnanimous purposes, tragically defeated by greed, licentiousness, and violence, yet opening to a utopian vision at last. At the end, "splendid empires" would spread over the beautiful world Columbus did not realize he had discovered, and "nations, and tongues, and languages" would fill its lands with renown, blessing his name "to the latest posterity!" The ideal of America, however darkened by corruption and loss, lived on in the transcendent future when it would at last be realized in all its glory. Walt Whitman, in turn, understood that the hero of Irving's biography was really not the troubled historical Columbus but the primal generative American—namely himself—defeated by "mocking disbelievers" but rising again, to transcendent faith: "By me earth's elder cloy'd and stifled lands uncloy'd, unloos'd, / By me the hemispheres rounded and tied, the unknown to the known." At the end of this poem "The Prayer of Columbus," the speaker's darkened vision is unsealed and he sees the future: "And on the distant waves sail countless ships, / And anthems in new tongues I hear saluting me." America would free the world; if in the beginning all the world had been America, in the end it would be again.[4]

Paine, Irving, and Whitman shared their vision with Humboldt, who imagined the Casiquiare, like the Delaware, teaming with ships laden with grain and wood and rubber, passing banks of "fertile fields" cultivated by free men, sailing out to ports on every ocean. J. N. Reynolds closed "Mocha Dick" with a panorama of American ships toiling "along the coast of Japan, up the straits of Mozambique," around Guinea to Africa to the Antarctic, to "the vast expanse of the two Pacifics with their countless summer isles." Such free-ranging daring and enterprise was not the result of forced labor, but "the natural result of the ardor of a free people; of a spirit of fearless independence, generated by free institutions." As Reynolds continued, "Under such institutions alone, can the human mind attain its fullest expansion, in the various departments of science, and the multiform pursuits of busy life." America was more than a nation among equals; it was that nation that would catalyze a worldwide transformation into a new global state. America, that is, is transitive, a genesis not an end, a passage not a place. A "Passage to India," as Whitman had it:

> Passage to India!
> Lo, soul, seest thou not God's purpose from the first?
> The earth to be spann'd, connected by network,
> The races, neighbors, to marry and be given in marriage,
> The oceans to be cross'd, the distant brought near,
> The lands to be welded together.

Who will do this work of crossing and welding? Everyone will have a hand, but after them all, after captains and engineers, "After the noble inventors, after the scientists, the chemist, the geologist, ethnologist, / Finally shall come the poet worthy that name, / The true son of God shall come singing his songs." At the end, the final transformative agent would be the poet, for only he can take up and hook and link all together:

> (He shall indeed pass the straits and conquer the mountains,
> He shall double the cape of Good Hope to some purpose,)
> Nature and Man shall be disjoin'd and diffused no more,
> The true son of God shall absolutely fuse them.

Humboldt's influence is visible everywhere in Whitman's work, and here as elsewhere Whitman chimed with Emerson, who in his essay "The Poet" obliquely identified Humboldt, in the figure of his iconic mountain, as the true poet, he who would "stand out of our low limitations, like a Chimborazo under the line, running up from the torrid base through all the climates of the globe, with belts of the herbage of every latitude on its high and mottled sides." Such Humboldtian poets would be "liberating gods," who "are free, and . . . make free" by rendering all objects transparent to thought, completing the circuit that links God to Nature to Man, fusing the trinity into unity through Beauty, "the creator of the Universe." Or as Whitman answered, the American poets "shall be kosmos"—as America is: when the work is done, all the world shall be Planet America. The Passage to India is the passage to Cosmos.[5]

America, then, catalyzed a difference that would irresistibly generate a multiform sameness, the Old World transformed into the New. This pattern connected genres traditionally thought to be separate: literature, political economy, history, science, exploration; canonical and noncanonical, self-conscious works of imagination and narratives of fact. Seeing all these works as compatriots in a common field reveals common features, with some illuminating differences.

The field as a whole is marked by contestation. Nations and empires battle to claim the hemisphere. Some battles lie far enough in the past to seem trivial and even hilarious, as in Irving's *History of New York* (1809), which uses the Dutch settlement of New Amsterdam as a vehicle to mock the Jefferson administration. Skirmishes become monumental: in Hawthorne's story "The May-pole of Merry Mount" (1835), nothing less than "the future complexion of New England was involved in this important quarrel" between the "grisly saints" and the "gay sinners," Puritan settlers and Anglican merrymakers, "Jollity and gloom . . . contending for an empire." In Cooper's brooding tragedy *Last of the Mohicans* (1826), France, England, and an emergent America heave like colliding tectonic plates across a landscape already thick with nations of Indians. Ultimately in Cooper's Leatherstocking novels the Indians are swallowed and eliminated, their traces dissolved back into the encroaching forest or blown away by prairie winds. Where were the American memorials of past battles, past glories? Humboldt had worried that colonial peoples were damaged beyond repair by their lack of "monuments" and memories from a shared and unifying past. Indigenous monuments might have served, he thought, had Europeans not destroyed them both physically and culturally, as Cooper destroyed his Indians. John Lloyd Stephens took up Humboldt's suggestion, finding and claiming indigenous monuments scorned by previous generations, turning archaeology into nationalism and thereby making himself the single most popular travel writer of nineteenth-century America. What Humboldt had done by "pocketing" the Aztecs and Incas, Stephens did for the Mayas, exploring the Yucatan, uncovering fabulous Mayan buildings and carvings, and purchasing them as an American patrimony.[6]

Humboldt's vivid portraits of ancient American civilizations ravaged by the Spanish conquerors also inspired William Hickling Prescott, who with Humboldt's help explored archives rather than jungles to write his haunting, beautiful, and magisterial works of tragic history, *The Conquest of Mexico* (1843) and *The Conquest of Peru* (1847). Prescott had had his fill of conquest—even as intellectually inclined soldiers packed copies of his books along the road to the Halls of Montezuma, he deplored the war as "mad and unprincipled." Francis Parkman looked north instead of south to develop, in the monumental detail of his multivolume *France and England in the New World*, the deep history of the wars, betrayals, and compromises that came to define North America as English rather than French or Indian. This is a universe, these works variously proclaim, that is unstable, shifting, uncertain. The solid ground of the familiar shakes underfoot and betrays our placid belief in continuance. Fissures open and swallow whole civilizations, centuries of power crumble and collapse overnight, new nations emerge like lifting mountain ranges, entire populations wash across the face of the planet, colliding, drowning in the undertow or floating to safe harbor.[7]

In this turbulent world, priority is given to the role of the witness. In historical writing there can of course be no direct witnessing, yet Irving,

Prescott, and Parkman all wrote in the tradition of German historicism articulated most famously by Wilhelm von Humboldt in his cornerstone essay "On the Historian's Task." Wilhelm insisted that the historian should become, in effect, a virtual witness, an observer who can see not just random events but their "inner causal nexus." History, said Humboldt, is "no less an art, free and self-contained, than are philosophy and literature." Like the poet, the historian must work the "scattered, disjointed, isolated" fragments into a whole, but unlike the poet, the historian cannot impose order but "subordinates his imagination to experience and the investigation of reality." Thus the historian's first task is the "exact, impartial, critical investigation of events," but he cannot stop there, for these events cannot be fused into a living whole by an operation of "mere intellect." The historian must go beyond intellect, to connect events through an intuitive, sympathetic imagination. Emphasizing facts alone will miss the truth; emphasizing imagination alone will falsify the truth.[8]

Parkman takes up this Humboldtian historical method in the introduction to *Pioneers of France in the New World*. The most seemingly fanciful incidents are based, he insists, on "authentic documents or personal observation," for faithfulness to the truth of history involves more than patient and scrupulous research into facts. Such facts may be "detailed with the most minute exactness," yet the whole may, as Humboldt says, miss the truth. Therefore, "The narrator must seek to imbue himself with the life and spirit of the time. He must study events in their bearings near and remote; in the character, habits, and manners of those who took part in them. He must himself be, as it were, a sharer or a spectator of the action he describes." The narrator must be, in other words, a virtual witness, using imagination to animate the past, to "clothe the skeleton with flesh." Alexander von Humboldt, who collaborated closely with Wilhelm, used similar terms to praise Irving's history of Columbus: Irving, he wrote, "has demonstrated that the power of imagination in the higher intellect" does not preclude devotion to "the more serious study of the historian." By combining both imagination and fact, Irving's work succeeded, Humboldt judged, where his own had failed, by virtue of the poet's ability to give imaginative life to the "skeleton" of Columbus. The key, Wilhelm had said, was to use the method of the artist to grasp the truth of nature organically from within, to study "the way in which the outward shape emerges from the idea and structure of the whole," recognizing, not imposing, an inner shape that is then "reborn through the imagination." This was the imaginative leap Alexander saw in Irving, whose deeply sympathetic portrait of Columbus invoked an organic whole that embraced not just Columbus himself but the iconic American, as Whitman recognized. Similarly, Parkman ransacked whole libraries of fact and incident to reach the inner truth of European empire in North America: New France was "all head," England was "a body without a head," and in the centuries of conflict, the one darkened into barren despair and absolutism while the other quickened into the productive life of liberty. "Each followed its natural laws of growth, and each came to its natural result."[9]

It is one thing to enter sympathetically into, say, the beautiful, rich, and exotic civilizations of the Aztecs and Incas, but could one sympathize with the Spanish conquistadors who destroyed them in such bewildering paroxysms of violence and greed? Sympathetic identification with the conquistadors had not been a problem for Humboldt, who condemned the Spanish as yet another violent manifestation of European feudalism. As for Irving, his sympathy for Columbus precluded deep understanding of the Spanish who betrayed his dream; they emerge in his narrative as an unreasoning embodiment of human vice, the snake in Eden replaying the overdetermined story of the Fall. Prescott, however, wrestles with the problem. How, he asks, can we judge the Spanish fairly? How can we find "a justification of the right of conquest, at all?" The solution is "to transport ourselves to the age when it happened," and try to inhabit imaginatively a world where religious infidelity was a sin "to be punished with fire and faggot in this world, and eternal suffering in the next." Before we recoil too far in horror, Prescott gently reminds his readers that this "monstrous" doctrine is at the foundation of our shared Christian heritage. We cannot simply and unthinkingly condemn the Spanish. Yet still, Christianity's assumption of the right of conquest and conversion didn't explain the excessive nature of the conquistadors' violence. To understand this additional dimension, Prescott offers the psychology of "fear armed with power" and passion "inflamed" by a sense of injury. Once again, he brings the monstrous acts of the Spanish uncomfortably close: look in the mirror, he suggests, and compare their conduct "with that of our own contemporaries under similar circumstances." The Spanish, the French, the British—all are equally guilty. "The wanton butchery, the ruin of property, and above all, those outrages worse than death . . . show a catalogue of enormities quite as black as those imputed to the Spaniards." The conquistadors are not monstrous aliens; they are us.

And we who judge the past, how would we wish to be judged by the future? Prescott refuses to vindicate the Spanish, but he can ask that we not judge them "by the lights of our own age. We must carry ourselves back to theirs, and take the point of view afforded by the civilization of their time." We must extend to them the same justice we will "have occasion to ask from Posterity" for ourselves, when it is we who are held up to critical scrutiny.

The result of this sympathetic form of virtual witnessing in all three American writers is historical works shaped by the aesthetic standards of organic unity together with a strong sense of historical relativism. These works blend rigorous attention to fact with both emotional and intellectual sympathy while striking a fine-tuned note of moral outrage at the present as well as the past. They are creative works of history that read with the coherence and power of epic novels.[10]

In other forms of writing—particularly fiction and the literature of exploration—the role of witnessing is given to a small band or community who find themselves projected out onto the raw edges of conflict where changes are most rapid. Often the witnesses form a dyad: Humboldt and Bonpland; Pym and Augustus, or Pym and Peters; Natty Bumppo and Chingachgook; Ishmael and Queequeg; Huck and Jim.[11] As with Humboldt's, these narratives are self-consciously directed homeward, to "witness" experience to those who never left, to ask what it means, what is happening to the larger human community. Humboldt voiced this responsibility when, in a statement suppressed in the United States, he wrote, "It is for the traveler who has been an eye-witness of the suffering and the degradation of human nature, to make the complaints of the unfortunate reach the ear of those by whom they can be relieved."[12] Often these witnesses are, like Humboldt, overwhelmed by historical and natural forces beyond their control. From the opening pages of *Last of the Mohicans*, for example, Natty and his band are in over their heads. The women, Cora and Alice, are, absurdly, traveling to the front lines of a war zone, and the incompetence of their military escort quickly precipitates a crisis which the combined resources of Natty, Chingachgook, and Uncas are inadequate to resolve. The result is catastrophic: the stupidity of British generals causes the fall of Fort McHenry; the cupidity of the French precipitates a massacre by their Indian allies of innocent civilians; the eventual deaths of both Cora and Uncas destroy both all hope for interracial marriage and the successful amalgamation of the white and Indian races, as well as the hope for a rebirth of the Delaware nation. In this precursor to J. R. R. Tolkien's *Return of the King*, it is as if Aragorn were killed by Saruman's forces and the true race of men died forever. The contest has been decided: the Indians will die, the French and the British will vacate, and the Americans will rule. Natty and Chingachgook, abandoned and alone, weep together over the graves of their children.

Humboldt on the Orinoco was similarly overwhelmed by natural and political forces beyond his control, and like Cooper, he looked to a more distant future for redemption. Irving repeats the pattern in his historical work *Astoria*: the daring entrepreneur John Jacob Astor projects a Pacific empire to be based in a trading post at the mouth of the Columbia, but Astor's designs are defeated by forces his men cannot predict or comprehend. While anchored off Vancouver Island, the *Tonquin* is overtaken by friendly Indians who without warning massacre the white crew, leaving four survivors who decoy the Indians back onto the ship, which they have rigged to explode. "Arms, legs, and mutilated bodies were blown into the air, and dreadful havoc was made in the surrounding canoes." Nothing is left but fragments, and for days the limbs and bodies of the slain wash up onto the beach. The British navy forces Astoria to surrender, Astor's plans are wrecked, yet off in the distance is hope: America, defeated, will return to claim Oregon for itself, north past Astoria to Puget Sound's safe harbors, clear to the 49th parallel. We may be overwhelmed, but we will prevail. The theme is theorized by Emerson in his 1860 essay "Fate." "We are incompetent to solve the times," he opens. We are hemmed in by the immovable limitations of "Fate, or laws of the world"; yet while the "ferocity" of nature will consume us we can fight the tyranny of Fate with our own ferocious resistance, "Power": "If Fate follows and limits power, power attends and antagonizes Fate." Where the laws of nature bind us, thought allows us to understand those laws and thereby deploy them for ourselves, seizing nature's power for human purpose. In various ways these works propose that we will be overwhelmed, but understanding that very fact will be our redemption. Knowledge will give us control.[13]

Unless, of course, you are Edgar Allan Poe. Emerson called on the power of inner, psychic resources; Poe shows those resources in the torment of collapse. Poe, a friend of Reynolds, has his alter ego Arthur Gordon Pym pursue his own private Wilkes expedition to the south polar seas; en route Pym falls asleep reading Lewis and Clark, and awakes to find himself trapped in the ship's hold by a mutinous crew of pirates. To survive, Pym descends through circle after circle of deception, piracy, and cannibalism. Rescued, his new ship descends "still farther south" to total calamity, as the black natives they colonize turn on them, massacre the crew, and burn the ship, which blows up in a gruesome episode lifted straight from Irving's account of the *Tonquin*: "The whole atmosphere was magically crowded, in a single instant, with a wild chaos of wood, and metal, and human limbs." Reaching their absolute polar nadir, Pym and the half-Indian Dirk Peters rush "into the embraces of the cataract" of Symmes' Hole, which extinguishes Pym, and his narrative, forever.[14] Less melodramatic but more disturbing is Melville's chilly anatomization of the psychically starved Captain Amasa Delano, portrayed as the representative American, literally "at sea" in a twilight world of historical forces he cannot comprehend. In a lonely harbor in Chile (not quite so far "south" as Pym, but far enough), Delano takes pity on a foundering slave ship, the *San Dominick*, which he attempts to rescue. His smug and benevolent racism blinds him to the reality of the situation: the crew has been massacred by the slaves, who control the ship and lure Delano on board in order to seize his ship as well. (The ship's name, were he paying attention, might have offered a clue.) Delano attributes the strange ship's obvious disarray to the incompetence of its Spanish captain, Benito Cereno, for he knows all Spaniards are effete and deceiving; but the slaves cannot be in power, for he knows all blacks are too stupid to be anything more than hapless victims. Delano's folly nearly destroys himself and his crew, through blindness that never is repaired by insight. What is in Poe's *Pym* psychic inversion (the world collapsed into the self) is in Melville's "Benito Cereno" a social and cultural eversion: the American projects his psyche onto the world, blinding himself to cosmopolitan reality, blundering ever onward with gifts in one hand and guns in the other. Civilization, and the weapons to defend it.

In both narrative instances—whether the witnesses are overwhelmed by physical forces or by the failure of psychic resources—the resulting narrative can take on the form of ideological critique. This is clearly the case with "Benito Cereno," where the critique is implicit in the action (and reinforced by the fact that the incident is drawn from history, as Melville makes clear when he reproduces the court deposition almost verbatim). In *Typee*, Melville's critique is by contrast explicit: the narrator, Tommo, escapes from the tyrannical and inhumane rule of the "civilized" ship to find a utopian life ashore amidst the "savages." Drawing on both his own experiences in the Marquesas and his reading on the fate of Tahiti and other South Sea Islands, Melville meditates on the cruelties of civilization and the happiness of the savages to conclude that rather than send missionaries to the Marquesans, the islanders might more usefully send missionaries to the United States. As Kant had suggested, it is we, not they, who are the true cannibals.[15] Melville arrived in 1842, just in time to witness the French military takeover of the Marquesas at gunpoint, and in the sequel *Omoo* he further dissected the consequences of British rule on Tahiti. Melville's early novels could usefully be read in the tradition of Enlightenment critique that traces back through Humboldt to the generation of Bougainville, Diderot, Georg Forster, and Bernardin de Saint Pierre.

Melville's cheerful South Sea novels are shadowed by our knowledge of the inevitable fate of the peaceful natives; Melville mourns the Islanders of the future, as Humboldt mourns the Guanches of the past. Washington Irving's version of this critique, like Humboldt's, looks backward rather than forward. In *Columbus*, our hero arrives at the New World in 1492 to find a perfect paradise, but in only twelve years paradise is annihilated. By 1504, "several hundred thousand" Indians have perished, "miserable victims to the grasping avarice of the white man." This devolution climaxes with the fierce resistance of two Indian holdouts, the beautiful, intelligent, and civilized Anacaona, the queen who had once welcomed the Spanish to her paradisiacal valley of Xaragua, and the heroic cacique Cotubanama, ruler of the rugged wild province of Higuey. Anacaona welcomes the Spanish, who stage a joust as courtly entertainment, but once horsed and armed, the Spanish turn on the weaponless natives and massacre them before the queen's horrified eyes. Her city burned, she herself is taken in chains to a kangaroo trial and execution. Cotubanama leads a long and bitter guerilla war which ends when the Spanish penetrate his people's mountain fastnesses, pursuing and butchering every last person, mothers and infants not excepted, torturing the men with unspeakable violence, and finally capturing and hanging Cotubanama in the public square "like a common culprit." The Europeans found a perfect paradise, says Irving, and filled it with "horror and desolation." He recoils from relating the "horrible details," for they might feed anti-Spanish prejudice, but "historical veracity" demands that he narrate the evidence of human cruelty. His history should stand as a warning, for "every nation"—not just Spain— "has in turn furnished proofs of this disgraceful truth."[16]

Irving likely was thinking of his own nation, for in his *Sketch Book* essays "Traits of Indian Character" and "Philip of Pokanoket," he paints an attractive picture of the Indians as an innocent people whose rights, then and now, are trampled by white men. Irving does not shrink from narrating the gory details of the "indiscriminate butchery" by the Puritans of the Pequods, whose stoic Roman resistance was reviled when it should have been honored. King Philip, leader of the resistance that became known as King Philip's War, becomes in Irving's treatment a heroic warrior, like Cotubanama defending his land and people against heartless invaders: "He was a patriot attached to his native soil . . . a soldier, daring in battle, firm in adversity, patient of fatigue, of hunger, of every variety of bodily suffering, and ready to perish in the cause he espoused." While Columbus's Indians are lost in history, King Philip, Irving's American patriot, "proud of heart, and with an untamable love of natural liberty," steps forward as the precursor to George Washington himself, who so successfully adapted Indian tactics to defeat the British. Much of this literature proposes that Americans became who they are by absorbing, like Natty Bumppo, the Indian identity into their own. To be American is to be marked by difference, and increasingly, that difference was identified as Indian. Irving spoke to his generation when he reassured them that, tragic as was the Indians' fate, it was redeemed by the Americans who, by eating the heart of the lion, had thereby acquired its courage—a prolepsis made credible only by the Indians' extermination.[17]

The New England Transcendentalist Margaret Fuller found this national narrative troubling. Like Irving, she could offer no practical alternative; unlike him, she could offer no absolution. In 1843 she ventured from her New England stronghold to experience the uncertainties and trials of the Great Lakes frontier, publishing her personal narrative as *Summer on the Lakes*. Like Nicollet, she had to learn how to see the beauty of the encircling vastness of the prairies, although where Nicollet witnessed the prairies instinct with Indian life, Fuller is haunted by their absence. "How happy the Indians must have been here!" she remarks. "It is not long since they were driven away, and the ground, above and below, is full of their traces." She finally encounters living Indians in Mackinaw, members of the Chippewa and Ottawa tribes encamped while waiting for government payments. She is surprised to find them not "taciturn" but talkative, "with much variety of gesture," and her heart rises at the hypocrisy which sends the few survivors into a Christian church: "Better their own dog-feasts and bloody rites than such a mockery of that other faith."[18]

Fuller's heart was moved to more than just angry words. Fuller saw herself, and the fate of American women in general, in the Indians she witnessed, and she returned from Mackinaw to New York, where she developed her career as a reformer and activist. In her mind differences of race, gender, and class had fused into a common cause: she expanded an earlier essay, "The Great Lawsuit," into *Woman in the Nineteenth Century*, the first manifesto of the American movement for women's rights, then traveled to Europe, settling in Italy, where she became a partisan in the Italian Revolution of 1848. It infuriated her when Americans, who should be the leaders of republican revolution worldwide, stood back and allowed the fledgling Italian republic to be reclaimed by despotism. In her last published work, her final dispatch to Horace Greeley's *New York Tribune*, the disgrace of defeat is redeemed by a prophetic vision of the new and coming revolution: "Every man who assumes an arbitrary lordship over fellow man, will be driven out. It will be an uncompromising revolution. England cannot reason nor ratify nor criticize it—France cannot betray it—Germany cannot bungle it—Italy cannot bubble it away—Russia cannot stamp it down nor hide it in Siberia. The New Era is no longer an embryo; it is born; it begins to walk."[19] Even as Humboldt despaired, the vision of a planetary America was being articulated afresh, by a new American voice, a woman's voice—which was tragically silenced when Fuller was drowned. In 1850 the ship that was bearing her home to America was wrecked on the rocks off New York's Fire Island.

It was Fuller's fate that started her friend Emerson writing about fate. Thoreau interrupted his reading of Humboldt to search, without success, for her corpse. Pacific beaches were not the only ones littered with body parts— so were those turned toward Europe. Years later, facing east from Cape Cod, Thoreau meditated darkly: "This gentle Ocean will toss and tear the rag of a man's body like the father of mad bulls, and his relatives may be seen seeking the remnants for weeks along the strand." Emerson had the answer: "The water drowns ship and sailor, like a grain of dust. But learn to swim, trim your bark, and the wave which drowned it, will be cloven by it, and carry it, like its own foam, a plume and a power."[20]

For of course not all these narratives carry the burden of social critique; nor do they all end in failure redeemed by prophecy. The comparison of *Moby-Dick* with its source text, "Mocha Dick," is instructive. Melville's novel ends in cosmic failure with a redemption dubious at best: when the American *Pequod* goes down, bearing with it representatives of all the nations of the earth, the sinking ship drags behind it a yawning gulf, "then all collapsed, and the great shroud of the sea rolled on as it rolled five thousand years ago." Thoreau's "gentle Ocean," the father of mad bulls, swallows (almost) all. But this was not how J. N. Reynolds ended "Mocha Dick." In Reynolds, the vindictive white whale meets his death like Cotubanama or King Philip, besieged by weaponry, fighting to the end, thrashing the sea into a crimson froth, but doomed: "He then turned slowly and heavily on his side, and lay a dead mass upon the sea through which he had so long ranged a conqueror." The conqueror, conquered: the whaler who assaulted the historical Moby Dick did not go down into the void, but lived to render no less than a hundred barrels of oil from the white whale's mighty corpse, and to regale appreciative audiences with the jolly tale of his exploits. Reynolds ends his story with the Humboldtian vision of a prosperous and free people ruling the waves, with no trace of Melville's Delanoesque suspicions; he points less to Melville than to Frémont, who cast himself as the "Pathfinder" leading a free people to their manifest continental destiny.[21]

The shift can be seen in another comparison. Both Humboldt and Frémont suffered the loss of their key scientific instrument, the crucial barometer, just as their missions approached their climax. For Humboldt, the loss was one more blow in a gathering series that pointed him not upriver to further glory, but downriver, back to civilization, recovery, and repair. There finally was no climax. But Frémont's story ended quite differently. His barometer broke just as they approached the Rockies: "The snowy peaks rose majestically before me, and the only means of giving them authentically to science, the object of my anxious solicitude by day and night, was destroyed." But defeat was not an option. The ingenious American managed, MacGyver-like, to repair the delicate instrument with his powder horn, a bit of wood, some animal skin, thread, and buffalo glue. His team triumphantly ascended the highest peak of them all and measured it with the rebuilt barometer: 13,570 feet. The name of the peak became a token of that triumph: Mt. Frémont. The gung-ho spirit of boyish adventure that animates Frémont's adventures with his mountain-man guide Kit Carson led directly to a host of gleeful narratives of jolly times on the wild frontier, the best of which are Parkman's haunting *The Oregon Trail* (1849) and Mark Twain's ironic *Roughing It* (1872). Few of their followers echoed, as they did, the brooding disquiet of Cooper in *The Prairie* (1827), where old Natty mourns the past and fears for the future: "What the world of America is coming to, and where the machinations and inventions of its people are to have an end, the Lord, he only knows. I have seen, in my day, the chief who, in his time, had beheld the first Christian that placed his wicked foot in the regions of York! How much has the beauty of the wilderness been deformed in two short lives!"[22]

Leading away from the polar limits of defeat or victory, overwhelming loss or triumphant gain, is a third path that departs in an altogether new direction. One sees it in young Humboldt's endless zest for new lands, new men, new thoughts; in Darwin's exuberance on the *Beagle* voyage; in Nicollet's celebration of the rolling prairie. These voices witness not destiny but the journey, not the goal but pleasure in the sheer ongoingness of life. The paradigmatic example is Melville's Ishmael, named after the Bible's half-caste wandering outsider son of the slave Hagar. Ishmael is Melville's foil to the murderous Ahab, whose monomaniacal focus on one transcendent goal, the death of the white whale, brings the ship and its crew to destruction. Where Ahab drives, Ishmael floats, open to all. It is of course Ishmael who pops like a cork out of the vortex of the sinking *Pequod*, clinging to Queequeg's coffin, the sole survivor and all the redemption Melville has to offer. In the book's opening sequence Ishmael recoils from horror at his cannibal bedfellow, only to decide, after a thoughtful look, "The man's a human being just as I am; he has as much reason to fear me, as I have to be afraid of him. Better sleep with a sober cannibal than a drunken Christian." He and Queequeg make a "cosy, loving pair" from then on, as Ishmael the easygoing cultural relativist absorbs everything from cannibal to captain, reflective, unresisting but ever watchful of the deepening currents that are drawing the *Pequod* to its destruction. There is something of this sensualist in Pym, too, the participant-observer who is drawn ever deeper into the vortex, and at least one critic has seen in Poe the figure of Humboldt the flaneur, "the strolling urban observer" who wanders and browses in a nonlinear and nondirected manner, observing intimate detail yet always able to step back and, with a cock of the theoretical eyebrow, comprehend the whole. Humboldt's is not the direct but the zigzag path, not the direct gaze but the side-of-the-eye, the oblique regard of Detective Dupin: the flaneur figure is also Thoreau, the peripatetic nonlinear wanderer who advised himself to look less and see more, or even not to look at all but see with "a true sauntering of the eye." As he admonished himself in his journal, "Man cannot afford to be a naturalist, to look at Nature directly, but only with the side of his eye." Three years later he added, "The poet will so get visions which no deliberate abandonment can secure. The philosopher is so forced to recognize principles which long study might not detect. And the naturalist even will stumble upon some new and unexpected flower or animal."[23]

As all these writers in various ways suggest, the motor behind the venture out to the margins was the drive to increase knowledge. These writers posit a vast and unknown world which can, and must, be grasped by an effort of intelligence, of (in the words of the Brothers Humboldt) intellect combined with imagination. Though themselves writers of books, they do not presume that reading books is the only or the true path to knowledge; from Humboldt to Natty Bumppo to Emerson's American Scholar, each asserts that knowledge is experiential. In this age of Humboldt, the first commandment is to "explore": leave the study and the library. Knowledge is found not within but out-of-doors. "Life is our dictionary," declared Emerson; why should we "grope among the dry bones of the past, or put the living generation into masquerade out of its faded wardrobe? The sun shines to-day also. There is more wool and flax in the fields. There are new lands, new men, new thoughts." The new lands to the west, south, and north made all lands seem new. The new men, from South Sea cannibals, to Guayaquerias and Chaymas, to Sioux and Chippewas, put the fundamental nature of humanity back into question. The immense and growing backwash of data and images from the margins to the center was forcing new thoughts. The administrative centers—Philadelphia, Washington, Boston; London, Paris, Berlin—could not simply tack on more facts. The floods of new species and phenomena, of artifacts and languages, were forcing the centers into wholesale reorganization.[24]

And so, as Emerson said, "Classification begins." The mind, "tyrannized over by its own unifying instinct, . . . goes on tying things together, diminishing anomalies, discovering roots running under ground, whereby contrary and remote things cohere, and flower out from one stem." Science begins, as it "reduces all strange constitutions, all new powers, to their class and their law, and goes on for ever to animate the last fibre of organization, the outskirts of nature, by insight." This project of animating the "outskirts of nature, by insight" was a collective effort by all intellectuals, not just physical and natural scientists. In the 1830s, when Emerson was publishing these thoughts, "science" was still conceived as knowledge in general, open to all who wished to investigate the world. Humboldt did not suggest that Gallatin return to school for a degree in anthropology; he told his friend to take up his hobby and turn it to account, and Gallatin did just that. Möllhausen, Bodmer, and Catlin followed Humboldt's and Gallatin's maps to meet the peoples themselves, to paint and record them, and in their search for American monuments, Emory and Stephens founded New World archaeology. Humboldt, then Nicollet, inscribed Indian names into their maps, in an attempt to stabilize spatial knowledge even as their techniques opened temporal depths, both historical and geological. Their maps are full of events, incidents, journeys, and measurements, names that are fossils now of once vital human actions on the land, every name once a story: Columbia, America, Humboldt River; Pike's Peak and Mt. Frémont; Mississippi, Mackinaw, Ohio and Hawaii; Seattle, Astoria, and Santa Fe.[25]

Classification had for a generation meant the Linnaean binomial system, the very hallmark of science. The burden of thousands of new species was already breaking down Linnaeus's original artificial system and forcing a reorganization into a revised "natural" system that reflected genealogical relationships which Darwin would recognize as evolutionary. Where binomial names had once been slots on a static grid, they were changing to capture developmental depth and interrelationships among species, a dynamic field in four dimensions. Writers like Melville and Cooper relished opportunities to mock the old static systems: *Moby-Dick*'s extract-collecting "poor devil of a Sub-Sub" belongs to "that hopeless, sallow tribe which no wine of this world will ever warm," and Melville's own playful systematization of whales categorizes them as books and chapters: Folio, Octavo, Duodecimo, as in his "type" species: "Boοκ I.(Folio), Chapter 1. (*Sperm Whale*)." In *The Prairie*, Cooper sallied out after old-fashioned natural historians under the figure of the eccentric Obed Bat, "M.D. and fellow of several cis-Atlantic learned societies." As Dr. Bat is introduced to the audience, he is complaining that in two days he hasn't found "even a blade of grass that is not enumerated and classed," but he has by starlight come across a most remarkable new species:

Vespertilio Horribilis, Americanus. Dimensions (by estimation)—*Greatest length*, eleven feet; *height*, six feet; *head*, erect; *nostrils*, expansive; *eyes*, expressive and fierce; *teeth*, serrated and abundant; *tail*, horizontal, waving,

and slightly feline; *feet*, large and hairy; *talons*, long, curvated, and dangerous; *ears*, inconspicuous; *horns*, elongated, diverging, and formidable; *color*,
plumbeous-ashy with fiery spots; *voice*, sonorous, martial, and appalling; *habits*, gregarious, carnivorous, fierce, and fearless.

The joke is complete when young Ellen points out that blind old Dr. Bat has
just classified his own ass.[26]

Humboldt was less interested in discovering new species than in tracing
new relationships among all natural phenomena. Yet he did insist that careful systematization was still the foundation of natural science. He himself
turned that task over to specialists, while he ventured to define new and collective forms of classification: climate zones, bioregions, plant communities.
To paraphrase Emerson, Humboldt's goal was to find the rhizomatic underground connections between phenomena, to identify how apparently disparate phenomena were actually the product of underlying law. Binomial classification was a necessary first step toward recognizing patterns and, beyond
them, a matrix of causal relationships: like the historian, the naturalist would
work from fact to truth. Thoreau took both poles of this process very seriously. In his later years he became an accomplished botanist who relished the
way scientific names allowed him "to know my neighbors, if possible,—to
get a little nearer to them," opening to his vision details and distinctions and
beings he had never noticed before. The key was to both focus and saunter,
to be part Ahab and part Ishmael, holding in balance the intense vision of
science with the open-ended creative synthesis of philosophy and poetry, to
sustain both fact and imagination and fuse them into a new whole. As he
wrote in his first natural history essay, "Let us not underrate the value of a
fact; it will one day flower in a truth."[27]

Science was, in other words, not blind—as Cooper had joked—but half
blind. Emerson criticized the "half-sight of science," and Thoreau noted with
mingled glee and concern that the botanist intent on finding grasses "does
not distinguish the grandest pasture oaks. He as it were tramples down oaks
unwittingly in his walk." Their solution was the Humboldtian one, to fuse
scientific fact with poetic imagination. This was necessary because, as both
Wilhelm and Alexander wrote, the world of history and the world of nature both composed, in Wilhelm's words, "an infinitude which the mind can
never press into one single form." That very plenitude incites the historian to
try again and again, and with each trial "to achieve it *in part*." For both the
Brothers Humboldt, history and science spoke truth, but never the whole
truth. Because human beings cannot see as God sees, omnisciently, there will
always be a provisional and aesthetic component shaping human knowl-

edge, and neither history nor science would ever come to an end. As Thoreau wrote, "The sun climbs to the zenith daily high over all literature and science . . . the sun of poetry & of each new child born into the planet has never been astronomized, nor brought nearer by a telescope. So it will be to the end of time. The end of the world is not yet."[28]

Yet as Thoreau also recognized, there was another side to science, especially in an age when Humboldtian science was government-sponsored and taxpayer-funded. The features of Humboldtian science—explore, collect, measure, connect—all sought to abstract from the chaos of physical nature quantities and qualities that could be rendered portable. One cannot move a mountain or a river basin, but one can move measurements, specimens, and descriptions. Back in the administrative centers, data mobilized by the activities of scientists can be reassembled into new and useful wholes: maps of mines, of navigable waters and future railroad routes, of zones suitable for agriculture or rangelands. "Science applies a finite rule to the infinite," Thoreau understood, "& is what you can weigh and measure and bring away." When he thinks like this, his enthusiasm for science dims: "Its sun no longer dazzles us and fills the universe with light." Such scientific reports do not throw open doors to new worlds of insight, but merely inventory the available resources, creating for the state "complete catalogues of its natural riches." Like cosmic clerks reporting back to the company boss, scientists tote up economic usefulness and military potential, a work of labor with no room for enthusiasm.[29] Humboldtian science could animate nature with insight, or collapse back into a tool of empire, a form of possession that led to dispossession: the extinction of wild animals, the dodo and bison and passenger pigeon; the "removal" of whole peoples, in the government-mandated Indian Removal Act of 1830 that began by evicting entire tribes off their ancestral lands and ended by herding their remnants onto barren "reservations," forced ghettoes that made Humboldt's much-maligned Spanish missions look utopian by comparison. Contestation is the keynote of this broad field of literature, and as science became a weapon for victory, it withdrew from its alliance with poetry.

Thus far I have traced a selection of texts that compose this field synchronically, without considering change across time. However, that temporal pattern is perhaps the most important feature of the Humboldtian era. When Humboldt visited America, various kinds of writing still operated across a common field: a political economist could turn his hand to ethnography; a poet could become a botanist, and a scientific explorer could write poetry;

a Harvard professor of languages could tutor his audience in the fine points of German science. Distinctions that today seem intuitively obvious and necessary are hard to make out: not only Humboldt, but everyone, was experimenting with "hybrid" texts. But by the time of Humboldt's death, those distinctions became crisp and well marked. Professional scientists increasingly wrote only for fellow specialists and held popular science in contempt. Travel writing took the science out of exploration and turned the strenuous discipline of scientific travel into tourism, a middle-class entertainment. Imaginative literature gathered to itself a new kind of exclusive, educated audience and took on an elite status, elevated above the lowbrow entertainment of the penny press.

This is, of course, an enormous story, far beyond the scope of the present project. Yet a look at one single episode will help suggest the shape of the whole. Let us return, therefore, to the Wilkes exploring expedition to the South Seas of 1838–42, and watch how the shape of American literature alters around this pivotal moment.

The Wilkes expedition was, as previously mentioned, inspired by the tireless lobbying of J. N. Reynolds, whose enormously influential speech on the floor of Congress fired the imagination of politicians, scientists, and poets alike, and whose sprightly sailor's yarn "Mocha Dick" inspired Melville to write what is, arguably, America's greatest novel. As Reynolds recruited the expedition's staff, a question arose: who would be the expedition's official historian? Reynolds himself had demonstrated considerable talent as a writer, but another and lesser-known competitor lobbied hard for the position: Nathaniel Hawthorne. In the spring of 1837, Hawthorne's Bowdoin College friend, the future president Franklin Pierce, sent Reynolds a sample of Hawthorne's writings (as Pierce noted, *Twice-told Tales* had just been published and favorably reviewed) and a letter commending Hawthorne's modesty, genius, and good qualities as both a diligent editor and a "companion and friend." Horatio Bridge, another college schoolfellow, threw in his own influence, promising that of "the whole Maine delegation" for good measure, and lobbied George Bancroft, the historian who would in a few years be prodding Humboldt over the Mexican question, to put in a good word for Hawthorne with the secretary of the navy. The effort came to nothing, and the next year the expedition sailed without Hawthorne—and without Reynolds, either. In the end, Lieutenant Wilkes appropriated to himself the role of historian, cobbling together the narrative of the expedition from the various journals of its participants to produce an incoherent text that William Stanton labels "a national disaster." What if, one wonders—what if Hawthorne's application had been successful? How might the shape of American literature have been different? For Wilkes's narrative had a surprisingly wide impact.[30]

Poe had no wish to join his friend's expedition, but he did "borrow" liberally (some would say plagiarize) from Reynolds's writing, as well as that of other explorers, to produce *The Narrative of Arthur Gordon Pym.* Through *Pym,* the Wilkes expedition entered canonical American literature. It also entered the literary canon through the novels of Herman Melville, whose cousin Henry Gansevoort sailed on it (briefly) as a midshipman. Melville used the royalties from *Typee* to buy a complete set of Wilkes's *Narrative* (it cost him $21.00), which he used as a source for *Typee, Omoo, Mardi,* and *Moby-Dick,* modeling the character of Queequeg after the New Zealand Maori chief Ko-towa-towa, and, perhaps, Ahab after the tyrannical Wilkes. James Fenimore Cooper was a friend of Lt. Wilkes's uncle, the banker Charles Wilkes, and he too drew on Wilkes's *Narrative* for his sea novels *The Crater* (1847) and *The Sea Lions* (1849); when Melville reviewed Cooper, he commented on how much he was reminded of the Wilkes expedition. And as Stanton notes, Wilkes's discovery of Antarctica made him a hero to American boys everywhere, including one Samuel Clemens of Hannibal, Missouri, who wrote in his autobiography that Wilkes's name was on everyone's lips: "Wilkes had discovered a new world and was another Columbus" who had seen dreamplaces with his own eyes. Mark Twain went on himself to become one of the age's most famous travelers.[31]

The Wilkes expedition also caught Emerson's attention; he reported in the *Dial* their triumphant return and extraordinary achievements, and soon after regaled his wife Lydian with the marvels on display in the Patent Office, with the suggestion she tell Henry Thoreau that

> I had seen stones & sand & volcanic scoriae from the Antarctic Continent, and all manner of corals sponges and "Neptune's goblets," and all manner of arms trinkets implements, & natural and artificial curiosities from the Feejees and Tonga and Navigators' & Sandwich Islands, from Japan & from Peru brought home by our recent Explorers: the most invigorating facts by far, coming from our friends the Feejees, tattooed heads & baked heads, and headdresses more striking than beautiful, and Feejee pillows simpler than any devised by Mr Alcott or Mr Lane, and Japan books & novels of Siam.

For his part, Thoreau, an inveterate reader of exploration narratives, professed skepticism. In *Walden* he turned the Wilkes expedition, in a manner suggestive of Poe, into a metaphor of the self: "What does Africa,—what does the West stand for? Is not our own interior white on the chart?" As Thoreau continues,

> What was the meaning of that South-Sea Exploring Expedition, with all its parade and expense, but an indirect recognition of the fact, that there are continents and seas in the moral world, to which every man is an isthmus or an inlet, yet unexplored by him, but that it is easier to sail many thousand miles through cold and storm and cannibals, in a government ship, with five hundred men and boys to assist one, than it is to explore the private sea, the Atlantic and Pacific Ocean of one's being alone.

Yet sail on, Thoreau concludes: "You may perhaps find some 'Symmes' Hole' by which to get at the inside at last."[32]

In short, there is a real sense in which, as the Wilkes expedition was preparing to sail, everyone was "on board": Reynolds made all literate America think of it as their path to destiny. Hawthorne wanted to join it, and in 1837 he saw no obvious barrier preventing him from working and writing alongside the scientists James Dwight Dana, Titian Peale, Charles Pickering, and Horatio Hale. Just as Whitman envisioned, the poet would join the scientist, the chemist, the geologist, and the ethnologist, doubling "the Cape of Good Hope to some purpose," fusing Nature and Man. In a sense Poe did join it, at least in imagination, penning *Pym* even as the squadron sailed, sending his doppelgänger on a parallel voyage to the southern polar seas to discover that his own interior was indeed, as Thoreau said, "white on the chart." Melville, who liked to say his life as a sailor was "my Harvard and my Yale," used the Wilkes narrative to supplement his own experience; and Cooper made it into the stuff of his adventure fiction. Yet by the mid-1840s, when the ships had all returned and the scientific results were being processed, catalogued, displayed and described, it was only the scientists who were on board. Everyone else was on shore, observing, cheering them on, but no longer welcome to participate. Scientific democracy was gone. This had not been a romp for the many, but a strenuous and disciplined mission for the few, and as the scientists labored they, unlike Wilkes, directed their writings not to the wide world of the merely literate but to the international circle of fellow scientific specialists. The federal government reinforced this limited sense of audience by refusing to print more than one hundred copies of each specialist's report. The movement they initiated, that had actually and imaginatively brought all America on board, ended by removing science from the public realm and into the domain of elite, trained professionals.

Meanwhile, Hawthorne would consolidate his emerging career in a very different direction, narrowing his own audience to those who could follow his densely allusive prose, which he developed into a finely tuned instrument of psychological exploration. Melville discovered greatness in Hawthorne's fictions, a dimension he thought was unsuspected by an uncomprehending

American readership, and he addressed his own later novels not to the broad popular audiences who were titillated by *Typee*, the story of his life among the cannibals, but to the few, and ever fewer, capable of reading symbolically complex and nonlinear narratives. Poe developed his own theory of art for art's sake that offered to sever what Thomas De Quincey, on the other side of the Atlantic, called "the literature of power" from its diminished cookbook cousin, "the literature of knowledge."[33]

At the end of this process of disciplinary formation there would be, not the one common intellectual culture of Humboldt's heyday, but the "Two Cultures" of science and literature anatomized so memorably in the twentieth century by the physicist and novelist C. P. Snow: scientists indifferent to poetry, literary intellectuals oblivious to even the most basic truths of nature. As science rigorously excluded the aesthetic and emotive, Humboldt dropped from the cutting edge to the cutting-room floor, edited out of the narrative except for whatever segments of "Humboldtian science" could be salvaged and refashioned into new disciplinary specialties. The exuberant literature of scientific exploration would sink into background material, to be read not for its own sake but for the learned footnotes it could provide into "real" literature, books that had once been its blood brothers. The histories of Irving, Prescott, and Parkman, works that had once reigned as the queens of literature, the poetics of the real, paled into padding for literary specialists, trotted out occasionally as ideological artifacts. "Literature" itself became an exclusive discipline, an elite art form demanding trained and refined reading skills which could discriminate, defend, and appreciate those linguistically self-conscious, organically unified, aesthetically complex texts deserving of canonization. Humboldt's baggy, rambling books, with their experimental hybridizations of science and aesthetics, narrative and description, fact and imagination, would disappear from the shelf, even from memory. But remarkably, a few of the most impressive Humboldtian texts would not only survive, but be canonized, long since cut loose from their now forgotten siblings: *Pym*, *Moby-Dick*, *Walden*, *Leaves of Grass*, foundational works of American literature.

Many factors account for the massive shift that parted literature from science and both from popular culture, a shift that flags the passage to twentieth-century modernism and its inverse twin, postmodernism, but one factor stands out as both particularly consequential and generally unremarked. In Humboldt's heyday, and in Humboldt's hands, science was often the vehicle for social critique. Naturalists returned from Tahiti and Mauritius and the Americas posing challenging questions about white European supremacy and the primacy of Western civilization. Humboldt would focus his early, socially oriented writing, *Vues des Cordillères, Political Essay on the Kingdom of New Spain* and the *Personal Narrative*, around a critique of colonialism, racism, and imperial capitalism; while in later years he was forced to subdue this voice, his writings from those years would emphasize instead his critique of scientific objectivity and his insistence that true knowledge must bridge mind and nature, subjective and objective. Like Emerson's and Whitman's "Poet," Humboldt sought to unify the growth or *Bildung* of the individual self with the progressive development of the Cosmos, pointing to a utopian project where different forms of knowledge would operate across multiple perspectives to generate a relational field that interlinked nature, culture, and society.

But this counter-Enlightenment project could not be sustained. By mid-century, it gave way to "science" as we know it today, a collective enterprise that had either been wholly captured as an arm of state power, as with Frémont's army or Maury's navy, or else had fought hard to resist such capture by declaring itself exempt from politics altogether, and exempt from subjectivity too—ideologically neutral, universal, and "objective." Literature, by drawing apart from this emergent form of science, inherited the function literature and science had once shared of social critique. Think of *Walden*, of *Uncle Tom's Cabin, Scarlet Letter, Little Women*, anything by Melville, Frederick Douglass or William Apess, Rebecca Harding Davis, Charlotte Perkins Gilman. However, this divorce came at a tremendous price: it cost literature its foundational grounding in natural knowledge, in what Jefferson had called, in the Declaration of Independence, "the laws of nature and of nature's God." Under the banner of "science," forms of racism proliferated and gathered strength even as the abolitionist and women's rights movements lost their footing in scientific evidence, which was turned with increasing complacency away from exploring the continuities and differences that characterized the global human community and toward demonstration of the natural and cultural superiority of the Anglo-Saxon race.

Even as racism was being naturalized, "nature" became the court of highest appeal: movements for liberation of the oppressed fought back by reclaiming their grounding in Jeffersonian or Enlightenment scientific critique, appealing over the top of nineteenth-century racial science back to foundational "natural" principles, what Thoreau and Emerson called "higher law," as declared by Jefferson's phrase "all men are created equal." This discourse gave Humboldt an important role to play in American abolitionism, but when it came to the environmental movement, Humboldt's vision was not so fortunate. As Lance Newman and Jeffrey Myers have recently demonstrated,

American environmental and nature writing was born as a literature of social protest; but even as social critique was carried forward vigorously in fiction, second-generation American nature writers starting with Muir echoed the new, modernizing form of science by removing nature away from politics, to a privileged realm apart from the polluting concerns of social life. Outdoor, physical "Nature" became, not the beloved and richly particularized backyard environment of Thoreau's Concord or Susan Fenimore Cooper's Cooperstown, laced with calls to social reform, but a remote and sacred wilderness so pure of human presence that the Indians who had called it home for thousands of years had to be purged. Humboldt, for one, would have missed them dearly.[34]

In sum, historians of science have given Humboldt's name to the very era that saw the growth of the United States as a world-class political power and to the very kind of science that gave that power its shape, scope, and texture. Literary historians who work in the vacuum created by the divorce of the "two cultures" have yet to read American literature through the distinctive kinds of activist knowledge that shaped their authors' natural, political, and intellectual worlds. They therefore pursue their literary critique oblivious to the deep contestation of the metaphysics on which they rest their critical projects. The very two-culture split inherited by the twentieth century, the split that defines scientific and humanistic knowledges against each other, was the unintended consequence of the immense body of natural and social knowledge generated by Humboldt and his followers. The methods of gathering such knowledge militarized science into an arm of the state, and the flood of knowledge generated by those methods forced science to reorganize into a highly specialized and disciplined profession. Literary artists were left to carry on the work of social critique, further setting them at odds with a science that seemed increasingly as complicit, or foreign, or monstrous as Mary Shelley's Frankenstein—so far at odds that literary works, so deeply involved in the Humboldtian science of their time, have traditionally, and mistakenly, been categorized as a "romantic reaction" against all science.

The terms of that reaction were set by American expansionism: instead of transforming the globe, the United States would racialize, nationalize, and militarize its sense of territorial destiny. Instead of catalyzing the world revolution that would recognize the freedom and independence of all peoples, it worked to contain and eliminate the threats posed by those peoples. The flood of new knowledge that opened to view so many new cultures, languages, and races resulted not in a Kantian cosmopolitanism, a flowering of pluralism and negotiated federations, but in the cancerous spread of scientific racism,

4

"All are alike designed for freedom": Humboldt on Race and Slavery

The principle of individual and political freedom is implanted in the ineradicable conviction of the equal rights of one sole human race.

ALEXANDER VON HUMBOLDT

There was, of course, a tragic flaw in the great dream of America: while America stood for freedom, the American economy was based on slavery. The Second Continental Congress of 1776, in deference to Georgia and South Carolina, had forced Jefferson to strike out his antislavery wording for the Declaration of Independence, leaving his declaration "that all men are created equal" with the "inalienable" right to freedom to preside, in a terrible irony, over a slave nation.[1] As the decades passed, the contradiction between ideal and reality grew ever more stark. Jefferson and Gallatin had planned to incorporate Indians peacefully into white society by educating them as farmers, imagining they would meld organically into Jefferson's agrarian state. But as conflicts deepened, the frontier Indian fighter Andrew Jackson ran, and won, the presidency on a platform of Indian removal, which he lost no time in implementing. As for slavery, many assumed, like Humboldt, that when the twenty-year moratorium mandated by the Constitutional Convention expired in 1807, the U.S. government would naturally begin the long and delicate process of emancipation. With abolitionism sweeping the British Empire and the Spanish government attempting reforms, it must have seemed, for a moment, possible, if not inevitable. But in the United States, slavery was in the ascendency. South Carolina had reopened the slave trade in 1803, and even after the federal ban on transatlantic slave trading passed in 1807, demand for slaves grew as the markets for sugar, cotton, and rice exploded. What swept the United States were not calls for equality but a growing need to rationalize inequality. As Reginald Horsman says, "If slavery was to continue then it became essential to demonstrate that the fault lay with the blacks, not with the whites." Enlightenment-era egalitarianism, with its faith in human improvability, sounded increasingly quaint as protoanthropologists declared racial differences biologically determined and nonwhite races inherently and permanently inferior. Education was deemed useless, for Indians and blacks were ruled incapable of civilization. The only options were to contain, remove, enslave, or exterminate.[2]

The history of race and antislavery movements in the United States is a huge field, but few seem aware of Humboldt's role in American abolitionism or the trajectory of racial science. Humboldt was the only major scientist during the nineteenth century to argue consistently, for six decades, that "race" was not a biological category and that, as he declared in *Cosmos*, "all are alike designed for freedom." Stephen Jay Gould states in his classic study of racial science that Humboldt should be the hero of modern racial egalitarians, for "he, more than any other scientist of his time, argued forcefully and at length against ranking on mental or aesthetic grounds." And it wasn't just Alexander: the Humboldt brothers together developed a historical anthropology that sought to appreciate every human group on its own terms, for none were in any meaningful sense "superior" or "inferior" to any other. Their arguments are important both for their own sake, and for the impact they had on modern conceptions of race.[3]

(De)Constructing Race

Humboldt's views on race and slavery were not forged in the New World but in the liberal intellectual circles of Berlin and in the halls of Göttingen, where he studied with Blumenbach. Though Blumenbach is frequently demonized for identifying five separate "races," with the white race as the original from which the other four had "degenerated," in the Latin of his original text the word "degeneris" means, not "degraded," but "removed from one's origin." As C. Loring Brace points out, it might more accurately have been translated as "evolved." Blumenbach himself explained it as a technical term intended to designate the mechanism by which all varieties had diverged from the original "primeval" type. He placed little importance on the apparent existence of basic racial types, since really there were not five but "innumerable varieties" of humankind all running into one another by "insensible degrees," making any attempt to identify principal varieties loose and to some degree arbitrary. Blumenbach believed that every race had the capacity for intellectual and moral improvement. As proof, he showed off a library of books written by African Negroes, whom he judged were, of all "savage" nations, the most distinguished by their capacity for "scientific culture."[4]

For Blumenbach in his Göttingen study, the question of human racial differences was deeply interesting but rather abstract: his notorious collection of skulls did not talk back to challenge his views (though he did find it "melancholy" that so many belonged to peoples destroyed by their conquerors). By contrast, his student Humboldt was plunged into the politics of racial difference the moment he stepped onto the shores of South America, from his decision to put off the colonial governor while visiting an Indian's home, to the Cumaná slave market under his window, to the "zambo" who attacked Bonpland on the beach, to his conversation at Mt. Vernon with George Washington's former slave Billy Lee. One of the last incidents he recorded in his *Personal Narrative* was the night the travelers landed on the coast near Cartagena to do a little moonlight botanizing. Out of the woods stepped "a young negro . . . quite naked, loaded with chains, and armed with a cutlass." He was courteous and offered to guide them in exchange for some clothes, but Humboldt and Bonpland were wary, particularly when he whispered something to companions concealed in the brush. These were, Humboldt guessed, "*maroon* negroes, slaves escaped from the prison where they were held in irons," an "unfortunate" class most to be feared: "They have the courage of despair, and a desire of vengeance, nourished by the rigor of the whites." He and Bonpland backed off and returned to the ship, troubled by the sight of a man clothed only in chains, then troubled still more by the reaction of the ship's crew, who wanted to seize the fugitives and "sell them secretly at Carthagena." Thus, Humboldt moralizes, does slavery deaden the ennobling instinct of pity. The Spanish colonies showed him slavery in its most softened form, yet his mind was unchanged: "I preserved, on quitting America, the same horror of slavery which I had felt in Europe."[5]

Nearly fifty years later, Humboldt summarized in *Cosmos* his lifetime's reflections on the subject of race. Experience teaches that mankind is distributed in varieties, "designated by the somewhat indefinite term *races*." How many races? His esteemed colleagues cannot agree, therefore "we fail to recognize any typical sharpness in definition, or any general or well-established principle in the division of these groups." Race was not, that is, a scientific category. The different so-called races, arbitrary and overlapping and vaguely defined as they are, clearly point to a community of descent, and this has one overriding consequence:

> While we maintain the unity of the human species, we at the same time repel the depressing assumption of superior and inferior races of men. There are nations more susceptible of cultivation, more highly civilized, more ennobled by mental cultivation than others, but none in themselves nobler than others. All are in like degree designed for freedom; a freedom which, in the ruder conditions of society, belongs only to the individual, but which, in social

states enjoying political institutions, appertains as a right to the whole body
of the community.

In the last volume of his *Personal Narrative*, Humboldt develops in detail
the "community of actions and efforts" that must be pursued to enforce the
political right of freedom for all. Aristotle himself, Humboldt noted in *Cosmos*, had laid down the "cheerless doctrine of the unequal rights of men to
freedom" and the notion that slavery was "in conformity with nature." From
then on, civilization had advanced without respect for the right of freedom.
This showed that slavery could, and did, maintain itself in the midst of elegance of manners, boasted progress of knowledge, and "all the charm of civilization." Its evils were not self-correcting but self-perpetuating and could be
ended only by concerted moral and political action.[6]

The unity of human races under the banner of freedom was hardly what
Humboldt witnessed in America, where one might have hypothesized that the
promiscuous mixing, over hundreds of years, of Indians, blacks, and whites
would have broken down racial barriers. On the contrary, Humboldt observed, it multiplied them. For example, in Mexico, Guatemala, Ecuador, and
Peru, the presence of over five and a half million Indians slowed the spread
of colonization: their "isolated position, partly forced and partly voluntary,
attachment to ancient habits, and mistrustful inflexibility of character, will
long prevent their participation in the progress of the public prosperity, notwithstanding the artifices employed to *disindianize* them." That is, Indians,
exactly insofar as they remained Indians, resisted the homogenizing influence
of European colonialism, a political reality that must be taken into account.
Fear of Indian uprisings was an important organizing force in establishing racial boundaries and enforcing colonial structures of power. Humboldt noted
that the 1780 uprising in Peru of Tupac Amaru, at the head of an army of
forty thousand mountain Indians, had shocked the Creoles and the Europeans into thinking "that the contest was between the Copper-coloured race
and the whites; between barbarism and civilization." Where Tupac Amaru
had begun by making common cause with the mixed-race mestizoes and the
Creoles, as events unfolded, Indians turned on all non-Indians, and "a rising
for independence became a cruel war between the different casts." This race
war had bonded whites together by forging "a feeling of common interest,"
and they focused on their united number *as whites* even as those who wished
for independence, and those loyal to Spain, continued to divide along party
lines. The result was stagnation, a political calm brought about by "the equilibrium established between the hostile forces"—an unstable equilibrium
founded on disunion and easily shaken.[7]

The prospect of a black uprising terrified whites of both parties, for although the African populations in South America tended to be small (Venezuela's, for instance, was only a fifteenth of the whole), they were of critical importance. Humboldt observed that blacks had the lesson of the Haitian Revolution well in mind, and he judged that in the struggle for independence they would join whichever side seemed most likely to advance their freedom. The abolition of slavery that resulted would be not a humanitarian effort but a strategy calculated to secure the support of "an intrepid race of men, habituated to privation, and fighting for their own cause." His shrewd and lengthy analysis showed that the balance of racial terror was deadlocking the prospects for South American independence, and when Bolívar broke that balance, his dream of a confederation of independent republics of free peoples foundered and wrecked on the very divisions Humboldt had described.[8]

As for Mexico, it was, Humboldt wrote, "the country of inequality. No where does there exist such a fearful difference in the distribution of fortune, civilization, cultivation of the soil, and population." In science, art, architecture, "the tone of society," luxuries and refinements, Mexico's major cities rivaled the great capitals of Europe, a high degree of civilization "to which the nakedness, ignorance, and vulgarity" of the lower classes form "the most striking contrast." Most miserable of all were the Indians, banished to the most barren districts, forced to pay tribute, deprived of the basic rights of citizens. The laws of New Spain treated Indians like children, put them "under the perpetual tutory of the whites," and denied them the right to sign contracts, placing "insurmountable barriers between the Indians and the other casts, with whom all intercourse is almost prohibited." White lawyers and Creole proprietors argued that Indians could not be given more liberty because "the whites would have everything to fear from the vindictive spirit and arrogance of the Indian race." Yet, Humboldt replied, the real danger lay in perpetuating the Indians' isolation, misery, and consequent motives for hatred. Humboldt pointed once again to the revolt of Tupac Amaru and concluded that it was of "the greatest importance" for whites to "interest themselves in the Indians, and rescue them from their present barbarous, abject, and miserable condition."[9]

As for Mexican blacks, Humboldt was struck by how few there were: "We may go through the whole city of Mexico without seeing a black countenance." The few thousand, free and enslaved, that did live in Mexico were concentrated on the coasts. There were, however, Indian slaves, thanks to a legal loophole that allowed frontier Indians, often Apaches, to be "dragged to Mexico," imprisoned, and sold to die in Vera Cruz or Cuba. And while Spanish law allowed Negro slaves to purchase their liberty on payment of a moderate sum, Humboldt nevertheless witnessed shocking abuses, like the woman who had stabbed her two slaves all over with scissors, pins, and knives and knocked out their teeth when they complained of a gum disease. Barbarity is the same everywhere, Humboldt grieves, when governments allow an order of things "contrary to the laws of nature, and, consequently, to the welfare of society."[10]

The miseries and abuses Humboldt documented were belied by the remarkable colonial genre of "casta paintings" which flourished in Mexico from 1760 to 1790. Composed of a set of, usually, sixteen panels, each painting portrayed a mixed-race family grouping of father, mother, and one or two children, carefully labeled to denominate the elaborate racial subcategories created by the proliferation of racial hybrids. Created for export to Spain, these paintings showed a hierarchical and racially harmonious society in a New World setting characterized by natural wealth and material abundance. Here a Spanish father goes shoe-shopping with his Mestiza wife, who holds her Castiza daughter; a Spaniard plays with his Morisca daughter while his Mulatto wife looks on; a China Cambujo father hands a toy bird to his Wolf daughter who is held by her Indian mother; a Camizo son helps his Castizo father roll cigarettes while his Mestiza wife studies the completed pile. The names proliferate beyond comprehension: Albino, Return Backwards, Hold Yourself in Midair, Albarazado, Barcino, Coyote, Zambaiga, all attempting to systematize a bewilderingly complex process of racial mixing. Although Humboldt does not mention casta paintings, he does note that the colonists have "enriched their language with terms for the finest shades of the colours" which result from intermarriage, and he details a similar list with regional variations, plus the curious information that Peruvian Indians can identify whites, blacks, and Indians by odor alone.[11]

In a society governed by whites, "naturally" a person's rank in society was determined by "the greater or less degree of whiteness of skin." As Benedict Anderson concluded, racism was associated with European domination because it welded together dynastic legitimacy—which in Europe had been class-based—and a national community by "generalizing a principle of innate, inherited superiority." In short, class-based hierarchies were racialized. Humboldt observed this process in action, noting, for instance, the way whiteness established a presumption of inequality among members of the same social class: in Colombia, he had been amused at overhearing cargueros "quarrelling in the midst of a forest, because one has refused the other, who pretends to have a whiter skin, the pompous title of *don*, or *su merced*." Whiteness also fostered a presumption of equality across class divisions: "When a common man disputes with one of the titled lords of the country,

he is frequently heard to say, 'Do you think me not so white as yourself?'"
This, Humboldt comments, may be taken to characterize "the state and the
source of the actual aristocracy." Thus "public vanity" demands very careful
calibration of the exact degree of white blood in each of the different casts,
and Humboldt reproduces a chart of the customary fractionation. The catch
was that in a population of such rampant hybridization, whiteness became a
matter of convention, a social construction ratified by law. Humboldt means
this quite literally: mixed-race families often repaired to the courts to have
themselves declared "white." "These declarations are not always corroborated
by the judgment of the senses," remarked Humboldt. "We see very swarthy
mulattoes who have had the address to get themselves *whitened* (this is the
vulgar expression)." When the court could not permit itself to go quite so
far, it rendered instead the somewhat equivocal judgment that the petition-
ers "may consider themselves as whites (*que se tengan por blancos*)." Perhaps
he wondered whether Peruvians might find employment in sniffing out the
truth.[12]

According to the historian David Brion Davis, in 1820, John Quincy Adams, then secretary of state under James Monroe, was shocked when the secretary of war John C. Calhoun confided in him "that one of the major benefits
of racial slavery was its effect on lower-class whites, who could now take pride
in their skin color and feel equal to the wealthiest and most powerful whites.
Thus slavery, in Calhoun's eyes, defused class conflict." Davis concludes that
"the great mission of proslavery theorists was to convince and command the
loyalty of nonslaveholding whites," who by virtue of their color could always
hope for upward mobility. The duty and burden of all whites to protect and
care for a so-called inferior and dependent race defined what was regarded as
"the moral superiority of the South," and the great majority of nonslaveholding whites were prepared to—and of course many did—defend this ideology
to the death.[13]

Humboldt did not witness the paternalism of the American South, but he
did put his finger on the false racial consciousness defended by Calhoun. In
the Orinoco jungle he was bemused by one "Don Ignacio," their host for a
night, "almost naked" and as dark as a zambo but insisting nevertheless that
he was "of the cast of Whites." The don had never bothered to erect even the
simplest of shelters for his family, and as the rain came down in torrents,
soaking them all to the skin, the tiger-hunter bragged of his exploits in killing
Indians and seizing their children for the missions, while assuring the travelers of their good fortune in finding themselves not among such savages but
"among Whites and persons of rank; *entre gente blana y de trato.* Wet as we
were," comments Humboldt dryly, "we could not easily persuade ourselves

of the advantages of our situation." More seriously, Humboldt concluded his long essay on race in Mexico by asking about the influence of racial division on "the general wellbeing of society." His answer was surprisingly personal: it hurts, so deeply and powerfully that the man of sensibility has no choice in a racist society but to opt out of society altogether. The social pathology of race, by forcing Americans of sensibility to withdraw to private life, creates "a want of sociability" and "the hatreds which divide the casts of greatest affinity." Whereas a wise government might succeed in extinguishing "the monstrous inequality of rights and fortunes" which so poisons American society, "it will find immense difficulties to overcome before rendering the inhabitants sociable, and teaching them to consider themselves mutually in the light of fellow citizens." The family so anxious to reap the social and legal benefits of whiteness has little in common with their colored neighbors, even if they don't go so far as the predatory Don Ignacio, hunting Indians like animals while bragging of his white superiority.[14]

The pathologies of race and inequality that Humboldt analyzed played out rather differently in the two Americas. As Calhoun's comment illustrates, in the Southern United States, not only did racial hatred blind whites to their moral obligation to free black slaves; to mitigate their republican guilt, they rationalized the power structure that favored them and used whiteness to paper over ominous class divisions. In the Spanish colonies, Humboldt's diagnosis was rather different: colonial government policies deliberately instilled and deepened divisions between the races as well as between loyalists and revolutionaries, in order to cause dissension and forestall united political action. As he concluded in his *Political Essay*, "The mother country foments incessantly the spirit of party and hatred among the casts and constituted authorities." Maintaining a balance of fear might disturb "the enjoyments of social life," but it kept the peace. However different the social evolution of North and South, though, Humboldt thought the sources of the pathology were the same:

> The *conquest*, on the continent of Spanish America, and the slave-trade in the West Indies, Brazil, and in the southern parts of the United States, have united the most heterogeneous elements of population. Now, this strange mixture of Indians, whites, negroes, métis, mulattoes, and *zambos*, appears accompanied with all the perils which the heat and disorder of the passions can gender, at such critical periods, when society, shaken to its very foundations, begins a new era. The odious principle of the *colonial system*, that of security, founded on the hostility of castes, and prepared during ages, then bursts forth with violence.

Humboldt did not quite predict the U.S. Civil War, but he did warn that if the peaceful abolition of slavery failed, apocalyptic violence would follow.[15]

In this revolutionary age, in short, the clock was ticking, the reckoning inevitable. Could a security based on fear last much longer? Could governments afford to neglect the remedy for the evil? Doubtful, thought Humboldt. Ironically, though, the very fear which paralyzed whites in power from taking action might also provoke a solution: look, Humboldt said, at Haiti. Humboldt was one of very few white intellectuals to defend the Haitian Revolution of 1791–1804, in which slaves in the French colony of Saint-Domingue, inspired by the French Revolution's call for "Liberty, Equality, Fraternity," had successfully thrown off their white masters and declared an independent nation. Even his heroes Washington and Jefferson had fought against Haitian freedom: George Washington had come to the defense of the white colonists, sending them $726,000 in aid, and Jefferson had attempted to isolate Haiti economically and diplomatically "to end the spread of black revolution." But all, said Humboldt, ended well. The Haitian storm "was appeased on the spot which gave it birth," and the new nation of free blacks, far from troubling the peace of its neighbors, was making progress "towards the softening of manners, and the establishment of good civil institutions." Free to turn to their own needs rather than those of their colonial masters, Haitians were planting food crops and their population was flourishing. And to Haiti's west and south, in Mexico, Guatemala, and Colombia, legislators took the correct lesson and instead of clamping down with further repressions labored ardently to extinguish slavery. Other governments should follow: "The fear of danger will force concessions that are claimed by the eternal principles of justice and humanity." Humboldt insisted that the circumstances of slavery and oppression, not the example of Haiti, were what would cause peoples of color to rise up and seize control of their own fate; but ironically, perhaps fear of Haiti's very success would finally force white leaders to enact the justice demanded by principle.[16]

Humboldt's analysis of New World racial politics, however valuable, did not lie at the core of his thinking on race. For all the breadth of his interests and his humanitarianism, he thought of himself above all as a natural scientist, and for Humboldt, the evidence of human variation was a pressing, and fascinating, scientific problem. While he assembled an extensive collection of ethnographic materials from the New World, it must be admitted that he did remarkably little with them, beyond publishing some of them in the *Personal*

Narrative. At least one commentator has regretted that Humboldt did not become the ethnologist he so clearly could have been.[17] Why not?

The answer is not hard to find, although pursuing it will mean digressing briefly into the career of Alexander's brother Wilhelm. In the human sciences, Alexander always deferred to Wilhelm, widely said to have one of the best minds in Europe. This was more than a matter of sibling rivalry: Alexander had concluded that his ability to gather reliable ethnographic materials was compromised by his ignorance of native languages and by his limited time in the field. The work had to be done, but it needed to be done by comparative linguists and ethnologists—now we would say anthropologists—who could learn the relevant native languages and immerse themselves in native cultures. While Alexander was coming to this conclusion, Wilhelm was pursuing a career in politics and diplomacy, until he was forced to retire from public service because of his liberal politics. He then returned to scholarship, pioneering the comparative study of non-Indo-European languages along anthropological lines. At his death in 1835 he was completing a monumental analysis of the languages of Southeast Asia (published posthumously by Alexander), which became famous as the first great work of general linguistics, foreshadowing twentieth-century structuralism.[18]

The Humboldt brothers worked as a team, applying shared ideas to different realms of knowledge. In his essay on history, Wilhelm turned to Alexander's natural science: "Even a simple depiction of nature cannot be merely an enumeration and depiction of parts or the measuring of sides and angles; there is also the breath of life in the whole and an inner character which speaks through it which can be neither measured nor merely described." And vice versa: when Alexander described his approach to science, he drew on Wilhelm's theory of history, as when in *Cosmos* he stated that his goal was to explore an alternative form of science analogous to "historical composition." As the historian of anthropology Matti Bunzl observes, "Alexander von Humboldt's approach to the phenomena of the world was that of a natural historian rather than a physicist," and both Humboldt brothers shared a suspicion "of the Enlightenment attempt to reduce the world to abstract principles along Newtonian lines." Bunzl concludes that Alexander's "natural historian's method essentially duplicated Wilhelm von Humboldt's hermeneutic methodology in his 'On the Task of the Writer of History.'" It was by devising a method of science based on history, rather than on deductive reasoning, that Alexander laid the groundwork for the study of ethnic difference: the Humboldt line of reasoning would seek to understand each people's unique historical development, rather than devise universalist abstract metanarratives of human racial evolution.[19]

Thus what distinguished the Humboldt brothers was their shared focus on individual phenomena as the starting point for any just apprehension of the whole. God really was in the details, what Alexander called the "accidental individualities," the astonishments of the particular, not the flaccid generalizations of the abstract. In his turn, Wilhelm wrote that there is an "idea" in every fact, every event, but "this idea can be recognized only in the events themselves." Whether one studies nature or history or languages, the scholar must take the greatest care "not to attribute to reality arbitrarily created ideas of his own, and not to sacrifice any of the living richness of the parts in his search for the coherent pattern of the whole." Understanding was rooted in lived experience, and to understand the past, or a culture, or a language, one must shed as much as possible one's own preconceived ideas and experience it on its own terms. For Wilhelm's study of languages, this meant setting aside the grammatical categories of Greek or Latin as a basis for comparison: "Every language must be understood from the point of view of the sense the nation has given it, not from any foreign viewpoint." Interestingly enough, Wilhelm noted that American Indian languages were particularly vulnerable to the "error" of imposing one language onto another, and he advised that the "incorrect views" of Spanish and Portuguese scholars be set aside to reveal their actual, not presupposed, structures.[20]

For the Humboldt brothers, to step inside a language was to step inside an entire culture and worldview. Language was both an expression of thought and the vehicle through which thought was developed. Hence every language embodied a people's unique worldview, its character or "genius," in a dialectic between society and nature that was always changing. And while every individual could think only in his native language, Wilhelm did not agree that one's nation or one's language determined how one thought. Every individual had the cognitive freedom to stretch and explore the resources of his language in his own way, making the relationship between national and individual character a dialectical process, with each forever "intertwined" with the other. Wilhelm flatly rejected the idea that any language was superior or inferior to any other. As Bunzl remarks, "He regarded all languages as functionally equivalent, capable of expressing any conceivable idea," an idea Alexander echoed in *Personal Narrative*. Wilhelm spent much of his career defending languages others had dismissed. For instance, some had relegated Chinese to a state of infancy, but after years of study Wilhelm concluded that on the contrary, the Chinese people had developed the unique resources of their language in an extraordinary way, resulting in a language with, in his words, "unparalleled excellence in the realm of ideas . . . on a level with the most perfect languages we know."[21]

As for American Indian languages, Wilhelm could find in them no trace of "primitivism" or corruption or degeneration. On the contrary, those he knew best "have a great regularity and very few anomalies in their structure." Their study held particular promise, and each one ought to be analyzed in careful detail. Mere word lists and random notes about grammatical peculiarities were not enough, for "the dialect of even the most primitive nation is too noble a work of nature to be presented for observation in such fragmentary form and with such a random choice of elements. Language is an organic entity and must be treated as such." It would be "most unjust" to designate American Indian languages "as coarser and more savage, however much they may differ in structure from fully developed languages," such as those with alphabetic script.[22]

It is important to note that Wilhelm could have treated languages as racial entities. He did not. Rather, languages were characteristic of "nations" or "peoples," and far from seeking racial or linguistic purity, he believed that intermixing was key to developing a language's richness and beauty. Languages, in their historical development and dialectical interplay, connected the whole human community. Accordingly, Wilhelm, too, insisted on the unity of the human species. Alexander closed his first volume of *Cosmos* with a tribute to his late brother, quoting from his great work on the Kawi language. If there is one idea that has spread throughout history, Wilhelm wrote, "it is that of establishing our common humanity—of striving to remove the barriers which prejudice and limited views of every kind have erected among men, and to treat all mankind, without reference to religion, nation, or color, as one fraternity, one great community." Thus "the recognition of the bond of humanity becomes one of the noblest leading principles in the history of mankind." In Wilhelm's view, the course of modernity was marked by the spread of "the benevolent view of the whole human race linked together," a linkage most visible in language: "More than anything else in man, language embraces the whole race. It is precisely in its characteristic of dividing peoples that it unites the variety of their individuality through the mutual understanding of the foreign tongue." Moreover, as Wilhelm adds, "It does this without endangering the particular qualities of each people." It was, then, through the common embrace of language, and an appreciation of all languages in their great diversity, that civilization would advance to the kind of pluralistic, collaborative cosmopolitanism envisioned by Kant and manifested in the humanistic science of both Alexander and Wilhelm von Humboldt.[23]

It was Wilhelm, then, who was to carry forward the humanistic side, the linguistic and anthropological studies that formed one side of the great

Humboldtian project. Rather than work them up himself, Alexander passed his ethnographic and linguistic materials on to his brother, who was writing a volume on the languages of America Indians for Alexander's magnum opus on his American travels; this project was the impetus behind Wilhelm's correspondence with John Pickering. However, Wilhelm died before his essay was completed. Although he worked on American Indian languages through the 1820s, he turned in later years to his massive work on the Kawi and related languages of the South Pacific, leaving unpublished a bulky, unfinished manuscript comparing twenty-four American Indian languages. Lost until the twentieth century, this manuscript is only now being recovered. Although Wilhelm's comparative approach was largely bypassed by the mainstream of German Indo-European linguistics, it was kept alive by his followers, to resurface with renewed energy in the work of Franz Boas.[24]

While Wilhelm's influence on German linguistics may have been limited, the Humboldts' combined influence on American linguistics, ethnology, anthropology, and archaeology was immediate and profound. Humboldt's spectacular recovery of Aztec and Inca objects and monuments and his speculations about the Asian origin and various migrations of American tribes stimulated, for instance, John Lloyd Stephens to rediscover the Mayas, William H. Emory to initiate the archaeology of Pueblo tribes, and Karl Bodmer and George Catlin to conduct salvage anthropology. Humboldt also played a decisive role in Mexico's rediscovery of its indigenous past. By publishing information on Mexican manuscripts, he showed their value for historical reconstruction while crediting the Aztecs as a scientifically advanced civilization, a history the Creole revolutionaries were happy to honor when they adopted the Aztec word "Mexico" for their new nation. Eloise Quiñones Keber demonstrates that through this work Humboldt became a key figure in establishing the "'canon' of major Aztec sculptural works" and in promoting a burgeoning interest in pre-Columbian art, which became not "the province of a small, specialized group of local antiquarians," but a "generalized awareness" still strong today. The art historian Halina Nelkin sees a continuity from the Humboldtian display of Pre-Columbian "primitive" art, so shocking to the Victorian public, to the radical changes it inspired among artists, changes "resulting in the rise of the modern art of our time." Humboldt himself found Pre-Columbian figures revolting in their lack of naturalism, but by displaying them respectfully nevertheless, he gave rise to a tradition of art history that could see a statue like the "Azteck Priestess" not only as a cultural artifact but as a sophisticated work of art, valued today by aesthetic standards not yet available to Humboldt himself.[25]

(Re)Constructing Race

Though Alexander von Humboldt may not have contributed directly to American linguistics and anthropology, he did influence those individuals who became its pioneers. As Horsman has argued, the United States in these years was the seedbed for racial thinking, as the new republic adjusted its ideals to embrace the enslavement of blacks and to legitimize the elimination of Indians who stood in the way of western expansion. Science increasingly was used to rationalize these political and strategic fronts. Hence this aspect of Humboldt's story is vexed, even tragic. Humboldt's conviction that the human race was one great family became a minority view even among his most loyal followers, who in the growing racism of nineteenth-century America twisted and reinvented Humboldt's ideas to fit their own ideology.

The man who carried Humboldt's ideas most directly into the American mainstream was his old friend Albert Gallatin, Jefferson's secretary of the treasury. Like Wilhelm von Humboldt, Gallatin had retired from public service to take up the study of American Indian languages, an interest dating back to 1780 when he spent his first summer in the United States on the Maine frontier next to two tribes of Indians. In 1823, at Alexander von Humboldt's request, Gallatin wrote an essay classifying North American Indians for a new edition of his *Political Essay on the Kingdom of New Spain*. Gallatin next set about deepening his knowledge of Indian vocabularies and grammars, designing a vocabulary questionnaire which he persuaded the secretary of war to circulate to Indian agents and missionaries (copies went along on the Wilkes expedition). The results were disappointing, but Gallatin was able to publish the first table and map showing tribal language groups as related by comparative methods. Then in 1826, diplomatic service called him away to Europe for a year, where he hobnobbed with his old friend Humboldt in Paris, and on his return Gallatin plunged back into his Indian studies with renewed enthusiasm. In 1832 the American Antiquarian Society asked to publish the essay he had written for Humboldt, which had never actually appeared in print: Humboldt had lent it out to Adriano Balbi, who mislaid it, after praising it and incorporating it into his *Atlas ethnographique du globe* (1826). So starting from scratch, Gallatin produced his "Synopsis of the Indian Tribes of North America" (1836), a Humboldtian synthesis that argued against the emerging view, given fresh currency by Ephraim Squier's studies of the Indian mounds of the eastern United States, that Indians were degenerate.

Gallatin's essay laid the foundation for American ethnology, and as ethnology was increasingly drawn into racial science, he and others agreed that a professional society was needed to set scientific standards and sift through competing claims. From 1842 until his death in 1849 the American Ethno-logical Society met regularly at Gallatin's house, where he gathered around himself the circle of fellow believers in "monogenesis," or the unity of the human race, against believers in "polygenesis" who held that human races had developed from separate creations. Gallatin's last writings extended and deepened his comparative linguistics to Mexico and Central American and to Emory's new findings in the American Southwest. The aging statesman and diplomat did not keep completely out of politics: in 1844 he protested at a proannexation rally, denouncing the annexation of Texas as nothing less than a declaration of "unjust war" against Mexico that violated trea-ties, discredited democracy, and would "disgrace the national character." Henry Adams thought this the single most courageous act of Gallatin's life. Through jeers and catcalls, the frail eighty-four-year-old persisted in finish-ing his speech, though his voice was so feeble few could hear him. Just before he died, Gallatin wrote a series of widely circulated pamphlets urging peace, which according to Adams played a significant role in public acceptance of the compromised Treaty of Guadalupe Hidalgo. His most direct influence on the twentieth century was through the work of John Wesley Powell, whose 1886 "Indian Linguistic Families of America North of Mexico" built directly on Gallatin, whom Powell called the Linnaeus of North American Indian philology.[26]

Gallatin perhaps more than anyone else took up and advanced Hum-boldtian thinking in the context of North American languages and cultures. Meanwhile a different line of approach took up Humboldt's suggestion in the *Essay on the Geography of Plants* that migrating humans have carried their garden and agricultural plants from one end of the globe to another: Greeks had carried wine grapes to Europe, Romans wheat, Arabs cotton. The Toltecs spread maize and potatoes throughout America; Europeans cultivated wal-nuts and peaches from Persia, apricots from Armenia, figs, pears, olives, and prunes from Syria, indigo from Africa. A century after its first appearance in Italy, the cherry was common in France, Germany, and England. "This is how man changes the surface of the globe to his liking," wrote Humboldt, "and gathers around him plants from the most remote climates." Associa-tions of agricultural plants recall to imagination, then, "the series of events which spread the human race across the whole surface of the globe," and in turn human migrations may be followed by identifying their characteristic food plants. For instance, the Guanches' cultivation of wheat shows them to have been an Old World people, while the New World replacement of wheat with maize shows they had independently developed agriculture, using only New World plants.[27]

One ethnologist who took up Humboldt's theory was the English Quaker James Cowles Prichard, who in the 1820s plunged into the new literature on biogeography, hoping to do for humans what Humboldt and others had done for plants. Prichard was especially taken by the correlations Humboldt drew between altitude and specific organisms, and he investigated "the distribution of plants and animals, correlating their habits to environmental circumstances" and inferring their likely places of origin. There were, Prichard concluded, not just one but several centers of creation around the globe, although human beings had just one original habitation from which they had migrated across the planet. Racial differences were adaptations to the globe's varying climates, making race not biological but historical. Just as Humboldt said, "race" was properly only a synonym for tribe or nation. Prichard identified dozens and dozens of these "races," all of them so variable and passing into one another by such "insensible gradations" that apparent racial divisions melted away. "We are entitled," Prichard declared at the end of seven-hundred-odd pages of explication and examples, "to draw confidently the conclusion that all human races are of one species and one family." This being so, the fact that Europeans were exterminating so many human varieties amounted to a call to arms: the story of Cain and Abel seemed "designed to be typical of the time when christianized Europeans shall have left on the earth no living relic of the numerous races who now inhabit distant regions." As his biographer notes, Prichard did not argue that some superior destiny justified the white European in his massacres; "he did not explain the extinction of human tribes as a 'normal' historical fact." It was for him not some abstract force of history but man, civilized Christian Europeans, who wreaked such havoc, bent on self-indulgence, national self-aggrandizement, and reckless exploitation. All one could do was salvage anthropology, documenting what was being so ruthlessly destroyed.[28]

Prichard used both biogeography and the affinities of languages to trace the migration and dispersal of peoples across the globe, and his work impressed Humboldt, who praised it in *Cosmos*. Prichard himself, however, was an armchair researcher who never traveled, and he obtained much of his information from the American Charles Pickering, back from the Wilkes expedition. Before completing his expedition report on the nature and geographical distribution of the races of man, Pickering had taken off to do more fieldwork in the Near East, and by the time he returned, the topic of race had become explosive. His global survey of human variety would be the most extensive ever undertaken, and in the superheated climate of the late 1840s, before the federal government would publish it he had to make sure all his punches were pulled. Leaning on his training as a zoologist, Pickering declared that his zoological eye brought him to the following conclusions: first, that there were eleven well-defined races, no more and no less; second, that Prichard was wrong, for each race was separate and distinct, "*adapted by Nature*" for its particular climate.[29] Thus Humboldtian biogeography was used to produce two contradictory answers to the problem of human variation: that race did not exist as a biological category, and that it did—and there were exactly eleven of them.

But were these eleven groups races, united by common descent? Or were they actually separate species? Pickering's answers gave aid and comfort to no one. On the Humboldtian view of the unity of the human species, a view that not incidentally was strongly sanctioned in the United States by biblical literalism, Pickering professed neutrality. If the majority preferred there be a single human species, "I am content to let them have their way." In any event he was inclined to believe that humans originated in one spot, probably in Africa; John Bachman, when he came to recruit Pickering for his own crusade in favor of monogenism, was pleased to find that Pickering's map had placed a circle in Ethiopia. But before the polygenists could claim Pickering for their side, they had to deal with the fact that he rejected their racist certainty in white superiority. Different races did have different "characters," but as for his own race, Pickering had some choice words for the character of that "race of plunderers" who so systematically destroyed whatever monuments had ever been erected, "from the soul-inspiring works of Greece to the Simple Grave of the American Indians." "Every race," Pickering concluded, "possessing its peculiar points of excellence, and at the same time counterbalancing defects, it may be, that union was required to attain the full measure of civilization." In the immense variety of nature, every kind of being had its place, and Pickering doubted that any one kind of human being, "existing singly should, up to the present day, have extended itself over the whole surface of the globe."[30]

As the Humboldt brothers, Gallatin, and Prichard made clear, their brand of monogenesis, based as it was on the flourishing of human diversity, carried with it a strong humanitarian commitment. But their subtle conclusions were increasingly out of fashion. Robert Knox, the Scottish physician and anthropologist, sets the tone: "Well-meaning, timid persons dread the question of race: they wish it left where Prichard did. . . . But this cannot be: the human mind is free to think, if not on the Rhine or the Thames, at least on the Ohio and the Missouri." Knox's reference to the Rhine was intended as a slam on Humboldt's "failure" to deal adequately with race in *Cosmos*. Humboldt might quail before the truth, blared Knox, but the truth must be told: race exists, and everyone knows it. "Race is everything: literature, science, art, in a word, civilization, depend on it." Admit the obvious, and everything falls into place. Racial differences are so crisp, so determinative, so systemic—in Emerson's phrase of such "appalling" importance—that the bold and honest American thinker must start with that assumption.[31]

Such at least was the view of the band of polygenists who became so dominant that they were named "the American School of Ethnology," as if there were no other. Their starting assumption was that only multiple origins could account for such extreme racial difference. Races were quite literally different species. The leader of this small but noisy band (as Robert Bieder suggests, what they lacked in size they made up for in belligerence) was Samuel G. Morton, a Philadelphia Quaker who had been inspired by Benjamin Rush to become a doctor, and who had trained in Edinburgh and Paris. Morton's innovation was to classify race by physical rather than ethnological features. Since skulls were both resistant to environmental influences and diagnostically important for classifying mammals, he began with skulls; where Blumenbach had collected eighty-one Morton's collection, by his death in 1851, had topped a thousand. He arranged them according to Blumenbach's five races and set about applying to those races the statistical methods he had learned in Quetelet's Paris. Specifically, since the brain was the organ of the mind, Morton assumed that brain size correlated with intelligence, and thus intelligence could easily be measured by calculating the volume of lead shot that each cranium could hold. Morton's measurements proved definitively that Indians were less intelligent than whites, and blacks were the least intelligent of all. The data seemed clean and objective; as Stephen Jay Gould asked, "How could sentimentalists and egalitarians stand against the dictates of nature?" Gould detected a pattern of "fudging and finagling" in Morton's data, subtle manipulations that had been, apparently, unconscious. They were also, of course, fatal: as Bieder says, "In Morton's hands statistics not only delineated the Indian but also judged him: the hand of God, not the environment, shaped the Indian's cranium." God and science had declared the Indian incapable of civilization, and the black as weak and dependent, if perhaps as loveable, as a child.[32]

Morton made a convert in the South Carolina physician Josiah Nott, who after imbibing the religious freethinking of Thomas Cooper at South Carolina College in Columbia (now the University of South Carolina), studied with Morton in Philadelphia and returned to Columbia to open a medical practice. On Cooper's suggestion Nott went on to further study in Paris, returning in 1836 to settle in Mobile, Alabama, where he took up the problem of hybridity: since blacks and whites were different species, he judged, intermarriage must eventually result in the extermination of both races.[33] Meanwhile Morton had begun a correspondence with one George Gliddon, the U.S. Consul at Cairo, who sent him over a hundred skulls culled from Egyptian tombs. Morton published his first results in the lavish *Crania Americana* (1839), which he followed up with *Crania Aegyptiaca* (1844): the former established the inferiority of Indians, the latter, of blacks. These works made the retiring Philadelphia Quaker into the darling of American science, and in 1846, when the famous zoologist Louis Agassiz came from Europe to lecture in the United States, he headed immediately to Philadelphia to pay Morton his respects. What happened in Philadelphia would change Agassiz's career, and American science, forever.

Agassiz was in the United States through the auspices of his mentor Humboldt, who had first taken him under his wing back when Agassiz was a brilliant but penniless student in Berlin. In 1846 Agassiz, though at the top of his field, found himself bankrupt and abandoned after his debts caught up with him and his wife left with their two children. Humboldt once again came to his rescue, arranging for a sizeable grant from the king to fund a trip to the United States and pulling strings with Charles Lyell in England, who arranged for Agassiz to deliver a lecture series at Boston's Lowell Institute. (Within a year, Agassiz had decided to stay in the United States and was soon installed as a professor at Harvard's new Lawrence Scientific School.) Agassiz arrived in October 1846, and finding himself with a few weeks to spare, he took off for a tour of the neighborhood, including a visit with Morton. To his mother he wrote that Morton's collection of skulls alone was "worth a journey to America": "Imagine a series of six hundred skulls, mostly Indian, of all the tribes who now inhabit or formerly inhabited America. Nothing like it exists elsewhere." In the same letter he recorded his first encounter with a Negro, in a Philadelphia hotel. The visceral revulsion Agassiz experienced shook all his previous beliefs in "the confraternity of the human type." But, as Knox had said, "truth before all":

> It is impossible for me to repress the feeling that they are not of the same blood as us. In seeing their black faces with their thick lips and grimacing teeth, the wool on their head, their bent knees, their elongated hands, their large curved nails, and especially the livid color of the palm of their hands, I could not take my eyes off their face in order to tell them to stay far away. . . . What unhappiness for the white race—to have tied their existence so closely with that of negroes in certain countries! God preserve us from such a contact!

Agassiz left Philadelphia a polygenist.[34]

Around the same time, Morton also converted to polygenism, and like Nott, began to write on hybridity and species crossings: if he could establish their frequency, he could remove the major objection to designating blacks and whites as separate species, since clearly they could successfully interbreed even though one of the standard zoological criteria for a species was reproductive isolation. While Morton worked out the empirical details, it was Agassiz who advanced the sweeping new theory that would give polygenesis the final stamp of scientific authority. Agassiz concluded that Morton's evidence confirmed a view he had come to in 1845: that God had created multiple natural provinces, or zones, each with its own distinctive array of plant and animal species. There were, then, multiple centers of creation, not just one, and each organism had been separately created to fit its particular zone. There were no migrations across continents, no adaptations across millennia, and absolutely, positively no evolutionary changes. (When Darwin published his *Origin of Species* in 1859, Agassiz became his most passionate American opponent, even leading an expedition to the Amazon to acquire zoological and anthropological evidence to prove Darwin wrong; along for the ride was one of his students, a young William James.)

Agassiz first announced his views at an address in Charleston in 1847, and he continued to refine them over the next several years. His breakthrough was a triumphant appearance at the 1850 AAAS meeting in Charleston, where Nott weighed in with a paper on the immutability and purity of the Jewish race, to which Agassiz gave his blessing. Agassiz detailed his findings in a series of three articles in the *Christian Examiner*. In the second he explained, to a national audience, the necessary consequences: while we must recognize "The Unity of Mankind," we must also recognize that that unity is spiritual and "figurative," not natural or "genital," and does not require common descent or "ties of blood." The creator has established a higher plan, a succession of creations region by region. The animals and plants inhabiting each region were created exactly where they occur, in the numbers God intended, and God's plan included human beings. The different human races each originated in the districts they currently occupy. Call them races, varieties, or species, what you will: "The chief point is to distinguish between the unity of mankind and the origin of the different races," and then "to settle the relative rank among these races, the relative value of the characters peculiar to each, in a scientific point of view." It is "mock-philanthropy and mock-philosophy to assume" that all races have the same powers, dispositions, and abilities, and that therefore "in consequence of this equality they are entitled to the same position in human society." This is not, Agassiz repeats, a political or religious question. Science has the right, and must have the courage, "to consider the questions growing out of men's physical relation as merely scientific questions, and to investigate them without reference to either politics or religion." Science had spoken. It was up to the politicians to "see what they can do with the results."[35]

And what did science say about the character of the different races? "The indomitable, courageous, proud Indian,—in how very different a light he stands by the side of the submissive, obsequious, imitative negro, or by the side of the tricky, cunning, and cowardly Mongolian!" Laying out such racial rankings would be the task taken up by Nott and Gliddon, with Agassiz's help, after the death of Morton in 1851. To honor their fallen leader the three teamed up to publish the bestselling *Types of Mankind* in 1854. The subscription list of 992 names included many of the United States' most prominent scientists and intellectuals, and the first edition sold out immediately; nine editions were issued by century's end. In his introductory essay Agassiz laid out his eight separate zoological provinces, complete with gross caricatures of the nonwhite races (with Agassiz's own race represented by a portrait of the great French zoologist Georges Cuvier, and the "indomitable" Indian by a profile copied from Bodmer) and a declaration that he was now convinced that the differences between human races were "of the same kind" as those between "different families, genera, and species of monkeys or other animals." The American School's doctrine of polygenesis was now complete.[36]

But not without a fight. In the United States, exactly one scientist dared to take on the team of Morton, Nott, Gliddon, and Agassiz: John Bachman, who so many years before had met Humboldt in Philadelphia, who had corresponded with Humboldt through the years, visited him in Berlin in 1838, and who now was the center of the Charleston circle of naturalists. Yet before turning to Bachman, it must be remarked as a curious fact that many of the major figures in this controversy saw themselves as loyal Humboldtians. Bachman called Humboldt to his side to defend the unity of the human species; yet Agassiz's polygenetic theory of natural provinces built directly on his mentor Humboldt's climate zones and biogeographical researches. So did the global studies of Prichard, which had dissolved race away, and the precisely eleven races of the equivocal, and well-traveled, Charles Pickering. And still another figure important to American racial science also claimed descent from Humboldt: Arnold Guyot, who had emigrated to the United States on the urging of his friend Agassiz to deliver, like Agassiz, a series of Lowell Lectures and had stayed to become a major voice in American science. Guyot, a geographer trained in the school of Humboldt and the German geographer Carl Ritter, used his lectures to display for Americans the results of the new Humboldtian biogeography, which he published first in *Earth and Man* (1849), then in a series of graded textbooks designed for American schools. In Guyot's hands, the Humboldtian dialectic between humans and their natural environment curdled into an argument for the geographical determination of permanent racial inferiority. The "privileged" white races owed to the "inferior races the blessings and comforts of civilization" and all the limited intellectual development "of which they are capable"; in turn, the inferior races owed the white race their labor. The people of the northern, temperate continents will, Guyot concluded, "always be the men of intelligence, of activity, the brain of humanity. . . . The people of the tropical continents will always be the hands, the workmen, the sons of toil." In his lectures to his American audience, Guyot found the formula that would both release the revolutionary energies of America, and contain those energies safely within his own race, gender, and class: the motto of the present age is "Viva la Révolution!" "*Liberty, Equality, Fraternity.* Yes, gentlemen, liberty to unfold all the living forces" of man, "the equality of rights lying in the moral nature of man, *but not that absolute, impossible equality which is contrary to nature and the course of Providence, and annihilates all progress*; that fraternity which is the law of the gospel" which unites all brethren into a "diversity in unity." Humboldt's science could not be refuted, but it could be appropriated. His humanitarianism, his egalitarianism, could not be openly defied, but it could be, and was, subverted into the ideology of slavery.[37]

The most paradoxical example of this is the Reverend John Bachman. Since his 1804 dinner with Humboldt, Bachman had risen through his collaboration with John James Audubon to become one of America's premier zoologists. While Audubon painted his magnificent plates, it was Bachman who wrote the scientific descriptions that turned Audubon's volume on the mammals of North America from a coffee-table wonder into a pioneering work of natural science. Bachman, who considered himself one of Morton's friends, first got wind of the turn American racial science was taking when he heard Agassiz speak in his adopted home town of Charleston in 1847. Bachman was outraged on two counts: first, Agassiz undermined the scriptural account of the Creation; and second, Agassiz's theory was plain bad science. For it was, as Lester Stephens has shown, science rather than scripture that directed the good reverend's arguments, first in his own address written to counter Agassiz's, then in his letters to his old friend Morton, in which he declared Morton's authorities unreliable but hoped the two scientists could remain partners "in the search of truth." Bachman was by then working on his own book, intended to refute the arguments of the polygenists. *The Doctrine of the Unity of the Human Race* came out in 1850, just in time for the pivotal AAAS meeting in Charleston. It was, however, not the truculent hometown Bachman who dominated the meeting, but the charismatic and worldly Agassiz. It didn't help when Bachman refused to debate Agassiz in public, instead referring everyone to his new book. Agassiz went on to spend a week with Bachman's colleague Lewis Gibbes, who showed Agassiz around the plantations of Charleston for a close-up look at actual slaves. Delighted, Agassiz had a series of photographs taken to document the degraded nature of the Negro race.[38]

Bachman kept writing. For years he was embroiled in the controversy, which got worse after Morton died and Nott took over, rather too gleefully, the nasty task of annihilating Bachman's scientific reputation. The torrent of belligerence, the relentless chorus of contempt orchestrated by Nott, eventually wore Bachman down, until the Civil War drew the combatants together in a common cause. Bachman's letters from Humboldt were, as earlier noted, one small casualty of that war; one wonders what comfort Humboldt offered to his friend in his distress. As scholars have noted, Bachman fought alone. Not one American scientist came publicly to his defense.[39]

Bachman's arguments may have made such a defense unpalatable. As Lester Stephens has shown, his science was impeccable: he was a first-rate taxonomist, and he easily demolished the argument for separate species by showing, in exacting zoological detail, that not one single racial variation came close to meeting the universal scientific criteria for defining separate species. Furthermore, he showed that Morton's argument that species and even genera were hybridizing madly in the wild was a tissue of fabrications, as was the polygenetic argument that the children of mixed-race couples were weak, deformed, and sterile—an argument belied, of course, by the racial diversity and vitality Humboldt documented so extensively in Mexico. The upshot was that taxonomy, reproductive biology, and behavior all established the unity of the human race. However, this most certainly did not mean that blacks were the equal of whites. In Bachman's thinking, blacks were "stamped with inferiority," childlike and "incapable of self-government," and whites had the duty to lead them by the hand, to protect, instruct, and console them in their weakness. Like Guyot, Bachman praises "the wise provision of heaven" in so constructing the various races such that "they in their several gradations can mutually benefit each other . . . by this means binding together the whole human family in one bond of universal dependence and brotherhood."[40]

Bachman, whose family had owned slaves in New York (where slavery was still legal when he left for South Carolina in 1814), thus reproduced the standard paternalistic argument used by Southern intellectuals to defend their "peculiar institution." William Gilmore Simms understood that democracy meant not "leveling" but more properly "the harmony of the moral world," a harmony that depended on men of unequal merits finding their proper place in society. As a Lutheran minister, Bachman knew this argument had deep roots in Protestant theology. In 1630, John Winthrop, future governor of the Massachusetts Bay Colony, had preached to his fellow Puritans on the troubling question of inequality: why did God make some rich and some poor, some "high and eminent in power and dignity; others mean and in subjection"? In order, said Winthrop, to develop in each the virtues necessary to their circumstances. Inequality was God's way of binding the society together by assuring "that every man might have need of other, and from hence they might all be knit more nearly together in the bonds of brotherly affection." Adapting this argument—which as Guyot shows was common to both South and North—allowed Bachman to tie his monogenism to slavery by arguing that polygenism, which might seem a logical fit with the ideology of slavery, actually undercut it by undoing the Christian structure of social harmony. As Bachman warned, Northerners would accuse the polygenists "of prejudice and selfishness, in desiring to degrade their servants below the level of those creatures of God to whom a revelation had been given and for whose salvation a Saviour died." In undercutting the scriptural basis for paternalism, the godless polygenists threatened to expose Southern slavery as mere greed, "an excuse for retaining them in servitude." This, not just crude biblical literalism, was the reason polygenesis finally had such a limited impact on the South. Without the claims of "brotherhood" conferred by a single creation, inequality became not the benevolent policy of a wise Providence but, as Humboldt insisted, nothing more than the raw product of European colonial power. [41]

In this climate, Bachman's use of Humboldt becomes doubly revealing. In a climactic passage, Bachman invokes the authority of Humboldt, quoting in full his long defense of the unity of the human race from *Cosmos*. Or more precisely, *almost* in full: in place of Humboldt's ringing declaration that "all are alike designed for freedom" and the lengthy justification that follows, Bachman gives his readers five well-spaced dots. It was one thing to cite Humboldt as scientific support for monogenism; it was quite another to cite his humanitarian support for universal freedom. Bachman knew that he was on thin ice: at the very end of his book, he defends Humboldt, the freethinking atheist, from charges of atheism. Unlike the polygenists, "he has perpetuated no doctrines that are opposed to the laws of nature—that are injurious to morals or subversive of Christianity." It was not the last time that loyal American Humboldtians would struggle to contain their hero's dangerous radicalism. Ironically, Bachman closes his book by borrowing a couplet from Humboldt:

> Auf den Bergen ist Freiheit; der Hauch der Grufte
> Steigt nicht hinauf in die reinen Lufte.[42]

"There is freedom in the mountains; the breath of the crypt / does not rise into the pure air." Humboldt would have demurred: the breath from the crypt of slavery surely tainted the air of Bachman's adopted home.

Bachman, the New Yorker, may have made his accommodation to the Southern ideology of slavery, but a few miles upstate his colleague Francis Lieber had not, and that refusal had consequences. A Prussian immigrant, yet another of Humboldt's liberal friends, Lieber had joined the faculty of South Carolina College, Nott's alma mater, in 1835, where he had built a reputation as an inspired teacher and one of America's great contributors to political science. Lieber served as acting president of the college from 1849 to 1851, and applied repeatedly for a permanent appointment to the presidency. When he was passed over once again, in 1855, the trustees made it clear that while the learned and beloved Lieber might have been the most qualified candidate, no one with his pro-Union, antislavery sympathies could lead South Carolina College. Lieber resigned the next day, moving to New York to join the faculty of Columbia University. Back in Columbia, South Carolina, his former students in the Euphradian Debating Society struck his name from their honor role and vandalized his statue. In a letter to Humboldt (one of the very few of his letters to survive the Civil War), Lieber sidestepped the loaded question of why he had resigned, but made sure his old friend knew of his new appointment and the success of his latest writings—on, of all things, civil liberty.[43]

The same polarization that was driving the old enemies Nott and Bachman together as fellow Southerners drove the fellow Humboldtians Lieber and Bachman apart. The deepening split would fracture not only Humboldt's intellectual children, but Lieber's own family: he left behind his eldest son Oscar, a promising young geologist who also traded letters with Humboldt. Oscar was killed in the Battle of Williamsburg, fighting as a Confederate officer against his brother Norman, who had joined the forces of the North.[44] Though Humboldt would not live to see it, the storm he feared had finally broken. The scientific battle over race that had splintered American Humboldtians into warring sides was eventually resolved by Humboldt's greatest follower, Charles Darwin. Meanwhile, the political battle had landed on his very doorstep.

Humboldt and American Slavery

Sixty years after the start of the Civil War, Franz Boas, the father of modern American anthropology, would point to the way slavery and its attendant racial ideology had twisted American science: "There is no fundamental difference in the way of thinking of primitive and civilized man. A close connection between race and personality has never been established. The concept of racial type as commonly used even in scientific literature is misleading and requires a logical as well as a biological redefinition." Boas went on to demolish racial thinking all over again, replaying many of the same arguments pioneered by first-generation Humboldtians a century before.[45]

Humboldt himself knew the battle for freedom was not faring well, and in letters to friends he worried constantly over the course the United States was taking.[46] Some of those private letters were caught up in the public debate. In 1863 William Lloyd Garrison's antislavery newspaper the *Liberator* published an 1845 letter to George Sumner (brother to Massachusetts senator Charles S. Sumner) in which Humboldt instructed the esteemed scholar to publish his travels, "on the condition that you remain most faithful to the sentiments of Liberty, to which many of your countrymen seem to be growing indifferent, either from the influence of social life in Europe, or under the pretext of exaggerated fears of radicalism. I desire, also, in your work, by the side of a love for the whites, some complaints at the ferocious legislation of the Slave States." Humboldt's brief 1858 letter to New York industrialist and inventor John Matthews was reproduced on a wallet-sized card distributed to "Sick and Wounded Soldiers in the U.S. Armies": "I have the warmest attachment to your beautiful and liberal city, New York, but have earnestly and deeply regretted that WEBSTER, whom I long respected, more than favored that *shameful* law which still persecuted colored men after they had regained by flight their natural inborn liberty, of which they had been robbed by Christians."[47]

Humboldt never passed up an opportunity to let fly his indignation over slavery, but he had mounted his most sustained and detailed critique in 1826, in the long essay on Cuba that concluded his *Personal Narrative*. It was published separately later that year as *Essai politique sur l'île de Cuba*, translated into Spanish the following year, and into English in 1829 in the final volume of Helen Maria Williams's translation. Like his other political essays, this one was written not just to document but to influence policy, and given the Cuban economy's growing dependence on slavery, the need for a policy change was urgent: "To remedy evil, to avoid public danger, to console the misfortunes of a race who suffer, and who are feared more than is acknowledged, the wound must be probed," wrote Humboldt. Fortunately, though the situation was dire, it was not hopeless, "for in the social body there is found, when directed by intelligence, as in organic bodies, a repairing force, which may be opposed to the most inveterate evils." Humboldt, acting as doctor to a sick body politic, would offer both diagnosis and cure.[48]

Cuba was of critical importance, in the first place because of its size and geographical centrality: the port of Havana was so situated as to command the entrance to "the Mexican Mediterranean," making it the New York of the Americas. Cuba had, accordingly, prospered, more than any other of the Spanish possessions, particularly after "the troubles of Saint Domingo" had driven to Cuba much of the sugar trade. By the time Humboldt published, the plantation colony was a major world producer of both sugar and coffee. This was a dangerous path, fretted Humboldt: the sugar market in particular was unstable and both capital- and labor-intensive. And that labor was, of course, slave labor.[49]

Humboldt's argument proceeded on two fronts, economic and political. The economic front deployed the "revolting calculations on the consumption of the human species" to demonstrate the fundamental inefficiency of a slave-based economy. A simple comparison with Europe proved this: in Germany, the sandy and infertile district of Brandenburg, though a third the size of Cuba, nourished a population nearly double. By contrast, Cuba produced almost none of the necessary provisions for life, forcing it to import nearly everything. Cuba was also wasteful of lives: even a modicum of care would have prevented the necessity of importing four hundred thousand blacks, "loaded with chains" and "dragged" to Cuba to replace the dead. "I have heard it coolly discussed," growled Humboldt, "whether it were better for the proprietor not to fatigue the slaves to excess by labour, and consequently to replace them less frequently, or to draw all the advantage possible from them in a few years, and replace them oftener. . . . Such are the reasonings of cupidity, when man employs man as a beast of burden!" Humboldt thus gives the lie to Bachman's patriarchal rationale. As he had seen with his own eyes, when economy trumped morality, the market logic of Simon Legree ruled. "I don't go for savin' niggers," Stowe had her plantation slavemaster say; "Use up, and buy more . . . makes you less trouble, and I'm quite sure it comes cheaper in the end." Humboldt listed some ways that paternalism could mitigate such violence: encourage the numbers of female slaves to increase by exempting them from labor during pregnancy; establish good child care and separate dwellings for families; provide good food in sufficient quantities, days of rest, moderate labor: "such are the means most capable of preventing the destruction of the blacks."[50]

But really, the argument over paternalism was beside the point. "The slave trade is not merely barbarous, it is also unreasonable, because it misses the end it would obtain." Those who claimed that slaves were essential for sugar cultivation were ignorant that of nearly 1.2 million slaves in the West Indies, less than half did all the labor for all colonial produce; and in Brazil, less than a quarter were occupied "in the labors of colonial productions, which we are gravely told render the slave trade a *necessary evil,* an *inevitable political crime!*" Knowing that all the economic statistics he could offer would make no difference to men who refused to listen to reason, Humboldt opened a second rhetorical front, a political one: if slaves were not given the rights due them by the laws of nature, they would revolt and seize those rights for themselves. "Who would venture to predict," speculated Humboldt, "the influence which may be exerted by 'An *African Confederation of the free states of the West Indies* . . . on the politics of the New World?'" Given his praise for the good progress made by Haiti, it is not clear that Humboldt was particularly disturbed by the prospect. What did disturb him was the complacency of the whites, who like Melville's Captain Delano were smugly certain "that their power is not to be shaken. All simultaneous action on the part of the blacks appears to them impossible; and every change, every concession granted to the captive population, a sign of weakness." They pretended that the Haitian Revolution was due to nothing worse than bad governance. Hardly, countered Humboldt: Haiti had played out the deepest laws of humanity and nature. If those laws would not move the white masters to justice, then perhaps fear of rebellion would. Humboldt deployed the grim satisfaction of being proved right: since he had left Venezuela, "civil dissensions have put arms into the hands of the slaves; and fatal experience has led the inhabitants of Venezuela to regret, that they refused to listen" to the voices recommending reform. But slave states did not have to end in violence. Humboldt envisioned a peaceful solution: free the slaves and make Cuba a self-governing republic, creating "a free, intelligent, and agricultural population" that would succeed "a slave population, destitute of foresight and industry." Placed in the hands of its cultivators, the face of Cuba would change. Soon a diversity of agricultural products, some for export and some for sustenance, would nourish "the progress of industry and national wealth," and above all, "the development of human intelligence. On these united powers the future destinies of the metropolis of the West Indies depend."[51]

After running through all the statistical and political arguments, Humboldt stepped forward in his own voice. Such a "minute investigation of facts" seemed necessary, he says, at a time when passion gave rise "to the most vague and erroneous statements." "I have abstained," he continues, from predicting the future, and examined solely the organization of human societies, emphasizing those "unequal rights" and "threatening dangers" which wise legislation might ward off. As a traveler to Cuba, an outsider who was only passing through, he bore a responsibility to bear witness to what he had seen. "It belongs to the traveller who has himself seen what torments or degrades human nature, to make the complaint of the unfortunate reach the ear of those by whom they can be relieved." It was the role of the ethical witness to step forward and speak where victims were silenced.[52]

How ferociously ironic, then, that Humboldt was silenced in the very country that most needed to hear what he had to say. His essay on Cuba was banned in Havana, of course, but he expected that. What he didn't expect was that it would face a still worse fate in the United States: here, of all places, he would be silenced in the most cool and calculating of ways. In 1856, with expansionist fever at its height, a proslavery Southerner named John Sidney Thrasher engineered an extraordinary propaganda coup by appropriating Humboldt's most important antislavery work, and by clever editing twisted it into a masterwork of proslavery, expansionist propaganda under the title *The Island of Cuba by Alexander von Humboldt*. As Vera Kutzinsky writes, "Thrasher's Humboldt is as much of an invention as the fictional slaves extolling the benefits of slavery." Thrasher had lived for many years in Cuba, where he mocked the Spanish government from the pages of his newspaper *El Faro Industrial* (The light of industry), and he was active in the Havana Club, a gathering of wealthy sugar planters and merchants who sought the annexation of Cuba to the United States by fair means or foul. In 1851 Thrasher was arrested for conspiring to seize Cuba, and sentenced to prison labor. The Fillmore administration got his sentence commuted, and in 1852 Thrasher returned to the United States to continue his campaign.[53]

Thrasher presented his book as nothing more than a modest updating of Humboldt's original, which he called "the best that has been written on the subject."[54] There was indeed much to update: in the three decades since Humboldt wrote, Cuba's sugar economy had exploded, the United States had become its major trading partner, and slaves were being imported at the rate of nearly eleven thousand a year. As the growth of the black population threatened to turn whites into a minority, racial tensions were soaring, playing into the Spanish government's policy of exploiting racial fears to maintain control; meanwhile, England was pressuring the government to free all the slaves, Jamaica-style, a prospect which terrified Southern slave owners. As Spain weakened, the future threatened the "Africanization" of one of the United States' key trading partners. Annexation of Cuba was the obvious answer, and the sooner the better.

In 1848, President Polk had tried to buy Cuba; when Spain rejected the offer, the conspiracies that had embroiled Thrasher erupted. In 1854 President Pierce renewed the purchase offer, which Spain again rejected. In retaliation, Pierce ordered the U.S. ambassadors to France, England, and Spain to devise a secret plan to seize Cuba by force. They did so at a meeting in Ostend, Belgium, and when this "Ostend Manifesto" was made public, it caused a firestorm of controversy. Horace Greeley's *New York Tribune* denounced it as the "Manifesto of the Brigands," and Humboldt called it "the most outrageous political document ever published," a "savage" policy; Pierce's claim that he had a right to conquer the island of Cuba was "barbarous." When one of those three ambassadors, James Buchanan, was elected president in 1856, Humboldt was incensed that Americans had elected not the free-soil Frémont but the "Cuba-mad [*Cubasüchtig*] Buchanan."[55]

What Humboldt thought made no difference to the equally Cuba-mad Thrasher. One might think that anyone could have detected his ruse by simply comparing his edited text to either of the two English translations, but Thrasher deflected such a possibility by claiming that Humboldt's Cuba essay had never before appeared in English. That was untrue, but since it was embedded at the very end of the multivolume *Personal Narrative*, it was easy to overlook, and reviewers largely took Thrasher's claim at face value. From that start Thrasher proceeded to recast Humboldt's argument by reframing it, distorting and undercutting it, and censoring it. To reframe it, Thrasher prefaced Humboldt's work with a long "Preliminary Essay" showing that abolition would not be America's salvation but would, in fact, lead to the destruction of American civilization. As he warned, compassion was a risky emotion on which to base national policy: slavery itself originated when misguided humanitarians sought to protect Indians by enslaving Africans instead. Thus did conscience and a "fallacious sentiment of humanity give life to the new social system in America." Beware: abolitionists acting out of "the same fervor and zeal, the same heedless inconsistency" were initiating a new reign of terror. Things must be left as they are. In Cuba's current social system "the inferior is subject to the superior race, to the manifest material and moral advantage of both." The "mutual dependence" between master and slave creates ties of "affection" unknown where the two races live in a state of civil equality. The result was the very material prosperity so fully documented by Humboldt. And fortunately, claimed Thrasher, the slave population of Cuba would stay docile under American rule, for it had never been polluted by what Thrasher called "the bloodthirsty teachings of European philanthropy."[56]

For confirmation of philanthropy's taste for blood, Thrasher pointed to the barbarism and heathenism of Haiti, and even more to the point, to the social collapse of Jamaican society. After the British had freed its slaves, "ignorance, irreligion, superstition, intoxication, profligacy, are hovering, like birds of prey" over Jamaica's schools and chapels. In Cuba, under British influence, insolent slaves salute the ladies and commend their beauty with rude remarks, marry whites, take up arms. "The [white] people of Cuba now stand alone in their resistance to this social revolution and ruin," writes Thrasher. British emancipation will result in the "Africanization" of Cuba, causing "a bloody revolution" and a "war of races" as whites rise up in defense. And why would the British choose to wreak such havoc? To keep Cuba out of the hands of America, the only power that can save Cuba, hence itself, from utter ruin. When the evil is done—when the English get their way—and Cuba has perished and America has descended into war, then "may England's states-men weep crocodile tears over our misfortunes." Consolidation of American power on our own continent—a Monroe Doctrine with teeth—is our only defense against Europe and the race war that European politicians hope to inflict on America.[57]

After reading Thrasher's introduction, Humboldt's mild speculation about "*An African Confederation of the free states of the West Indies*" sounds considerably more ominous. But Thrasher did more than reframe Hum-boldt's text with his own racist prognostications; at several points, he trans-lated Humboldt's language with an alarmist twist. For example, Williams's phrase "the *troubles* of St. Domingo," became "the *disasters* of St. Domingo." He scissored out Humboldt's moralistic punch lines, and relocated sentences to change their contexts. While such subtle distortions merely muted Hum-boldt's argument, Thrasher's running commentary at several points actively undercut it, most notably when Humboldt projected his vision of a "free, intelligent, and agricultural" Cuban population. Thrasher let Humboldt's words stand, but chided him for the "error" of such a foolish social theory, as proved "by the sad experience of Jamaica." Humboldt's noble but naïve hopes for liberation and peaceful progress were simply wrong. This was the lesson of the *realpolitik* of the 1850s. It was slavery, not abolition, that civi-lized, and abolition, not slavery, that would destroy the "vital principle" of organic American society.[58]

Yet up to this point, Thrasher had left enough of Humboldt's argument intact that an alert reader—one able to hear Humboldt's voice over Thrash-er's static—could have detected the inefficiency and even moral bankruptcy of slavery. This section had formed the bulk of Humboldt's original essay. What Thrasher censored out was the conclusion, Humboldt's passionate moral peroration in which he stepped back from detached analysis to take on the more personal voice of the witness. Never, before or since, had Humboldt allowed himself to be as politically engaged or as specific in his proposals. By silently eliminating his conclusion, Thrasher attempted to remove from American hearing the climax of Humboldt's humanitarian thinking.

Humboldt's argument in the section Thrasher eliminated had two inter-
locking phases. First, Humboldt reiterated that, thanks to Spanish reforms,
he had witnessed slavery not in its harshest but in its mildest forms, yet still
he came away horrified. No degree of melioration could soften the absolute
moral evil of slavery, nor could it be veiled by "ingenious fictions of language" like "*negro peasants of the West Indies, black vassalage,* and *patriarchal
protection.*" Do those who offer such "captious sophisms" really think that
by comparing the state of blacks with that of serfs they "have acquired the
right of dispensing with commiseration"? Such reasoning gives us "useless
arms" at a time of revolution. "Slavery is no doubt the greatest of all the evils
that afflict humanity," declares Humboldt, in an oft-quoted phrase. There
may, he acknowledges, be degrees of suffering—how different the lives of a
slave in a rich man's Havana house, and a slave worked to death in the sugar
mills!—but true philanthropy must do more than give "a little more salt-fish,
and some strokes of the whip less." It must end slavery altogether.[59]

This reasoning leads to the second phase of Humboldt's argument. Slavery, being an absolute moral evil, must be absolutely ended. It is not, that is,
a natural feature of civilization but a historical development, with its roots
in the European conquest and in the European colonial system. This history
is sorry enough in the Spanish colonies, yet Spanish policies are a model of
enlightenment compared with those in the Southern United States, where
whites in the struggle with England "established liberty for their own profit"
while augmenting the slave population. The thought that the United States
might become the center of world civilization makes Humboldt cringe:
"What a spectacle would that centre of civilization offer, where, in the sanctuary of liberty, we could attend a *sale of negroes after death* [of their owners],
and hear the sobbings of parents who are separated from their children!"
Humboldt hopes that the principles animating Northern states will extend
south and west. But he makes clear that principles alone will solve nothing.
Legislators who leave the problem for "time" to take care of are traveling a
fool's road, for time will only deepen the simmering anger of the oppressed.
Nor will "civilization" cure the evil. Since ancient Rome, civilization has been
content to maintain slavery in its midst. The only way to end slavery is for
good men to step forward and act: local authorities must exert firm will, with
"the concurrence of wealthy and enlightened citizens" and a plan of action.
Only "this community of actions and efforts" will end slavery, a community
willing to break through the illusion of legitimacy conferred by long habit.
Humboldt has no faith in "the influence of knowledge, intellectual improvement, and the softening of manners" alone; this man who consorted with
kings, presidents, and diplomats insisted that "only the directing action of

governments and of *legislatures*" can bring about "a peaceable change." This was Humboldt's own battle-scarred version of *realpolitik*: history had taught him that mere enlightenment was helpless unless backed by the power of law and authority. For Humboldt, the ethical was political. No wonder he engaged so closely with American politics, with Jefferson and Adams, Polk and Buchanan, Webster and Frémont. Humboldt would have scoffed at the notion of American exceptionalism. He knew that without real and sustained enlightened political leadership, America would sink to the level of every other corrupt imperial power.[60]

Thrasher sent Humboldt an advance copy of his "translation" with a fawning letter apologizing for differing from the great man "on some of the general principles of social economy in this continent." Humboldt was furious. He denounced Thrasher's version of his book in a July 1856 letter to a Berlin newspaper that was translated and republished in newspapers across the United States. Humboldt complains bitterly that while his "frank and open remarks" had circulated freely in Spanish and English translations, in the *American* edition his entire chapter on slavery has been "arbitrarily omitted. To this very portion of my work I attach greater importance than to any astronomical observations, experiments of magnetic intensity, or statistical statements." To Humboldt, none of his science counted so much as his moral argument against slavery. Knowing his American audience would read in his letter what they could not read in his mutilated book, he took advantage of the publicity to summarize his argument, then concluded that he would never have complained if he had been attacked for what he actually said; "but I do think I am entitled to demand that in the Free States of the continent of America, people should be allowed to read what has been permitted to circulate from the first year of its appearance in a Spanish translation."[61]

Thrasher came out swinging. In a disingenuous reply to the *New York Times*, he claimed he omitted Humboldt's chapter "On Slavery" simply because it did not pertain directly to Cuba. Of his other omissions and distortions he said nothing. The Prussian ambassador Friedrich von Gerolt hastened to send Humboldt a copy of Thrasher's reply, or rather "excuse," which he agreed was "exceedingly lame." Meanwhile the controversy escalated, as Southerners reacted angrily to Humboldt's assertion, repeated in his letter, that Spanish legislation "is less inhuman and atrocious" than in the slave states of the United States. While the controversy ran on, Humboldt scribbled in the margins of his own copy the comment that Thrasher's mutilations had been made "doubtless for political reasons."[62]

Eighteen months later Humboldt was still steaming. In a letter to Julius Fröbel (a German "Forty-Eighter" or political refugee from the failed

Revolution of 1848) that was also widely reprinted in American newspapers, Humboldt grumbled once again that "my book against slavery . . . is not prohibited in Madrid, but cannot be purchased in the United States, which you call 'The Republic of distinguished people,' except with the omission of everything that relates to the sufferings of our colored fellow-men, who, according to my political views, are entitled to the enjoyment of the same freedom with ourselves." But thanks to the controversy it generated, Thrasher's act of silencing actually helped Humboldt's voice ring loud and clear across the United States, and from then on he was honored by antislavery activists as their ally and sympathizer. Humboldt even found himself projected into the midst of an American political campaign: the loathed "Cuba-mad" Buchanan was running against none other than John C. Frémont, the scientific explorer trained by Nicollet and so extravagantly admired by Humboldt. In 1850, when the future Republican party's nominee was stepping into his brief term as California's first senator, Humboldt had arranged for the king of Prussia to award Frémont "the grand golden medal" honoring progress in the sciences. Humboldt closed his letter notifying Frémont of the award with an expression of pleasure at the prospect of Fremont's new political office: "California, which has so nobly resisted the introduction of slavery, will be worthily represented by a friend of liberty and of the progress of intelligence." At the height of the Thrasher controversy Frémont echoed this letter in a reply to Humboldt expressing his gratitude "that in the struggle, in which the friends to liberal progress in this country find themselves engaged, we shall have with us the strength of your name." The flurry of Frémont campaign literature did indeed borrow the strength of Humboldt's name: at least one campaign biography of Frémont was dedicated to Humboldt, and his passages praising Frémont (from his 1850 letter and *Views of Nature*) were reprinted, together with a glowing description of the "grand golden medal," in a Republican campaign flyer that was widely distributed to German Americans. Von Gerolt wrote Humboldt that in New York City, many thousands of German supporters had held "a mass meeting" and "a splendid torchlight procession" for Frémont, and the historian Philip Foner reports that in the 1856 election, Frémont won overwhelming support from German Americans.[63]

Frémont lost, of course—though he won all the Northern free states, he received only 114 electoral votes to Buchanan's 174. The election polarized the nation along sectional lines, helping to align the North and West against the South, and in 1860, the next Republican candidate, Abraham Lincoln (who had campaigned vigorously for Frémont) would succeed where Frémont had failed. Yet Humboldt did not live to see this, and his reaction to the 1856 election was bitter. Although in a visit with Bayard Taylor he cheered himself with the fact that Frémont had won well over a million popular votes, in a private letter to Varnhagen he muttered, "And the disgraceful party which sells negro children . . . who prove that all white workmen should rather be slaves than free—have succeeded. What a crime!" When an American visitor tried to comfort him by relaying the promising gains made by the Free Soil party in the 1858 election, Humboldt would have none of it: " 'Ah,' said he, shaking his head, 'you have done the same thing twice already!' " Unable to influence American policy on slavery, he turned to Prussia, the one government where he still had a degree of power, and in March 1857 he pushed through a law declaring that every slave who stepped on Prussian soil would become free.[64]

Abolitionists were circulating not only Humboldt's praise for Frémont, but also his condemnations of Massachusetts senator Daniel Webster for supporting the Fugitive Slave Act of 1850. No Prussian Medal of Freedom for Webster: when Professor Cornelius Felton, the future president of Harvard, paid his respects to Humboldt in Berlin in 1853 and gave Humboldt a copy of a memorial to Webster, the old Prussian let the American have it. While Humboldt was heaping abuse on America in general, where "you make men slaves," and on Webster in particular, Felton angrily bit his tongue: Just how free, he wondered while Humboldt ranted, was Prussia? What would Humboldt's precious king do to him if he spoke his opinions openly? But rather than "bandy words" with the eminent old man, Felton tried to divert him by remarking that one of the last works Webster had studied was *Cosmos*. Humboldt replied courteously that he was "*much glorified*" that so eminent an intellect took an interest in his book, then, not to be distracted, gave Felton a sharp rap: "But I am deeply concerned about slavery; and I am sorry Mr. Webster had anything to do with it. *I would not have put my name to it.* Whatever political necessity may require, the individual need not, unless he chooses, connect himself with it." Prussia may not be free—as Humboldt of all people knew perfectly well—but the individual always retained the freedom not to perpetuate further injustice, a view that echoed Henry David Thoreau's recent essay "Civil Disobedience."[65]

Two years later Garrison's newspaper *the Liberator* reprinted another of Humboldt's anti-Webster diatribes to an American visitor, contributed by an anonymous correspondent: "For thirty years—for thirty years—(and he counted them on his fingers)—you have made no progress about slavery. You have gone backwards." The law of 1850 really should be called "the *Webster* law. I always before liked Mr. Webster. He was a great man. I knew him, and always till then liked him. But ever after that I hated him. He was the man who made it. If he wished to prevent it, he could have done it." But instead, Webster had carried those who ought to have known better along with him. Thirty years before, Humboldt had declared that abolishing slavery would require the combined action of an entire community, and no action could succeed without leadership from its legislators. Decades before, Jefferson had let him down by refusing to exert that leadership. By Humboldt's standards, Webster's betrayal was unforgivable.[66]

Such newspaper notices circulated far more widely than any of Humboldt's books. Many thousands of Americans who knew little of Humboldt's contributions to science thought of him instead as one of the age's great humanitarians. In March 1858 Anne Warren Weston, a member of the Boston Female Antislavery Society, sent a letter to Humboldt with a copy of the society's annual publication *the Liberty Bell*, explaining that she was encouraged to do so by his 1856 letter against slavery. "The noble declarations contained in that note have rendered your name dear to the hearts of the American Abolitionists," she wrote. "Surrounded as they are by every obstacle that the cupidity and selfishness of a great nation can place in their way," she wished to send Humboldt their "most profound gratitude." She concluded: "The present age is full of the glory of your scientific reputation but in the coming future when human rights & human brotherhood are better understood, it shall not be merely the savant but the philanthropist that the nations shall honor." Sadly, a little over a year later, none of the speakers at New York's memorial to Humboldt's death would even mention his antislavery stand, and as Foner points out, George Bancroft would repeat his assertion that Humboldt wished Cuba to come to the United States, a judgment flatly contradicted by every one of Humboldt's published statements. Yet the famous historian, with his expansionist views, would do far more to construct Humboldt's future reputation than the unknown female Boston abolitionist.[67]

The most eloquent tribute to Humboldt's antislavery views was given by Theodore Parker, the Boston Unitarian freethinking minister. Parker was scheduled to speak at the July 1858 American Anti-Slavery Society Independence Day celebration at Framingham, Massachusetts, together with such luminaries as William Lloyd Garrison and Wendell Phillips, but Parker's failing health kept him home. He offered in absentia a resolution that the society send formal thanks to Humboldt, and after it was read, "a loud and unanimous AYE rang through the Grove, and testified to the respect in which the vast assembly held the venerable man, who, in his old age, had rebuked so honestly and faithfully the servility of the press and people of the United States, in regard to the great subject of human slavery." Several days later, Garrison published Parker's tribute to Humboldt in the *Liberator*. In it,

Parker honored Humboldt's services to science, declaring that no man since Aristotle "has done so much to widen the bounds of human knowledge," but on this day he wished rather to speak of Humboldt's "Humanity": "He is the Friend of Mankind, always on the side of Progress, of Humanity. He takes the side of the Indian in North and South America, against his conqueror. He recognizes the natural right of the African, and proclaims it in his early books and his most recent letters." Parker—quite possibly the most well-read man of his time—quoted long passages from the *Political Essay on the Kingdom of New Spain* and the *Personal Narrative*, and after reviewing the Thrasher controversy, he read to his audience from the pages Thrasher had censored. From Humboldt's "noble words," Parker turned "with a sickening feeling of disgust . . . to the vulgar and inhuman rant of the noisy American official who pretends to translate this book," who thinks freedom will never prosper with black slaves because "they have no great desire for it," never having been exposed to "the *blood-thirsty teachings of European philanthropy*." Parker drove home the point that those "*blood-thirsty teachings*" were none other than the "self-evident truths" of Jefferson's Declaration of Independence, truths on which America was founded, and truths they were gathered together that very day to celebrate.[68]

Humboldt may have felt that with nearly all his old friends gone, no one was left to hear his words, yet his willingness to keep speaking, to keep witnessing, nevertheless, gave courage and rhetorical tools to a new generation that would finally break slavery's hold on Humboldt's beloved republic. To Humboldt, science was humanistic to the core, and the humanities—literature, art, the social sciences—were hollow unless they embraced the results of science. As Parker wrote to his Framingham audience, Humboldt's antislavery activism formed part of his great vision of human knowledge creating one great whole, "which he names KOSMOS—both Order and Beauty—in one word, THE WORLD." In his last work, Humboldt braided together all the facets of his many-sided interests—in natural and physical science, art, history, linguistics and anthropology, politics and social justice—into one interrelated but pluralistic universe, a grand orchestral whole of many voices, unified by the ability of the human mind to recognize order and give wonder. At the end of all his passages lay his crowning work: *Cosmos*.

The Community of Cosmos

Understand and venerate the object.
Penetrate this secret orb
and let its flower be your amulet.

JORGE CARRERA ANDRADE

A word is dead, when it is said
Some say—
I say it just begins to live
That day

EMILY DICKINSON

Franz Boas, Cosmographer

The Humboldt River—the one that, as Humboldt joked, was "full of fish" though rather lacking in timber—rises in the mountains just south of Idaho and flows west and south, wide and strong, for some two hundred fifty miles. For generations it provided an easy highway to westering immigrants seeking the wealth of California, the latest New World. But then, mysteriously, the river vanishes, its waters sinking into the sands at the foot of the formidable Sierra Nevada mountains. Humboldt's river never finds its way to the sea.[1]

Humboldt too seems to sink away, disappearing at the foot of those equally formidable obstacles, the Civil War and Darwinism. Yet this is an illusion, an artifact of how Americans came to tell their history. Like groundwater, the "Humboldt Current" sank below the horizon to arise again in springs and oases across the American landscape—how fitting that the "Azteck Priestess" who reigns over his work turns out to have been Chalchiuhtlicue, the goddess of groundwater. From these springs and oases, Humboldt's thought, revitalized and reshaped, pursued new channels into the twentieth and twenty-first centuries. Darwin himself was one such channel, and another was Peter Kropotkin, the exiled Russian geographer and anarchist who countered "social Darwinism" by arguing that altruism and voluntary cooperation were far more important in "the free evolution of society" than competition and warfare. As he wrote, "Man is part of nature, and . . . the life of his 'spirit,' personal as well as social, is just as much a phenomenon of nature as is the growth of a flower or the evolution of social life amongst the ants and the bees." In the United States, writers and poets transmuted Humboldt's ideas into an American idiom that became the signature of a new "American" style of painting and poetry. But one of the most vital and important carriers of Humboldt's ideas is seldom recognized as such: Franz Boas, the German anthropologist who brought Humboldt's ideas back to the United States and used them to remake American anthropology, becoming in the process "one of the most influential figures in the history of the social sciences."[2]

Boas grew up in a home where, as he said, "the ideals of the revolution of 1848" were a living force. His family was Jewish, and as his recent biographer has written, they with many other liberal German Jews had adopted Wilhelm von Humboldt's Enlightenment concept of *Bildung*, or mental cultivation, as their own ideal. Boas, like Alexander von Humboldt, recalled that his favorite childhood book was *Robinson Crusoe*, and he too grew up longing to bring his fascination for natural history to faraway places. His scientific interests pulled him first toward physics—his early research was in the optical properties of seawater—but this study raised, in his mind, troubling questions about perception that drew him to explore "the relation between the objective and the subjective worlds." Humboldt's field of physical geography seemed the best place to explore such questions, and Boas's first field excursion was a year spent living with the Inuit on Baffin Island, where he had expected to derive a clear explanation of exactly how environment determined human behavior. What he discovered instead precipitated the collapse of his intellectual universe: the physical sciences, in which he had been so exactingly trained, were next to useless in the face of living, culturally complex human beings. He wrote up his results, which were, he thought, "a thorough disappointment," too "shallow" to be of any use. The Inuit turned out to be not at all "primitive," as evolutionary science had decreed, but culturally sophisticated, at ease in their apparently stark environment, and just as richly varied in their *Herzensbildung*, or inner character, as the Germans Boas had grown up with.[3]

The result was intellectual crisis. If the method of the exact sciences failed when applied to human beings, which should he sacrifice—science?—and lose his career? Or human beings? It was at this moment that he returned to his early reading in Humboldt; growing up, he had first approached *Ansichten der Natur* "with respectful awe," then dared, though "fearfully," to dip into "that 'sublime work,' *Kosmos*." It was in Humboldt that he found his solution. As he wrote in an 1887 article that initiated his career as an anthropologist, science exists in not one but two forms, each equally valid, but locked in conflict. The better-known and more widely respected physical scientist compares a series of facts in order to isolate from them some phenomenon

they have in common, then flings them aside to pursue the general law. Newton, for example, was interested in planets only as they led him to the law of
gravitation; once the law was described, "Newton's work was at an end." Or
take Comte, in whose system all sciences have the sole aim of deducing laws
from phenomena: "The single phenomenon itself is insignificant." But Boas
insists that Comte's system is radically incomplete. Its necessary but excluded
other is Alexander von Humboldt. As Boas writes, "To this system of sciences
Humboldt's 'Cosmos' is opposed in principle. Cosmography, as we may call
this science, considers every phenomenon as worthy of being studied for its
own sake. Its mere existence entitles it to a full share of our attention." Cosmography studies the *history* of phenomena, what they are and how they
came to be just that way. It cherishes the very particulars that science uses
then tosses away.[4]

Which one, Boas asks, is more valuable? The answer is, Neither. Each approach has its own value, and the two approaches arise "in two different desires of the human mind, its aesthetic wants, and its interest in the individual
phenomena." The physical scientist wants order and clarity, to ascend to the
highest summit and see from its height "the vast field of phenomena. Joyfully
he sees that every process and every phenomenon . . . is a link in a long chain.
Losing sight of the single facts, he sees only the beautiful order of the world."
By contrast, "the cosmographer" cozies up to the object of his study, whether
its rank in the system be high or low, "and lovingly tries to penetrate into its
secrets until every feature is plain and clear." The choice is personal, and will
be made subjectively. Which one chooses is "a confession . . . as to which is
dearer to him,—his personal feeling towards the phenomenon surrounding
him, or his inclination for abstractions."[5]

Boas made his choice and spent the rest of his life fighting for it. To him,
Humboldt's cosmography arose not from logical demands but from "the
personal feeling of man towards the world," an "affective" impulse expressed
by Goethe: "'It seems to me that every phenomenon, every fact, itself is the
really interesting object. Whoever explains it . . . usually only amuses himself
or makes sport of us. . . . But a single action or event is interesting, not because it is explainable, but because it is true.'" Boas might well have quoted
Thoreau, too: "The philosopher for whom rainbows, etc., can be explained
away never saw them." The point for Boas is that each event is worth study
"because we are affected by it." Knowledge starts in the object's emotional
impress on us, and we respond to that impress by romancing it, gratifying
our love for it—and through it, our love for the land, the planet, nature,
the universe, whose unity may be only "subjective" but which is not the less
real for all that. For mind, since it is affected by the universe, is part of all it

studies. Cosmography, Boas concludes, approaches the domain of art. "Understand and venerate the object," writes the Ecuadorian poet Jorge Carrera Andrade; "Penetrate this secret orb / and let its flower be your amulet."[6]

Boas was reacting against two trends in the science of his day: the demand that all the phenomena of the universe be ranked and ordered in some grand system, reducible to a general and far-reaching law; and the fragmentation of the universe of knowledge into an array of specialties: biology, psychology, meteorology, geology, history. While this compartmentalization of knowledge is useful as far as it goes, it ignores another kind of unity—and here Boas borrowed one of Humboldt's favorite expressions—"the physiognomy of the earth," or, the "face" of nature. This form of unity, the perceptual impression of the whole, may seem merely subjective and impressionistic, but Boas, like Humboldt, insisted that it is nevertheless real and important, particularly in the human sciences. For example, by Boas's day, Wilhelm von Humboldt's defense of non-Western language had become passé, and it was standard practice to array human languages on an evolutionary scale from debased and primitive (American Indian) to complex and powerful (English). Boas challenged this as a miscarriage of the ordering impulse of science and argued, consciously taking up Wilhelm von Humboldt's legacy, for the beauty, power, and sophistication of *all* languages, none of which is in any sense "primitive," all of which embody a coherent and distinctive perceptual whole and are historically unique and worthy of study for their own sake. Repeating a point Alexander von Humboldt had made long before—today it seems obvious—it was Boas who insisted that to have any understanding of a culture one must immerse oneself in the language of its people.

While in the throes of intellectual crisis, Boas was also struggling with the direction of his career. He found German academic life stifling, while America seemed full of opportunity. In 1886 he tested the American waters with a trip to New York that coincided, happily, with the annual meeting of the AAAS. Here young Boas struck up a conversation with none other than old Horatio Hale, the philologist and ethnologist of the Wilkes expedition that had sailed nearly fifty years before. Hale had grown up imbued with Humboldtian ideas, which by the 1880s ran counter to the all-pervading jingoistic celebrations of America's evolutionary triumph, and he found in Boas a fellow spirit. Hale became a fatherly mentor to Boas, who pursued new field studies among the Indians of British Columbia and rose quickly in the intellectual community of the United States. It was Boas who reinvigorated Gallatin's languishing American Ethnological Society from his base at Columbia University, where he taught starting in 1899. His students form a roll call of twentieth-century anthropology: Alfred Kroeber, Edward Sapir, Ruth

Benedict, Margaret Mead, Zora Neale Hurston, to name just a few. Many were themselves members of marginalized communities—Jews, women, a black woman—and all learned their calling under the gaze of the bust of Humboldt that presided over Boas's study.

It was Boas who took up Wilhelm von Humboldt's unfinished work on American Indian languages and in explicitly Humboldtian terms brought it to realization. As historian of anthropology Matti Bunzl says, "Boas' anthropology may be viewed as uniting the intellectual currents emanating from both Wilhelm and Alexander von Humboldt." Just at the point when Germany was abandoning its liberal tradition of anthropology, Boas imported its tenets to the United States and made them the cornerstone of modern anthropology, which in his hands would lead "away from racialist prejudice and toward a more pluralistic, democratic society." For Boas, like Alexander, championed progressive causes and fought against racism, biological reductionism, and the reactionary politics of his own day, which reached their nadir in Nazism. He ended his most popular book with these words:

> Freedom of judgment can be attained only when we learn to estimate an individual according to his own ability and character. Then we shall find, if we were to select the best of mankind, that all races and all nationalities would be represented. Then we shall treasure and cultivate the variety of forms that human thought and activity has taken, and abhor, as leading to complete stagnation, all attempts to impress one pattern of thought upon whole nations or even upon the whole world.

U.S. Americans may have forgotten that Humboldt was their teacher, but Humboldt's legacy is inscribed so deeply into the face of the twentieth century that without it, we would not recognize ourselves.[7]

Introducing Humboldt's Cosmos

In 1845, Humboldt was known in the United States mostly as an explorer and recorder of the exotic American tropics and one of Europe's leading scientists. It had been nearly twenty years since a major new work by Humboldt had appeared in English.[8] Then he began to publish *Kosmos*, and everything changed. A raft of book reviews alerted the intelligentsia that something new was afoot. Suddenly translations issued by both British and American publishers flooded the market: three competing versions of *Kosmos*, one cheap and pirated, one elegant and expensive, and one for the mass market; two competing translations of *Ansichten*; a new and updated translation of *Personal Narrative*, conveniently packaged in three trim volumes; Thrasher's expurgated *Island of Cuba*; and two new biographies, all in the space of a decade, attended with a torrent of reviews and notices. Major new works on or by Humboldt continued to appear for another fifteen years or so, capped by the authoritative life-and-letters biography of 1873. In the United States, the 1850s were the decade of Humboldt, and his popularity approached cult status.

Humboldt's *Kosmos* did important cultural work for America. Though the multivolume book published in English as *Cosmos* is known today (if it is known at all) as a popular science book about stars, that's a little like saying Darwin's *Origin of Species* is a book about breeding pigeons. Such a view miscalculates the broad impact Humboldt had on American literature and art. In memorializing his friend, the political scientist Francis Lieber caught the developing tension when he protested that, while "high authority" stated that the works of Humboldt presented "Nature in her totality, unconnected with Man," really Humboldt comprehended nature "in *connexion* with man and the movements of society, with language, economy and exchange, institutions and architecture." Humboldt's status as an icon was so great that for some years after his death a high-stakes game was played over how to define his legacy, and "high authority" made sure that what survived of Humboldt was safe, nonthreatening, and obsolete: his science, stripped of all its human connections. Cutting Humboldt down to size was a necessary move in the modernization of scientific knowledge, but it obscured the social and aesthetic dimensions of Humboldt's thinking, and it invited later generations to assume that his turn to Cosmos was a retreat from the tortuous and repressive politics he found himself helpless to influence.[9]

Humboldt did think of the study of nature as a kind of escape, to a zone of freedom where local discords were ultimately resolved into harmony. But this utopian vision was not a retreat but an advance that drove his social critique, which went on unabated in the pages of *Cosmos*. For the study of nature was, for Humboldt, inseparable from the study of the mind in its material, social, and cultural context. This reflexivity of mind, society, and nature became his overriding argument in *Cosmos*, which was, for Humboldt, the culmination of one of his oldest ideas—indeed, the idea that had propelled him from war-torn Europe to witness the harmonies of nature in the New World, where European empires were dying and new nations being born. As he put it as soon as he returned, such studies "make us live in both past and present times, gathering around us all that Nature has produced in the various climes, bringing us into communication with all the peoples on Earth." And from past and present such studies project us into the future as well, by enabling us to "erect forever the laws to which Nature submits. It is in

undertaking these researches that we prepare ourselves for an intellectual delight, a moral freedom that strengthens us against the blows of destiny, and
which no external power could possibly destroy."[10] In short, the study of nature creates and bonds the human community and gives us the strength to
resist the social pathologies that would tear us apart.

That "moral freedom" was written into the very fabric of nature encouraged Americans, busy inventing themselves as "Nature's Nation," to think of
themselves as the privileged inheritors of nature's sublime power and beauty,
which they cast, almost universally, into religious terms: in the United States,
Humboldt's Cosmos was made to glow with a penumbra of Providential
national destiny, of prophetic vision, giving it a supernal beauty that wove
science and poetry and art, religion and morality, together as expositors of
the New World, God's most exceptional Creation. Humboldt's Cosmos
seemed made for America, and Americans adopted it into their founding my-
thology.

Cosmos was born, Humboldt said, on the slopes of the Andes and first
took shape in the 1805 *Essay on the Geography of Plants* that he dedicated to
Goethe; but the idea had been with him since those formative years at Jena
and Weimar and even before, in those heady conversations on the Rhine
with Georg Forster. After this long foreground, the catalyst was a moment of
crisis. Humboldt loved Paris, the center of the scientific world, where he was
free to work and live as he pleased within a wide circle of liberal friends. But
in 1827, his freedom ended when, having spent himself into poverty publishing his great scientific works, his king reminded him of his debt and recalled
him to Berlin. In Paris he had recently given a series of private lectures to
friends exploring the "reflective influence" nature exerted on the mind; when
he arrived in Berlin (then, according to Alfred Dove, quite the cultural dust-
bin), he announced he would give a course of lectures on physical geography.
From November 1827 through April 1828 he delivered a series of sixty-one
lectures at the University of Berlin, speaking extemporaneously from a loose
outline to a room so crowded that he soon announced a second series, which
was held in a music hall before an audience of thousands, free to all comers.
The reception was ecstatic. Initially skeptics had grumbled that Humboldt
was so Frenchified he had probably forgotten his native German; for his part,
Humboldt aimed his lectures at the heart of German provincialism, particularly what he saw as the corrupting flatulence of Hegel and the ignorant me-
diocrity of Schlegel. Goethe loved it: "The mighty conqueror of the world of
science is perhaps the greatest orator," he wrote; to a friend he mentioned
"the great pleasure which Humboldt's magnificently rich colloquium on the
miracles of nature gave me." To another friend he reflected that Humboldt

made him feel "like an ancient mariner" who had spent his life skipping from isle to isle, but who now sees "that the immeasurable abyss has been fathomed . . . that the great work, beyond all belief, has been truly done." Historians credit Humboldt's lectures with jump-starting German science, which went on to surpass even the French in brilliance, and with demonstrating that the true value of science lay not in its coterie appeal to an elite few, but in its power to raise and educate the many.[11]

Publishers pressed Humboldt with offers, but he resisted, demurring that his off-the-cuff lectures were hardly fit for print. By 1834, though, he was ready: as he wrote to Varnhagen, "I begin the printing of my work (the work of my life). I have the extravagant idea of describing in one and the same work the whole material world—all that we know to-day of celestial bodies and of life upon the earth—from the nebular stars to the mosses on the granite rocks—and to make this work instructive to the mind, and at the same time attractive, by its vivid language." The long introductory essay was finished and to his friend he outlined the rest, regretting that he could not concentrate the whole in one single "magnificent" volume. There was, though, the problem of what to call it. Already twenty years before, in the introduction to his *Personal Narrative*, he had worried over what name to give his new science: Natural history of the world? Theory of the earth? Physical geography? None quite fit. With Varnhagen he fretted over the problem once again: all the obvious possibilities—"Physical Description of the Earth," "The Book of Nature," "Physical Geography"—were too vague or too narrow. He considered, and rejected, "Gaea" (now spelled "Gaia," as in James Lovelock's "Gaia hypothesis"), which had been recently used by another author. So, taking a deep breath, he declared, "The title shall be 'Kosmos.'" Yes, it sounded pretentious, but the ancient Greek word gave him what he needed, heaven and earth *together*, and Wilhelm, with his deep learning in classical languages, approved. So there it was. Would Varnhagen do him the kindness of a preliminary reading, kind but tough-minded? "And do also ease my mind as to the title." Evidently Varnhagen did, for *Kosmos* it remained.[12]

The work grew as Humboldt wrote. It would gather together two generations of scientific research and discovery: into more than a dozen boxes he sorted his notes on scraps of paper, which he pasted together by their corners to form what his nineteenth-century biographer called "the most wonderful serpent-like structure of erudition." Information came in from his hundreds of correspondents, from all over the globe: this was nothing less than the great ingathering of the Humboldt network, dispersed across the planet but united by Humboldt's concentered vision. The German scholar Petra Werner has recently "disaggregated and interrogated" the portion of Humboldt's vast correspondence that went into *Cosmos*, showing "the extraordinary extent" to which he relied on friends and admirers; "In a real sense," adds Nicolaas Rupke, "Humboldt, in writing the description of the physical universe, acted as the editor of a large, international, collaborative team," cajoling their co-operation by offering each of his correspondents in return "his much valued praise and patronage." Humboldt was again at the center of the world, in his walk-up flat in the heart of Berlin, stacked with boxes, books, papers, maps, and mementoes (fig. 13). "Glorious old man!" exclaimed Richard Stoddard: "We love to think of thee and thy immortal task." Next to blind old Milton dictating *Paradise Lost*, "we know of no grander spectacle than the white-haired Humboldt writing 'Kosmos' at midnight!" Stoddard wrote that Humboldt finished the last page of the fifth and last volume on 14 September 14 1858, his eighty-ninth birthday. "His friends assembled at his house and congratulated him," laughing off Humboldt's presentiment that he would not outlive the following spring.[13]

Cosmos was the scientific bestseller of the age. In 1845, the first edition of the first volume sold out in two months; by 1851, Humboldt estimated that eighty thousand copies had been shipped. He himself superintended the French translation, and by 1846 it had also been translated into English, Dutch, and Italian. His publisher wrote in 1847 that the demand for the second volume was "epoch-making": "Book parcels destined for London and St. Petersburg were torn out of our hands by agents who wanted their orders filled for the bookstores in Vienna and Hamburg. Regular battles were fought over possession of this edition, and bribes offered for priorities." As Cedric Hentschel notes, our post-Darwinian perspective obscures the intense enthusiasm with which the first two volumes were received. Scores of thoughtful and laudatory reviews of the five successive volumes of *Kosmos* circulated in England and America, stoking the fires of enthusiasm, and in 1851 Humboldt's bust was given a place of honor at London's Crystal Palace. King Friedrich Wilhelm IV had a commemorative medal struck featuring a profile of Humboldt on one side and a sphinx on the other, under the legend "ΚΟΣΜΟΣ." Humboldt himself was rather bewildered by it all. "How has it happened that Kosmos is so popular beyond expectation?" he wondered to Varnhagen. It must be, he thought, in the imagination of the reader, or the fortuitous richness of the German language—as if the age itself were writing through Humboldt, making him its instrument.[14]

Humboldt had worked hard to capture the public imagination, and in his opening pages he mounts a defense of popular science, directed both to the general reader and to his colleagues in science who needed to be shown how to bind their research to society. The great Romantic traveler offers to lead

FIGURE 13. Eduard Hildebrandt, *Alexander von Humboldt in His Library*, c. 1855. Royal Geographical Society.

his readers on a "journey," not to a far-distant land, but "through the vast range of creation." As we set out, we may "distrust . . . our own strength, and that of the guide we have chosen," and indeed, he worries he may lose us in the jungle of dry details, or strand us on a mountaintop of abstractions. Like all guides who delight in leading others "to the summits of lofty mountains," he fears he may have erred "in describing the path before us as more smooth and pleasant than it really is," praising the view when all we can see is clouds. If so, blame, then, he begs, not the landscape of the sciences but "the un-skillfulness of the guide who has imprudently ventured to ascend these lofty summits." For the journey is worth the risk: "Nature is a free domain" that can be truly delineated only by "exalted forms of speech, worthy of bearing witness to the majesty and greatness of the creation."[15]

To skeptics who doubted that the ignorant public could ascend to the heights of science, Humboldt answered that while they may not catch every detail, the journey itself will "enrich the intellect, enlarge the sphere of ideas, and nourish and vivify the imagination." After all, detail was like scaffolding—it must be removed if the edifice is to have "a striking effect." And it was the "effect" Humboldt was after. The reader struck with awe or moved by beauty will want to learn more. This had political consequences. Francis Bacon had said that "in human societies, knowledge is power. Both must rise and sink together." But to Humboldt this meant not that the powerful must claim knowledge but that knowledge must be "the common property of mankind." Societies that shared knowledge across "*all* classes of society" would rise and prosper, strong and invigorated by their arts and sciences even if poor in natural resources. Conversely, societies that did not value public education would "diminish," even though erected on mines of gold. For, Humboldt reiterated, "the knowledge that results from the free action of thought is at once the delight and the indestructible prerogative of man."[16]

Humboldt had a well-honed sense of how words and thought act on each other and how both in turn interact with wider society. Reintroducing an archaic word into the modern lexicon was an ambitious act, a deliberate intervention intended to change the intellectual and emotional landscape of modern knowledge. It would take Humboldt five volumes to fully define "Cosmos," but he dropped a hint in his second paragraph, calling it a "harmoniously ordered whole." When a few pages later he was ready to pull back the veil a little more, he addressed himself to citizens of the planet itself: his "*science of the Cosmos* recalls to the mind of the inhabitant of the earth" that his horizon is much wider than any nation or region: it embraces "the assemblage of all things with which space is filled, from the remotest nebulae to the climatic distribution of those delicate tissues of vegetable matter which spread a variegated covering over the surface of our rocks." From the farthest nebulae to the lowliest lichens—in this way his "picture of the world may, with a few strokes, be made to include the realms of infinity no less than the minute microscopic animal and vegetable organisms." No other existing word—universe, earth, *monde*, world—captured the reach of the harmonies observed from the heavens to the earth under our feet. Thus he would reintroduce the word that originated with Pythagoras and Aristotle: Cosmos "is the assemblage of all things in heaven and earth, the universality of created things constituting the perceptible world."[17]

As Humboldt soon adds, *Cosmos* signifies both the "*order of the world, and adornment* of this universal order." Herein lies his distinctive use of the word: there are two aspects of the Cosmos, "order" and "adornment." The

first speaks to the observed fact that the physical universe, quite independently of us, exhibits regularities and patterns that we can identify as laws.
Beauty, "adornment," however, is perceptual, literally in the mind of the beholder. This is the double side of Humboldt's Cosmos: first, the physical universe exists quite apart from us; as Margarita Bowen says, "The concept of the
preexisting universe is essential to Humboldt's philosophy." But that is not
the complete story: it exists as a *Cosmos*, both ordered *and* beautiful, through
the human mind. Humboldt's Cosmos is thus fundamentally developmental
and dynamic. It emerges and grows as human conceptions of nature and
the depth of human feeling about nature enlarge and deepen. As a narrative,
Cosmos is still being written. Or, in Humboldt's favorite metaphor, Cosmos
is a "picture" which comes into being as we paint it and view it. Without
art—taking the word in its broadest sense to include science, technology,
exploration, literature, and the visual arts, gardening, and the painting of
landscapes—there may be a perfectly fine universe, but there will never be a
Cosmos.[18]

To represent this double-sided aspect of Cosmos, Humboldt divided his
book into two parts, to be read stereoscopically, each in light of the other.
The first volume (following his lengthy introduction) shoots into the out-
ermost reaches of deep space, then leads the reader gradually back to earth,
visible now in the most profound way as a *planet*, one small globe spinning
in pulsing, swirling, limitless space. The journey continues across the face
of the planet and into its superheated interior, to show the earth, too, pulsing and swirling with energy; then concludes with the life on its surface, in
all its astonishing multiplicity, including the races of men who, like all else,
are united by virtue of their very diversity into the "one great whole . . . animated by the breath of life." Once he has concluded this outer or "objective"
journey through the external world of the senses, Humboldt next takes us,
in the second volume, on an inner or "subjective" journey through mind,
"the inner, reflected intellectual world." He means not psychological exploration, but something more Wordsworthian, the emergence and growth of
mind-in-nature, "the reflection of the image impressed by the senses upon
the inner man, that is, upon his ideas and feelings." Where the first volume
journeys through space, the second journeys through time—historical time,
from the earliest civilizations nestled in the Mediterranean basin, through the
ramifying globalization of the sphere of the mind as nations launch onto the
oceans or caravan across continents to meet, merge, mingle, separate, each
era and each people contributing something to the growing Cosmos: words,
poetry, gardens, concepts, the compass, rice and sandalwood, telescopes and
paintings, ships, treatises, and scriptures. Born in awe and wonder before

nature's power and beauty, the human mind reaches out from itself to grasp nature through words and tools, and through the millennia nature and mind develop each other in an ever-diversifying historical process.[19]

This was not quite like anything that had ever been done before, and Humboldt patiently clarifies what he is not doing. He is not, first off, writing "a mere encyclopedic aggregation" of the results of science. He is, rather, trying to find their unifying thread, "to show the simultaneous action and the connecting links of the forces that pervade the universe." A contemporary reviewer caught the spirit of Humboldt's work when he wrote that *Cosmos* was like "a burning-glass" that reflected the investigations of science "on the mind of the reader in a cleared state and united in an organic whole." Nor is he writing "science" in the strict sense—that is, he has no wish "to reduce all sensible phenomena to a small number of abstract principles, based on reason only." This is the distinction that Franz Boas would seize on for his own intellectual program: whereas science seeks laws, cosmography romances the phenomena. At one point Humboldt put it this way: the spirit of his project, "physical cosmography," "arises from the sublime consciousness of striving toward the infinite, and of grasping all that is revealed to us amid the boundless and inexhaustible fullness of creation, development, and being." "Grasping" the fullness of creation aspires to connection, to touch and hold and explore, and Humboldt worries about those who, in striving toward the infinite, abandon the physical world. This impulse has deluded seekers through the ages into thinking they have reached their goal, found the single great commanding principle that unifies all. But that, Humboldt ventures, will never happen. The sheer complexity and diversity of nature will always outrun the scientist: as Thoreau would say, "The universe is wider than our views of it."[20]

Thus—sounding a little defensive—Humboldt fends off his reductionist friends: "Devoid of the profoundness of a purely speculative philosophy, my essay on the *Cosmos* treats of the contemplation of the universe, and is based upon a rational empiricism," or "facts registered by science."[21] Humboldt is drawing a crucial distinction here. Both science and cosmography, as Boas recognized, are empirical, based on facts; both share a conviction that the universe forms "one great whole," and that human reason can aspire to understand the truths of that whole—but exactly where science then leaps beyond the physical to the abstractions of law and principle, Humboldt instead turns back to "contemplate" the staggering beauty of the Cosmos. His motion is thus parabolic, not linear. He wants his feet to burn on the steaming rocks of volcanoes, his soul to be riven by the bleak, fierce llanos, his mouth to taste the Otomacs' unctuous, earthen baked clay, his heart to be moved by the song of the capirote. Here is his harmony, his song: not in the austerity of the laws of science, but in the way all these things seem to him to be one great thing, infinite and infinitely interconnected. In such a view, oneness cannot be seized by a law—only sung by a poet.

The poetry of Humboldt's prose will not be evident to readers of English until better translations become available, and even then, Humboldt apologized repeatedly for his faults of style. Yet if he seldom managed to write like a poet, he always thought like one. The American nature writer John Burroughs recognized this when he wrote that Humboldt's "poetic soul, shines out in all his works and gives them a value above and beyond their scientific worth. . . . His 'Cosmos' is an attempt at an artistic creation, a harmonious representation of the universe that should satisfy the aesthetic sense as well as the understanding." To accept Humboldt as a poet means to view poetics, as Humboldt did, in the Aristotelian sense as *poiesis*, "making," emphasizing the process of making over the finished product; and in the Romantic sense as original creation rather than imitation, an art that grows organically from an inner impulse that arises ultimately from nature. In this work, which he called in his subtitle a "sketch" of the physical universe, he first painted his own picture of the Cosmos (recalling his skill as a portrait painter and landscape artist), then outlined a historical narrative showing humans and nature "making" a Cosmos. In effect, Humboldt did in language what he had done long ago in his Chimborazo cutaway, his thumbnail Cosmos—he used his double vision to give an aesthetically pleasing image of nature framed, literally, with the supporting reams of scientific data. In *Cosmos*, the imagination of the viewer—whether poet, artist, or scientist—fused information into a new and beautiful whole. As Kropotkin exclaimed, out of the confusion of facts and the fog of guesses, a "stately picture" emerges from their mist like an Alpine chain glittering under the sun; Kropotkin thought a copy of *Cosmos* should be given to every schoolchild. Humboldt too wanted this exhilaration not for the few but for the many. The "contemplation of nature" makes self, nation, and nature into a coherent world, and is a necessary part of the *Bildung*, or growth and integration, of the self in the world. [22]

Making that world, for Humboldt, had one very specific requirement: it must be based on accurate causal connections as established by modern science. The abuse of sheer reason or "speculative philosophy" had, he believed, misled the "noble but ill-judging youth" of Germany "into the saturnalia of a purely ideal science of nature . . . signalized by the intoxication of pretended conquests" and a "scholastic rationalism, more contracted in its views than any known to the Middle Ages." His disgust with idealists who abandoned actual nature to spin metaphysical fantasies brought him closer to the empirical materialism of his French colleagues. Yet while he approved and drew heavily on the results of their science, he was after something else, a "cosmical presentation awakened in me by the aspect of nature in my journeyings by sea and land." One could call what he was after *grounded imagination.*[23]

For instance, one of the keenest passages of the *Personal Narrative* describes the working of imagination: barely six weeks after landing in the New World, entering the mountains above Cumaná, Humboldt is watching night fall and the southern stars come out. "The tree under which we were seated, the luminous insects flying in the air, the constellations that shone toward the south; every object seemed to tell us, that we were far from our native soil." Then the ringing of a cowbell or roaring of a bull from the valley below would suddenly awaken a remembrance of home: "They were like distant voices resounding from beyond the ocean, and with magical power transporting us from one hemisphere to the other. Strange mobility of the imagination of man, eternal source of our enjoyments, and our pains!" While the body is grounded, it is the imagination that connects, that can be in many places at once. Humboldt attests that the mind of "the most savage nations" is moved to awe at nature's power and beauty, forming "a bond of union, linking together" the visible world with the higher world of spirit. The deep emotions thus awakened, the sense of "intimate communion with nature," lead first to worship and deification; then, awakening reflection allows the developing mind to separate ideas from feelings. "Vague presentiments" of nature's harmonious union are no longer enough. The mind moved by awe and wonder succeeds to reflection, to understanding, to doubt and investigation—to science.[24]

Does science then turn on and kill the imagination that gave it birth? On the contrary, argues Humboldt: the prejudice that science must kill the feelings is wrong. The excitement of discoveries, of "mysteries to be unfolded" and the "inextricable net-work of organisms," carries thought forward. Wonder and the pleasure of discovery feed the desire to know, and knowledge leads back to wonder, in an ascending spiral fed by imagination at every turn that ever enlarges and will never end. Nature is without limit, "ever growing and ever unfolding itself in new forms," and even "when thousands and thousands of years have passed away," the surface of the earth, its interior, its oceans, its atmosphere, will forever "open to the scientific observer untrodden paths of discovery." As Thoreau wrote in the full flush of his excitement after reading Humboldt, "the sun of poetry & of each new child born into the planet has never been astronomized, nor brought nearer by a telescope. So it will be to the end of time. The end of the world is not yet."[25]

It seemed to Humboldt that our feeling for nature has two different dimensions. First is the impact of the whole. Nature in its sheer allness, "the image of infinity" revealed by "the starry vault of heaven," the "far-stretching plain," or "the vast expanse of ocean," awakens us to an intuition of "the order and harmony pervading the whole universe." But there is also nature in its eachness, its individuality. Here Humboldt revels in his memories. His language turns to poetry, as he recollects

> the calm sublimity of a tropical night, when the stars, not yet sparkling, as in our northern skies, shed their soft and planetary light over the gently heaving ocean; or . . . the deep valleys of the Cordilleras, where the tall and slender palms pierce the leafy vail around them, and waving on high their feathery and arrow-like branches, form, as it were, 'a forest above a forest;' or I would describe the summit of the Peak of Teneriffe, when a horizontal layer of clouds, dazzling in whiteness, has separated the cone of cinders from the plain below, and suddenly the ascending current pierces the cloudy vail, so that the eye of the traveler may range from the brink of the crater, along the vine-clad slopes of Orotava, to the orange gardens and banana groves that skirt the shore.

In such scenes, the heart is moved not by nature's general charm but by "the peculiar physiognomy and conformation of the land, the features of the landscape, the ever-varying outline of the clouds," by the irreplaceable uniqueness of each place, each different way of being. In a deep insight into Humboldt's language, David Kenosian observes that to him, "South American ecosystems were not so much unconnected sentences . . . as they were poems."[26]

It is this "physiognomy" that most entrances Humboldt, for it is here, at this threshold, that nature interpenetrates mind. Painters casually refer to "'Swiss scenery' or 'Italian sky,'" but Humboldt takes such phrases seriously and tries to identify the individual elements that compose these painterly impressions: "The azure of the sky, the effects of light and shade, the haze floating on the distant horizon, the forms of animals, the succulence of plants, the bright glossy surface of the leaves, the outlines of mountains, all combine to produce the elements on which depends the impression of any one region." Yet how is it that such elements of sky and light and form "impress" themselves on nations, peoples, individuals? Why do we rejoice at the simple appearance of fields and woods? How does the look of vegetation influence "the taste and imagination of people," and more, how does it impress "the soul of those who contemplate it"? Humboldt thinks that nature moves us, shapes us, creates us, in ways that art seeks to capture and repeat but that we have never really thought about. In his *Essay on the Geography of Plants* this insight opens a flood of questions:

> What is the moral cause of these sensations? Are they produced by Nature, by the grandeur of masses, the contour of forms, or the haven of plants? How

can this haven, this view of Nature more or less rich, more or less pleasant, influence the mores and, primarily, the sensitivities of peoples? Of what consists the character of the vegetation of the tropics? What difference in physiognomy distinguishes plants from Africa from those of the New Continent? What analogy of forms unites Andean alpine plants with those found on the summits of the Pyrennées? These are questions little broached to at present, and doubtless deserved to occupy the physicist.

Doubtless they do, but as he added some years later, while science can measure and tabulate and compare, it cannot by these means communicate the character of nature. "What speaks to the soul, what causes such profound and various emotions, escapes our measurements, as it does the forms of language."[27]

Here, then, where it matters the most, science must necessarily fail. Thus Humboldt turns to art. Where science must weigh and measure, abstract and bring away, art can make present to the senses and the imagination the fundamental experience of contemplating nature in its wholeness, generating a similar emotional impact. As Joan Steigerwald explains, Humboldt used his measurements "as instruments of judgment" which translated the phenomena of the natural world into a total impression, seeking, in reconciling "the opposing requirements of scientific precision and painterly effect," a new figurative vocabulary, "a new graphic form of representation." When it came to fine art, he encouraged the artist to become a cosmographer, analyzing the whole into its component parts: "In the sphere of natural investigation, as in poetry and painting, the delineation of that which appeals most strongly to the imagination, derives its collective interest from the vivid truthfulness with which the individual features are portrayed." Here is the heart of Humboldt's aesthetics: art can incorporate and surpass science in conveying the perceptual truth of the whole, but only if the artist paints the truth of particulars. By truth Humboldt means natural historical truth. The artist cannot paint just "plants," but must become botanist and know each species, its growth and habits; clouds are not puffs of pigment but studies in meteorology; mountains are visual embodiments of geological principles, water of hydrology. Landscapes become not static portraits but dynamic historical ecologies. Literature, too, acquires life as it approaches individual truthfulness: just as a painter must first sketch in the field, Humboldt felt his Orinoco journal, "written while the objects we describe are before our eyes," had a more vital "character of truth." In *Views of Nature* he further generalized his literary aesthetics: "Speech acquires life from everything which bears the true impress of nature, whether it be by the definition of sensuous impressions received from the external world, or by the expression of thoughts and feelings that emanate from our inner being." Write what you know, he might have said, whether your gaze turn outward or inward. In his artistic and literary realism, accuracy and imagination will not cancel out but reinforce each other, resonating together, and the higher the subject, the simpler and more truthful must be the writing. As he concludes, the poet who "knows how to represent with the simplicity of individualizing truth that which he has received from his own contemplation, will not fail in producing the impression he seeks to convey; for, in describing the boundlessness of nature, and not the limited circuit of his own mind, he is enabled to leave to others unfettered freedom of feeling."[28]

To the physical scientist, the face of nature is superficial and shallow, a mask that, like Ahab, he must strike through to reach the hidden truth. But for Humboldt the face of nature did not hide but revealed the depths within. Those depths were temporal, for all objects encode their past: "We cannot form a just conception of their nature without looking back on the mode of their formation." This is most obvious in the case of sedimentary rocks, which, laid down in successive ages, entomb the remains of organic worlds long since destroyed: "The different superimposed strata thus display to us the faunas and floras of different epochs." The notion that one could "read" strata like the leaves of a book was already a commonplace, but Humboldt's notion is rather different: "In tracing the physical delineation of the globe, we behold the present and the past reciprocally incorporated, as it were, with one another; for the domain of nature is like that of languages, in which etymological research reveals a successive development." Just as language reveals to the linguist a long history of linguistic evolution, so does the face of nature speak, like a language, of its own past, "reciprocally incorporated" into the present, which, as Lyell recognized, itself bespeaks the very mechanisms of change.[29]

In this deepening narrative of origins, Humboldt links up the dispersal and diversification of languages and cultures with geological time, the upthrust of continents and mountains, the draining of valleys, the formation of deserts, all the way back to the first formation of rocks and landforms by volcanic activity. Wherever Humboldt looked, whatever he looked at, he saw spatial patterns of distribution and change pointing back in an unbroken continuum through human history to the deep geological and even deeper astronomical past. This was the lesson he urged across the pages of his last volumes. All this could be "read" in the present by those who knew the language of nature, that most ancient of storytellers. The face of the land told its own story, its features "animate the scenery by the associations of the past which they awaken, acting upon the imagination of the enlightened observer like traditional records of an earlier world. Their form is their history."[30]

Wilhelm von Humboldt had seen his historical method as a logical extension of his brother's scientific method. Now Alexander turns the tables by calling his own method a kind of "historical composition," citing his brother Wilhelm. Where the physical scientist can discount "accidental individualities, and the essential variations of the actual" in his attempt to reduce all to a "*rational* foundation," the historian/cosmographer must turn to and treasure exactly those individualities and variations, as Boas had recognized. But there is yet another dimension to the telling of history: it is, in Humboldt's word, a "composition," not a transparent transcription of truth but a selection made from a particular and necessarily limited point of view. As Humboldt realized in his moment of bedazzled frustration on the sides of Tenerife's volcano, what lies around us is a jumbled chaos of apparently isolated facts. Some facts and connections are easy to see, others are cryptic or obscured, others still are lost forever and cannot be traced. None of them assemble themselves into a narrative in any obvious way: this is the work of the teller of stories. As the historian Michel-Rolph Trouillot observes, "history" has in this sense a double meaning: "In vernacular use, history means both the facts of the matter and a narrative of those facts, both 'what happened' and 'that which is said to have happened.'"[31] Human beings, not nature, write histories, and human authors cannot simply transcribe revolutions—planetary or political—onto the page. They must select, foregrounding some elements, silencing others, teasing out of the chaos a meaningful pattern, a causal narrative.

If histories are "compositions," does this mean they are fabrications, mere fictions? As Trouillot points out, this claim itself has a long history, but Humboldt would have rejected it. He required his cosmography to follow strict rules of evidence: just as landscape artists should not make up fantastic or stereotyped plants but portray real ones, so must the cosmographer be constrained by truth. Not everything his correspondents sent him went into *Cosmos*, only what Humboldt judged would deepen and refine our understanding of the causal interconnectivity of the universe. And while no one was more fascinated by the marvelous, Humboldt wanted his marvels to be real, not isolated, pointless curiosities but causally connected to the great whole: a man who gives milk, a people who can live on dirt, Amazon warrior women, eels that electrify their prey, all were woven into his causal narrative, "accidental individualities" that yet evidenced something important. He applied the same standards to historical questions: Had this valley once been a lake? Had these fossil seashells high in the Andes been deposited on the ocean floor? Had the Indians of the Americas migrated from Asia? The standards of science constrained his conclusions, and he took care to make those constraints visible: yes, I think so, and here are the reasons why. Constrained, then, by truthfulness, he nevertheless assumed the freedom to tell his narrative in his own distinctive way, and his frequent self-conscious reflections on the way he is constructing his narrative show he understood his alliance with *poiesis*: truth is made, not transcribed. And this includes the scientist.

As Trouillot points out, no one actually remembers history, for no one living now experienced it: "The collective subjects who supposedly remember did not exist as such at the time of the events they claim to remember." Sadly, no one can really remember the Alamo, or Wounded Knee, or slavery, and soon, no one will remember the Holocaust. What we do remember is what we were told: we believe Columbus discovered America because that is the narrative we hear from childhood onward. Insofar as we all do believe that narrative, we constitute ourselves as a collectivity, a people who share a common past and identity and sentiment of destiny. As Trouillot continues, "the past" is thus a creation of the present, and "their constitution as subjects goes hand in hand with the continuous creation of the past. As such, they do not succeed such a past: they are its contemporaries."[32]

If Trouillot is correct, then the writing of the past is not innocent: it has tremendous consequences. If resurrecting the ancient word "cosmos" was calculated to make the concept of Cosmos thinkable, actionable—real—then telling the history of Cosmos was another calculated intervention. The Cartesian dualism that separates spirit from body, mind from matter, humans from nature, was in Humboldt's day as now so dominant as to seem intuitive and inevitable. Humboldt's history, though, integrates mind and material nature by showing how humans and nature together create the Cosmos. In effect, Humboldt is trying to rewrite history as ecological history—or as historical ecology. If science had thrown out history, as Boas would argue, Humboldt would bring it back. How we tell our story about nature constitutes who we are as a people. If we tell it as dualistic, violent, and exploitative, we revalidate those qualities as our essential truth. But if we can learn to tell it as integrated, cooperative, and sustainable, we will advance our rough-edged and imperfect civilization to a higher level.[33]

In Humboldt's experience, the edges were rough indeed: he lived with warfare, reactionary politics, stark exploitation, and brutal inequalities, even as he had grown up with utopian ideals of peace, liberty, equality, and cosmopolitan brotherhood. He pours those ideals into his telling of the Cosmos: "Nature is a free domain," he asserts, and "the view of nature ought to be grand and free, uninfluenced by motives of proximity, social sympathy,

or relative utility"—it must start, in other words, from the viewpoint of the
stars and the planet earth, not our petty "human interests." Where humans
constrain and limit, that which is most wild is most free, like the capirote,
"which no effort has been able to tame, so sacred to his soul is liberty." In a
society that assumes—as we still do—that only humans speak while nature
is silent, Humboldt revels in the voices of nature, whether the heartrending
song of the capirote, the raucous jungle chorus that awakes him at night, or
the silence that only seems:

> Yet, amid this apparent silence, when we lend an attentive ear to the most
> feeble sounds transmitted by the air, we hear a dull vibration, a continual
> murmur, a hum of insects, that fill . . . all the lower strata of the air. Nothing
> is better fitted to make man feel the extent and power of organic life. Myriads
> of insects creep upon the soil, and flutter round the plants parched by the
> ardour of the Sun. A confused noise issues from every bush, from the decayed
> trunks of trees, from the clefts of the rock, and from the ground undermined
> by the lizards, millepedes, and *cecilias*. These are so many voices proclaiming
> to us, that all nature breathes; and that, under a thousand different forms, life
> is diffused throughout the cracked and dusty soil, as well as in the bosom of
> the waters, and in the air that circulates around us.

To those who deny animals intelligence or even agency, as if they were no
more than wound-up machines, Humboldt depicts the deep intelligence of
mules: "When the mules feel themselves in danger, they stop, turning their
heads to the right and to the left; the motion of their ears seems to indicate,
that they reflect on the decision they ought to take. Their resolution is slow,
but always just, if it be free; that is to say, if it be not crossed or hastened
by the imprudence of the traveller." The mule-drivers, unconfused by metaphysics, know this well: "Thus the mountaineers are heard to say, 'I will not
give you the mule whose step is the easiest, but him who reasons best;' *la mas
racional.* This popular expression, dictated by long experience, combats the
system of animated machines, better perhaps than the arguments of speculative philosophy."[34]
    Where humans believe themselves to be separate from or above nature,
Humboldt constantly uses metaphors of permeation: "The mind is *penetrated* by the grandeur of nature"; natural scenery leaves an "impression" on
the mind; the physical world "is *reflected* on the inner susceptible world of the
mind," in a "mysterious *communion* with the spiritual life of man"; the physical "influences" the moral world, in "a mysterious *reaction* of the sensuous
on the ideal"; clarity and serenity of mind "*correspond* with the transparency
of the surrounding atmosphere"; we are "moved" with emotion; majestic

scenes of nature "*mingle* with all our feelings of what is grand and beauti-ful." The body of the world has stolen through our senses into the deepest recesses of our mind, where it shapes our thought and language at the most fundamental level.[35]

The variety of landscapes and seascapes under ever-varying clouds opens the imagination's creative powers in a free play of mind: "Impressions change with the varying movements of the mind, and we are led by a happy illusion to believe that we receive from the external world that with which we have ourselves invested it." In Coleridge this insight provoked a mood of despair: "O Lady! we receive but what we give, / And in our life alone does Nature live." But this is not quite what Humboldt is saying. Where Coleridge col-lapses nature into mind, imprisoning himself in an unhappy illusion that all nature is no more than a screen on which he projects his despairs and desires, Humboldt escapes this trap by insisting on *difference*, on the very fact that nature is *not* human. It is that very independence that allows nature to "im-press" us so deeply. Only difference can "impress," like the bite of the print-ers' plate pressing the receptive paper. Humboldt's mentor Georg Forster put it this way, in recalling their travels on the Rhine: "The object, whatever it may be, that exists without the cooperation of man, that is, and was, and ever will be independent of him, impresses itself deeply upon the mind with a clear and sharply defined image." In some moods Humboldt found this power, this difference, shading into an indifference to the human. When he turned away from the tombs of the extinct Atures Indians, his dark thoughts gave him a curious kind of hope: "Yet when every emanation of the human mind has faded—when in the storms of time the monuments of man's cre-ative art are scattered to the dust—an ever new life springs from the bosom of the earth. Unceasingly prolific nature unfolds her germs,—regardless though sinful man, ever at war with himself, tramples beneath his foot the ripening fruit!"[36]

By asserting that nature is independent of humans in a difference that is profoundly generative, Humboldt is trying to bridge the impasse reached by Kant, who had deepened the Cartesian dualism of mind and nature into an unbridgeable abyss by arguing that the nonhuman or "noumenal" world could never be reached or conceived. We could see only its phenom-enal shadow, the mask, what little was open to the human senses. As Mar-garita Bowen details, Humboldt bridged this Kantian impasse by showing how humans developed their concepts over time, in a historical process by which they "are generated, tested and incorporated into the sphere of ideas." Through this historical process, ideas forged in the crucible of physical nature made the world of thought part of the process of nature. As Bowen observes,

Humboldt sees the very gulf between mind and nature "as the locus of the sciences."[37]

We are back, many pages later, standing on Humboldt's bridge, spanning the gulf between mind and nature. Humboldt wrote that while nature may be opposed to intellect, "as if the latter were not comprised within the limits of the former," and while nature may be opposed to art "when the latter is defined as a manifestation of the intellectual power of man," these "contrasts," though reflected "in most cultivated languages," must not be allowed to stand. We must not "separate the sphere of nature from that of mind, since such a separation would reduce the physical sciences of the world to a mere aggregation of empirical specialties." This was the very point at which Boas felt science had in fact arrived, by 1887: fragmentation into a variety of scientific specialties had abandoned the crucial problem of their union, which to Boas was not reductive but perceptual: we *see* that nature forms a whole, though science in its fragmentation no longer allows us ways to approach that sense of wholeness. Boas's solution was to resurrect Humboldt's "cosmography," because Humboldt started from the very perspective that science had eliminated in order to professionalize and specialize: the subjectivity of the observer. Humboldt thus ran exactly counter to the developing ideology of science, the objectivity which sought to purify science by removing subjectivity altogether. Ironically, in Humboldt's view, eliminating subjectivity to render nature as pure object actually rendered true science impossible: "Science begins where the mind takes hold of matter and attempts to subject the mass of experiences to a rational understanding; it is mind directed toward nature." One simply cannot take mind out of the scientific equation. The external world exists for us only in the image reflected to mind by the senses, and so, "as intelligence and forms of speech, thought and its verbal symbols, are united by secret and indissoluble links, so does the external world blend unconsciously to ourselves with our ideas and feelings."[38]

Humboldt has returned here to his analogy with language: thought is to language as mind is to nature. Language both interprets thought and paints the objects of the world, mediating intellect and nature. At the same time, "it reacts . . . upon thought, and animates it, as it were, with the breath of life. It is this mutual reaction which makes words more than mere signs and forms of thought." Words push back on wordless thought, animating it with unexpected energies; nature too pushes back on mind, enabling and limiting it, shaping and "animating" it even as we invest it with qualities that we then imagine were there all along. The process is recursive, reactive, reciprocal—as Charles Sanders Peirce would discover when he put down his Humboldt and took up semiotics. Humboldt's point is simpler: thought, feelings, language, nature, are all inseparable. Language is the medium of thought, as nature is the medium of mind. The teller is part of tale she tells, a truth from which the scientist is not exempt. We are all standing on Humboldt's bridge, fishing into the gulf below, casting thought into words, casting words out into the world and reeling back what they have captured, even if it is only, as Thoreau says, "some horned pout squeaking and squirming to the upper air." Humboldt himself, rather less poetically, recruits a phrase from his nemesis Hegel: "External phenomena . . . are in some degree translated in our inner representations." Or as Humboldt continues, "The objective world, conceived and reflected within us by thought, is subjected to the eternal and necessary conditions of our intellectual being." Science might cast out subjectivity in its search for truth, but cosmography would bridge and build exactly on the abyss that science had created.[39]

Humboldt felt he had not finished this work—had really only begun, offering "a first imperfect attempt" that would incite rather than satisfy. But he did hope that his science of the Cosmos would someday unite "both spheres of the one cosmos—the external world, perceived by the senses, and the inner, reflected intellectual world." This has not happened, for the path which Humboldt indicated remains to be explored. When it is, we will better understand Humboldt's home truth, that the products of our "spiritual labor belong as essentially to the domain of the Cosmos as do the phenomena of the external world." Humboldt is reaching here to a concept we are just beginning to grasp: as Ian Hacking observes, "Philosopher and historian alike are part of the ecosystem that has been transformed by bearers of that vision in their interactions with nature as they saw it." Gregory Bateson wrote of this insight as the "ecology of mind," and Margarita Bowen calls it the "ecology of knowledge," which acknowledges that "all knowledge occurs not only in society but in the dynamic space-time context of the earth ecosystem; the observer—and the scientist is no exception—thus must be considered always as part of the system being observed, the ideas and actions that issue from such observation are themselves incorporated in the dynamic system, to become part of the future environment." As Bowen concludes, if thought is really part of the Cosmos, as Humboldt says, it too has a carbon footprint: "The act of reading these words can be considered then as part of the carbon cycle," the complex system of energy flow that links the organic and inorganic worlds. (The man who paused in his discussion of meteorology to point out that the air every animal breathes also carries speech and thus thought, "maintaining social intercourse," would have gleefully agreed.) Bowen continues, "Humboldt's concept of nature incorporated human thought and culture; continuing that tradition, the ecology of knowledge affirms that all knowledge occurs within the functioning ecosystem and itself forms an integral part of that system, at least while mankind survives."[40]

We might not survive—as Humboldt reflected when he turned away from the graves of the vanished Atures. The Cosmos offers no guarantees. What humans have helped build, they may wholly destroy. High in the Andes, in the privacy of his journal, Humboldt felt a moment of despair: "Mountains of the moon and of Venus! When will we undertake that journey, propagating our culture over other planets—that is, our combination of vices and prejudices—devastating them as the Europeans have depopulated and sacked both Indies?"[41] Cosmos was not an accomplishment but a prospect—a viewpoint from which Humboldt could sustain a critique of a Western civilization that had, for good or ill, inherited the legacies of hundreds of nations across the millennia of cosmic progress. To reach that prospect, the Cosmos needed every one of Humboldt's many readers.

Behold the Earth

These may seem like abstract and difficult questions, but for Humboldt they really weren't. His great goal was simply to get his readers outdoors, into nature, to join the community of Cosmos. This did not mean just grand colonial expeditions to exotic and faraway places, but excursions to local parks and gardens, to art galleries, even libraries. His "incitements" to the contemplation of nature included travel to distant lands, but closer to home, also landscape painting, nature writing, and gardening. Near or far, turn your gaze outward. Why was this so important? Cosmos was a communal, participatory, collective project, a "communion" that created an ever-widening community open to all—and it was no less than the hope of humanity, which could progress only if it expanded and deepened the bonds among diverse peoples and with the nonhuman world. Exiling nature to somewhere "out there," to be exploited as raw resource, was a symptom of the colonial imperialism he tried to condemn and redirect. Nature was everywhere, in rivers and skies and streetscapes no less than oceans and jungles and high peaks; it was even deep in the human soul, a generative otherness within that was part of our fabric of being. In no other way could Humboldt imagine how to explain our susceptibilities to nature's power and beauty, our ability to be "penetrated" by grandeur, "impressed" by local scenery, or "moved" to reflection by such revelations as the infinity of deep space or the vast age of the earth.

Of course most people did not understand this, as Humboldt had known ever since he grew up mocked as the "Little Apothecary" for his love of wild nature. But not to look outward, beyond human artifice and convention, was to be imprisoned in a hall of mirrors and subject to social and political tyranny. Emerson's breakaway essay *Nature* had opened with a call for an "original relation to the universe," direct and unmediated by stale tradition, and this was Humboldt's call as well. The only way to recover from the corrupting pathologies of social convention was to grow beyond them, as a self and as a community or society, free and self-directed, using the resources of a native language, art and poetry and scientific knowledge. Humboldt thought of progress in the same way as Kant, as a steady, slow, progressive evolution of the entire human species, and he thought he saw progress all around him: "all classes of society" were "embellishing" their life with ideas, fully refuting "the vague accusations advanced against the age in which we live, showing that other interests, besides the material wants of life, occupy the minds of men." Cosmos as a collective project seemed well underway, and sketching its emergence allowed Humboldt to mount a sweeping critique of Western civilization even as he celebrated what in it was hopeful and redemptive. As nature was "a free domain," the study of nature cultivated "a moral freedom that strengthens us against the blows of destiny." Leading by example, then, Humboldt issued in *Cosmos* a gentle call to intellectual arms.[42]

Certain persistent misconceptions, however, must be addressed, for they have dulled and obscured Humboldt's message. The problem with originality is it is hard to hear—it is much easier to fold the truly new back into the familiar, and hear it as validation of the ready-made. To begin with, in Anglo-America, the ready-made view—almost the only permissible view—saw the universe as God's Creation, unified not as Humboldt said by human perception, but by God's creative act and sustaining energy. Virtually every Western scientist outside the circle of French influence operated within the assumptions of natural theology, which held that the physical universe, in its divine balance and ingenious mechanisms, revealed the nature and attributes of the God who created it. This made science, at least until Darwin, an arm of Christian religion. Humboldt, raised in the French Enlightenment and happy only in Paris, was not a natural theologian. Indeed, it is hard to tell whether he was a believer at all. While he did not, like Tom Paine and Constantin-François Volney, rail against priestly corruption and declare his own mind his only church—indeed, he had some favorable things to say about the influence of Christianity—he did exempt God from his Cosmos, both as an explanatory principle and as a spiritual overlord.

This fact greatly alarmed many of his English reviewers, who took him to task for irreligion even as others, such as the ministers John Bachman and Theodore Parker, defended him from charges of atheism, and atheists such as Karl Heinzen and Robert Ingersoll cheerfully declared Humboldt their intellectual hero. Humboldt himself, ever the diplomat, was careful not to offend, and he let stand a handful of references to the "breath" of life that could be interpreted as Christian allusions. But he nowhere refers directly to God, at the time an extraordinary omission. His Cosmos is entirely secular, his interest in religion comparative and even anthropological. Yet curiously it has become a commonplace in some academic circles that Humboldt, as a recent art historian put it, "subsumed all that he discovered under a comprehensive system of religious belief. For him God was the originator and central point of all life, and everything in creation could be taken as evidence of the divine order of the universe." This erroneous claim distorts Humboldt's philosophy and disables any understanding of what was innovative about his project, or of the peculiar pressures that Americans, in particular, exerted on its reception. Claims such as this serve to tame Humboldt by boxing him into the stable of pre-Darwinian scientists who are presumed to have been ignorant of deep geological time, organic evolution, and violence in nature. Actually Humboldt, who after all trained as a geologist, was a pioneer in the use of fossils to date geological strata (he contributed the word "Jurassic" to modern languages), and he took evolution for granted even as he was baffled by its cause.[43] As for the many commentators who repeat the bromide that Humboldt believed in a "static" universe, they are, rather like Rick stranded in the colonial desert of *Casablanca*, misinformed.

But the question of whether Humboldt's nature, like Tennyson's, was "red in tooth and claw" is more vexed and more interesting. In a widely cited and perceptive essay, the evolutionary theorist Stephen Jay Gould himself maintains that Humboldt was driven into "oblivion" by Darwin's theory of natural selection. According to Gould, Humboldt's "notion of pervasive harmony" was exterminated by Darwin's view of nature as "a scene of competition and struggle" with no progressive direction, propelled by mindless forces that are capricious and random. In one sense, Gould is exactly right: *Cosmos* opens with, and is dominated by, the view that nature is "a unity in diversity of phenomena; a harmony, blending together all created things, however dissimilar in form and attributes." It is because of this very unity that the observer can pass from any aspect of nature to any other, "by means of the mutual dependence and connection existing between them." Humboldt's "mutual dependence" became the foundation for Darwin's natural selection, and when Darwin was ready to sum up the universe conjured up by his theory, his image is astonishingly Humboldtian:

It is interesting to contemplate an entangled bank, clothed with many plants of many kinds, with birds singing on the bushes, with various insects flitting about, and with worms crawling through the damp earth, and to reflect that these elaborately constructed forms, so different from each other, and dependent on each other in so complex a manner, have all been produced by laws acting around us. . . . There is grandeur in this view of life, with its several powers, having been originally breathed into a few forms or into one; and that, whilst this planet has gone cycling on according to the fixed law of gravity, from so simple a beginning endless forms most beautiful and most wonderful have been, and are being, evolved.

Yet there is also great pain in Darwin's work, for his "exalted" view of nature's grandeur springs out of an eternal "struggle for existence" forever seething beneath nature's smiling face. My ellipses in the passage above conceals its non-Humboldtian element: the beauty and wonder Darwin describes follow from "the war of nature, from famine and death."[44]

Humboldt knew this side of nature. As he observed on the Orinoco, Eden was an illusion: "In this Paradise of the American forests, as everywhere else, sad and long experience has taught all beings, that benignity is seldom found in alliance with strength." Scenes of strife and violence and war are everywhere in his *Personal Narrative*, which Darwin carried with him on the *Beagle*. So how different were Humboldt and Darwin, really? As Robert J. Richards shows, Humboldt's influence on Darwin was profound: from the very beginning of his career, Humboldt's writings had inculcated in Darwin the belief in nature as a source of "moral and aesthetic value." Both saw "grandeur" in the whole of nature, both appealed to an ultimate harmony beyond the struggle and warfare of daily life, and both believed that cooperation across the continuity of all life was the engine of social progress. There is, in fact, more of Humboldt in Darwin—and vice versa—than Gould acknowledged.[45]

The difference, for there surely is one, lies in their different purposes. Darwin was proposing a scientific theory, Humboldt a humanistic worldview. Darwin in *Origin* needed the basic mechanism of natural selection to drive evolution toward the ever increasing diversity of forms that so delighted Humboldt; Humboldt in *Cosmos* needed the basic truth of natural harmony to drive the Cosmos away from the Hobbesian war of all against all, Tennyson's "nature red in tooth and claw." What later generations came to call "Darwinian" nature actually predated Darwin by many years. This means that Humboldt's readers as well as Darwin's were conditioned to see, under nature's smiling face, brutality and violence. Some of Darwin's readers took that fact much farther than Darwin ever intended. Were not, as Darwin observed, the stronger always extirpating the weaker? Then since men were part of nature, why shouldn't they do what comes naturally, and eliminate the "weak" and "unfit"? Humboldt provided his readers no such opening, for

Europeans' manic drive to eliminate the "weak" and "unfit" was the very fact he was writing against. Natural predation did not excuse political exploitation. Since Humboldt, unlike Darwin, felt no need to sever subject from object, he could insist upon "harmony" as both an objective truth about nature, visible to reason if not always to emotion, and a subjective truth of human perception, a way to look at nature. There is thus a willed quality to Humboldt's harmony, a need, going back to his first nature writings and even further back to Kant's honoring of his Jewish friend Moses Mendelssohn, to face down the worst by challenging it with hope. It was a strategy fundamental to American Transcendentalists as well, many of whom were thoughtful readers of Humboldt: the "higher law" of nature challenged the corrupt lower laws of the state, opening a space for social reforms that included women's rights and the abolition of slavery. Humboldtian harmony was largely exterminated, but, *pace* Gould, the "exterminating angel" was not Darwin.

As for the near-universal assumption that Humboldtian harmony is balanced and therefore static and unchanging, it helps to be reminded that Humboldt's science arose in late Enlightenment vitalism, in which, in Peter Hanns Reill's words, "harmony" meant "the joining of opposites within an expanded middle generated by reciprocal interaction . . . leading to ever-changing harmonic combinations." There are times when Humboldtian harmony can sound rather static, as when, just before launching into the Orinoco river system, he meditates on the "wonderful uniformity of succession" in the equatorial tropics, where "the active powers of nature limit and balance each other." Yet standing on the volcanic dome above Caracas, he was struck not by nature's reciprocity but by its violence: volcanoes, earthquakes, floods, and tempests wipe away in hours the human labors of generations; not calm succession but "the conflict of the elements" seems to characterize nature in the New World. In his nature essay "Steppes and Deserts," calm does exist, but far from the sublunary zone marked by the endless "strife of nations." He who seeks "intellectual repose" must turn to "the silent life of plants" or even better, "towards the celestial orbs, which, in undisturbed harmony, pursue their ancient and eternal course." As in the opening of *Cosmos*, Humboldt sometimes looks to "the calm of nature" that becomes evident when the expanding realm of scientific law dispels the apparent discord of the elements. This was the hope, at least, of an aging man, as he confessed to a friend: "In the disorder of our days and with the discord of our times . . . I seek refuge . . . as often as my position allows, in the unending cosmos, seeking and finding, in the investigation of its phenomena and laws, the peace that is so necessary to me in the evening of an extremely eventful life."[46]

Yet in *Cosmos*, when he actually launches out beyond the moon and the planets, out to the farthest reaches of interstellar space, the view is anything but calm and peaceful. Stars that appear calm and stable and restful on earth resolve into swirling nebulae, some like disks, alone, or in pairs, or connected with a thread of light, or "elongated, or variously branched, or fan-shaped, or . . . like well-defined rings, enclosing a dark interior." All, thousands of them, are stellar systems growing in "the great garden of the universe." Looking into the heavens is like looking into our forests, where "we see the same kind of tree in all the various stages of its growth, and are thus enabled to form an idea of progressive, vital development." The telescope teaches us to see organic change: in this great stellar garden, the seedbed of worlds, we do not merely conjecture evolution, we actually see it, "going on before our eyes." More, our telescopes are time machines that "penetrate alike through the boundaries of time and space," showing us that the Cosmos lives in space-time. The light we see took thousands, even millions, of years to reach us. Looking into deep space opens our eyes to deep time, even to deepest time: "The light of remote heavenly bodies presents us with the most ancient perceptible evidence of the existence of matter," where—he is quoting from a sonnet by brother Wilhelm—"myriads of worlds are bursting into life like the grass of the night!"[47]

Swirling cosmical vapor coalesces into stars and planets, whole systems, to one of which "our cosmical island belongs," our own sun just one among an array of single stars and double ones, comets and asteroids and aerolites and shooting stars that drop stellar matter out of the sky into our very arms. Standing on the earth, holding star-stuff in our hands, gazing through a telescope at nebulae and suns, we see a universe not calm and quiet but throbbing with energy: "The apparent rest that reigns in space would suddenly disappear." We see stars pulsing in and out of existence, and "the countless host of fixed stars moving in thronged groups in different directions; nebulae wandering through space, and becoming condensed and dissolved like cosmical clouds; the vale of the Milky Way separated and broken up in many parts, and *motion* ruling supreme in every portion of the vault of heaven, even as on the Earth's surface, where we see it unfolded in the germ, the leaf, and the blossom, the organisms of the vegetable world." All things, Humboldt concludes, "that have been, that are, and will be, are alike dependent on motion."[48]

Our own earth, whirled into existence out of condensing and superheating gas, throbs and pulses still, its light crust of rock heaving and cracking atop the molten masses below our feet; yet we hardly see this, since beyond a few mine shafts and the depths of the sea, our earth is "as much unknown to us as the interior of the other planets of our solar system." Yet look more closely: molten rock rushes "impetuously" upward, or oozes slowly, forming new rocks before our eyes, while deep in the oceans silt and the remains of animals filter to the depths, fusing into records of former worlds. The earth is not done and finished; this is all still happening today, this moment. To look around is to look deep into the past. Here in this truth, that the geological forces that shaped the earth's surface can be seen today, lies the key to the past: Humboldt's follower Lyell used this insight to shock the world into seeing deep time. Chemistry, by analyzing the earth's strong structures, will continue to reveal "the vast laboratory of nature" where subterranean forces form and transform the rocks under our feet. As Thoreau wrote, after peering for himself into "the laboratory of the Artist who made the world and me": "The earth is not a mere fragment of dead history, stratum upon stratum like the leaves of a book, to be studied by geologists and antiquaries chiefly, but living poetry like the leaves of a tree."[49]

If the telescope propels us outward into space-time, so geology propels us into earth-time. For the rocks deposited so slowly embed within themselves a clock which measures the true time depth of the earth, emancipating us "from the sway of Semitic doctrines." Freed from biblical literalism, paleontology imparts "grace and diversity" to the study of the earth, allowing us to "ascend the stream of time" as we descend deeper and deeper into the layers of rocks, until we behold past worlds. Humboldt marvels over Mary Anning's discovery of coprolites, the fossilized excrement of fish which reveal what they ate, throwing open a window onto a whole lost ecology; and at the "ink bag of sepia" from an ancient cuttlefish, "so wonderfully preserved, that the material, which myriads of years ago might have served the animal to conceal itself from its enemies, still yields the color with which its image may be drawn." Just so does nature inscribe its history. Humboldt pushes us back and back to the very dawn of life—then stops in bewilderment. Science can see no farther, nor solve the problem of ultimate origins, of life or of humans. Had he lived just one more year, Darwin would have allowed him to deepen the sight of science even this far.[50]

Life, Humboldt exulted, was everywhere, on the rocky crests of the Andes, in the ice of the poles, thronging in the clearest ocean water, teeming in the "ethereal regions of the air." Life, dynamism, change, was the condition of the universe. The stars that look so fixed wander, dissolve, flare, and migrate like buffalo herds across the firmament, and the plants that repose so restfully reveal, under the microscope, "the most varied currents, either rotating, as-

cending and descending, ramifying, and ever changing their direction"; glob-
ules gyrate, self-unrolling filaments spiral and twist. The very molecules quiver
against each other in a visible Brownian motion. Even the solid-seeming continents must heave and sink and reconfigure and drift, for we can see that
South America and Africa have slid apart. Climate too is on the move: ancient
tropical plants show us that the climate has cooled, for causes uncertain, and,
who knows? may warm again. Upheaving mountain chains divide continents
and excite human activity. All is "instability." Earthquakes agitate, lands thrust
upward or sink away, volcanoes erupt out of quiet plains. We live on an "un-
quiet" planet. There is stability nowhere, for all nature surges and billows, not
a place or a state but a process, "the creative force of the universe."[51]

Amidst all this turbulence and ferment stand the human beings who have
arisen so inscrutably to watch and contemplate. Here on this threshold of the
physical universe a new world emerges, ruled by other and more "mysterious"
laws. This is the world of "the human species in all its varied conformation, its
creative intellectual power," its multifarious languages—and here Humboldt
ends the first volume, "at the point where the sphere of intellect begins, and a
new world of mind is opened to our view. It marks the limit, but does not pass
it."[52] Science may stop at this limit, but not Humboldt. Having soared into
deepest space and tendered us safely home again, his task is only half complete; from the sphere of nature, he takes us into the sphere of the mind.

If there is one lesson to the first volume of *Cosmos*, it is, as Carl Sagan would
tell us over a hundred years later, that we are "star-stuff." In mind's eye
Humboldt saw earth as Sagan's generation learned to see it: a blue globe alive,
alone, an astonishment in the black abyss of space. To see that globe is to see
ourselves, reflected back into the abyss, and know that somehow on that blue
planet star-stuff concentered itself still further, condensing until it went su-
pernova and learned how to look back, to see not just the universe but itself,
seeing. What Emerson called "the half-sight of science" sees only the physical
world, but Humboldt wanted science to open both eyes, to see itself seeing.
This was his genius, and the bold goal of the second volume of *Cosmos*, which
is, as Gould recognizes, "an astounding *tour de force* that reads with as much
beauty and relevance today" as it did in the 1850s. Perhaps, one ventures to
say, even more, for we have traveled another 150 years down that road of half
sight. Except that now, as we finally realize, we are breaking our blue planet,
rending apart the very harmonies that gave Humboldt hope in a dark time.
Without us there will always, Humboldt knew, be a universe. Birds will sing,

insects will hum, the planets will cycle and the clouds will go on, nevertheless, in their direction. But once we have broken our world, there may never again be a Cosmos.[53]

Humboldt begins his second volume by discussing the three principle ways humans express their "pure love of nature": literature, art, and gardening. Each activity generates or "incites" more love, in the creator and in the reader or viewer, igniting the Humboldtian spiral of pleasure and knowledge, each yielding to the other in a widening circle of understanding. As the circle widens it draws in more and more people, for although science may be the province of a few intellectuals, everyone in every part of the globe can rejoice in natural beauty. For as Humboldt insisted, Edmund Burke got it exactly wrong when he said that "it is our ignorance of natural things that causes all our admiration, and chiefly excites our passions." On the contrary, Humboldt replies, knowledge, not ignorance, gives pleasure, and popular pleasures can lead to the cultivation of knowledge, to culture, which is not trivial or extraneous to science but its deepest motivator. Nor is science opposed to literature, at least not in periods "characterized by general mental cultivation, [when] the severer forms of science and the more delicate emanations of fancy have reciprocally striven to infuse their spirit into one another." Humboldt would have read C. P. Snow's "Two Cultures" as symptomatic of a society in deep trouble. Nor, finally, is this spiral of pleasure and knowledge arrested at the national border. As the bulk of this volume details, through trade and travel and, yes, conquest, nations since the beginning of recorded history have met, mingled, and enriched each other with new ideas, new words, new plants and animals; the entire globe is a contact zone. This process of mixing and exchange is the engine of Cosmos, how it gets made, for this, too, is a history, symmetrical with volume 1. In the course of these pages Humboldt moves from space-time, to earth-time, to historical time, to tell the story of the birth of Cosmos and its first steps.[54]

For one cannot understand nature without tracing "its image, reflected in the mind of man." Such images begin with childhood impressions, as in his own youthful reading of Georg Forster, his viewing of the exotic landscapes painted by William Hodges, and the sight of "a colossal dragon tree" in Berlin's botanical garden. These were the three sources of "my early and fixed desire to visit the land of the tropics," which determined the direction of his career. Representations of nature thus have tremendous power, and to show this, Humboldt gives a sweeping survey of nature in literature, from Greeks to Romans to Christians; Indian, Persian, Hebrew and Arabian; modern European, from Dante to Columbus to Shakespeare and Milton, Rousseau, Buffon, and Saint Pierre, Forster, Darwin on the *Beagle*, and Humboldt's beloved friend Goethe. Only quite recently has there been a name for this peculiar enterprise: ecocriticism. Humboldt endeavors, in his words, to show the influence of "the aspect of animate and inanimate nature at different periods of time, on the thoughts and mode of feeling of different races," extracting "from the history of literature the characteristic expressions of the love of nature." Or in the words of Cheryll Glotfelty, one of its founders, "Ecocriticism is the study of the relationship between literature and the physical environment. . . . ecocriticism takes an earth-centered approach to literary studies." Which is, of course, precisely what Humboldt is doing, in this, ecocriticism's foundational work. Humboldt shows considerable ability as a literary ecocritic, though he occasionally calls in the aid of friends: Ludwig Tieck on Shakespeare, the Brothers Grimm, Goethe, Schiller, Schlegel. As Clarence Glacken comments in his own classic of ecocriticism, *Traces on the Rhodian Shore*, Humboldt's "path-breaking historical chapters" inspired a legion of imitators, beginning a tradition of histories of the feeling for nature that extended well into the twentieth century.[55]

As always, Humboldt's project was comparative. How did various societies relate to the natural world? Did they see nature as central, or merely as background? How close or remote were they from "wild" nature? How original or how artificial were their constructions of nature? His history builds from his central thesis that "unconsciously to himself, the external aspect of the surrounding scenery impresses itself on the soul of man," interweaving itself "with the deep original tendencies and the free natural disposition of his mental powers." Some individuals, and some societies, are more "susceptible" to these influences than others, and Humboldt is particularly interested in the factors that assist or hinder nature's ability to "impress" the human soul. Thus, while the Greeks show an acute sensual awareness of nature, they are so wholly taken up with the drama of human life that they throw nature into the background. The Romans tended to imitate the Greeks, developing an artificial lexicon that, together with their practical, administrative interests, limited their response to nature; Humboldt is aghast at how generations of Roman travelers could have crossed the Alps without remarking their "romantic beauty"—"all these travellers think only of complaining of the wretchedness of the roads." The peoples of India show a deep sensitivity to nature, perhaps because their thoughts were awakened when they migrated from the cooler north to the lush tropics, where they found an unknown and luxuriant vegetation. Humboldt cites his brother Wilhelm, together with other Sanskrit scholars, on "the veneration and praise of nature" in the Vedas, and the sharply drawn portraits of particular places in Indian heroic poetry. By contrast, the Persians drew their images only from gardens, and are thus stiff and artificial, relying on clever wordplay rather than original experience. The Hebrews and, later, the Christians greatly advanced nature poetry because of their monotheism, which "embraces the universe in its unity, comprising both terrestrial life and the luminous realms of space," dwelling less on "individual beauty" than on the great masses, nature subjected to a higher power.[56]

In Humboldt's view, the keys to good nature writing, then, are a society's physical closeness with nature, preferably wild, and best of all tropical; the association of nature with the divine, as in India and the Hebrews, which opens the mind to see larger forces at work in individual particulars; and finally, the enlargement of views by migration or travel. So powerful is the effect of travel that it can open the eyes even of the uncultivated. The recently discovered journal of Columbus, "a wholly unlettered seaman," shows his depth of feeling for nature, for "the earth and the new heaven opened to his eyes." This observation leads Humboldt to restate the most fundamental axiom of his aesthetics: "Individuality of observation can alone lead to a truthful representation of nature." Humboldt's standard is, Have you been there, have you experienced it for yourself, formed your own original impressions? Buffon disappoints because he has not: "We instinctively feel that he could never have left Central Europe," and he lacks any connection between representations of nature and awakened feelings. By contrast, Bernardin de Saint Pierre, whose novel *Paul and Virginia* Humboldt and Bonpland read to each other on the Orinoco, throbs with natural life: "We felt ourselves penetrated by the marvelous truth with which tropical nature is described, with all its peculiarity of character." Though to Humboldt travel writing had lost "unity of composition" in recent years as the focus turned outward to observations of nature (he was criticizing himself here), it had compensated for this loss by widening the reader's field of view, of ideas and feelings. Humboldt postulates that this ability to project oneself through literature to distant regions, to see new worlds as their inhabitants see them, is uniquely modern, and he predicts a great new age of nature writing just around the corner as ease of travel by sea and land "renders the whole earth more accessible, and facilitates the comparison of the most widely-separated parts." Literature was going global, linking up the remotest corners into the great community of Cosmos, imbuing readers everywhere with "the unfettered freedom of feeling" achieved when imaginative power is joined with sensitive observation.[57]

Humboldt found all this verified by the long history of landscape painting, an "inspired branch of the poetry of nature," which, like nature writing, both teaches and inspires "a free communion with nature" and was about to come into full flower. Before Cook's second voyage, few artists paid attention to scenery beyond Europe, but the atmospheric and exotic paintings of Cook's expedition artist William Hodges—on which Humboldt had dreamed as a child—induced a whole generation of artists to move out across the oceans. Humboldt points to Karl Bodmer's paintings of North American Indians as masterworks in this tradition, but for the most part, his view is prospective: all across the tropics,

> an inexhaustible treasure remains still unopened by the landscape painter. . . . Are we not justified in hoping that landscape painting will flourish with a new and hitherto unknown brilliancy when artists of merit shall more frequently pass the narrow limits of the Mediterranean, and when they shall be enabled, far in the interior of continents, in the humid mountain valleys of the tropical world, to seize, with the genuine freshness of a pure and youthful spirit, on the true image of the varied forms of nature?

The locus of this new vision would be the tropics because nature there is most fully developed, without restraint, in all possible diversity. Humboldt predicts that once Spanish and Portuguese colonies win independence and cultivation grows in India and Africa, "a new impetus and a high tone of feeling" unique to those who call the tropics home will enrich both the sciences and the arts. Only imagine what the new nations of the Andes might produce: "What may we not, therefore, expect from a picturesque study of nature, if, after the settlement of social discord and the establishment of free institutions a feeling of art shall at length be awakened in those elevated regions?" Although the locus of creative energy would be centered in the tropics, new art would arise wherever the human imagination confronted nature directly, "independent of place," for every region has its own attractions and unique character. New technologies would help too: Humboldt looked forward to the day when photography would bring truthful representations of nature to the masses, and when museums and the "magic circle" of life-size panoramas would open the diversity of the earth's regions to the public.[58]

Gardens were also important—Humboldt never forgot the great Dragon Tree and fan palms he saw in Berlin as a child, and of course Tegel itself, where he grew up, was a royal botanical garden where New World plants were acclimatized for dispersal to parks all across Europe. Yet he found them problematic. On the one hand, the hothouse can create the powerful illusion that one is transported to the tropics, exciting "a stronger effect on the imagination" than even "the most perfect painting." However, the illusion is disturbed by the fact that these plants are not wild but artificially cultivated. "Perfect development and freedom are inseparably connected with nature," and in some ways dried plants collected from the wild are better than tamed living plants confined to a European hothouse: "Cultivation blots out some of the original characters of nature, and checks the free development of the several parts of the exotic organization." But what could one do? Civilized man had removed himself from wild nature, and there was no going back. However artificial, at least gardens compensate man "for the loss of those charms of which he is deprived by his removal from a free communion with nature, his proper and most delightful place of abode." Thus the cultivation of plants has its place, too, with landscape painting and "the inspired power of language." All three do the noble work of substituting for real, but difficult, journeys in the wild a lesser but still vital sense of the love for nature.[59] One can hear him sighing after the urban middle classes, confined on weekdays to pavement but bringing the children on weekend outings to London's Kew Gardens, or New York's Central Park or Museum of Natural History, the Washington Zoo or the Smithsonian Museum, or the National Gallery of Art, where they can stand in awe before the paintings of Church and Bierstadt and Moran, or buying Sierra Club books with the photographs of Ansel Adams and Eliot Porter, or making the pilgrimage to Walden Pond, a copy of Thoreau in their back pocket. He had faith—touching, if misplaced—that nature would become uplifting middle-class entertainment that would lead, not to commodification and commercial desecration, but to the shared communion of Cosmos.

After this excursion into ecocriticism, the volume shifts to intellectual history. It is all very well to examine century upon century of images of nature, but Humboldt's deeper question remains unanswered: Why and how have these images come into being? If, as he claims, they have grown and spread in response to the growth of scientific knowledge, then one must look to "the history of our world of thought" as it has learned to embrace the sensual reality of nature and the inner forces that set it in motion. This history has three intertwined elements: scientific knowledge, exploration of the world, and the invention of instruments of vision, "new organs" such as the compass and the telescope. Humboldt adds the all-important role of language as a means of communication across great distances, and of individual languages themselves, which when compared "as objects of the natural history of the mind" become "a rich source of historical knowledge." What a delightful overreach! Humboldt imagines that a single narrative can combine the origin and development of scientific knowledge through all recorded history, across all Eurasia, together with a history of scientific instrumentation and exploration both overland and oceanic, taking into full account the growth, diffusion, and interchange of languages across the globe and across the millennia. No wonder he looked back on this volume disappointed. Yet he managed to pull off an extraordinarily broad and variegated narrative, held together by one theme: "the development of the idea of the Cosmos" as the ultimate human achievement, a collective labor that spanned all human history, gathering speed and power over the centuries until it was poised on the verge of reaching critical mass, of reorganizing into a new form that Humboldt could sense but not articulate.[60]

Where to begin this intellectual excursion? The ecocritical section was relatively easy to formulate, since the written, canonical record sat on shelves or hung on museum walls, waiting to be read and analyzed. But there was no canonical record for this intellectual history. It was diffuse, global in scope, moving and shifting across time, much of it lost to view. And there was no one source, no predetermined starting point. So Humboldt sticks with his principles: if the teller is part of the tale, then the teller must make that fact explicit. Foregrounding his own identity as a Western European, Humboldt explains how this identity governs his starting point: "We will select for this purpose that sea basin around which have dwelt those nations whose knowledge has formed the basis of our western civilization." There is, he acknowledges, a problem with this. It implies a single point of origin, which is untrue. There is "no *primitive race*, no one primitive seat of civilization." Instead "we see many luminous points, or centers of civilization, simultaneously blending their rays": Egypt, Babylon, Nineveh, Kashmir, Iran, China, just for starters, but even these are like "the largest of the sparkling stars of the firmament." We know their brightness, but not their relative distances. Stars less bright might actually be closer, but we do not see them so well, and so we don't enroll them into our narrative.[61]

Turning, then, to the waters of the Mediterranean, Humboldt reminds us (in an approach reminiscent of the French Annales school led by Fernand Braudel) that this ocean basin is composed of confluences from many streams, but again, "amid the diversity of these streams we are unable to trace one primitive source." Here he takes up his governing metaphor: streams and currents, rivers and tributaries, images that have structured his narratives ever since he first launched into the great currents of the Atlantic Ocean. Knowledge runs in rivulets, collects in pools, connects through channels. Streams divide, and merge, on their way to the sea. Some, like his own Humboldt River, never make it, sinking into underground aquifers. We are again threading the maze of the Orinoco-Amazon river systems, where rivers can flow both ways through channels that connect watersheds. Humboldt seeks to understand not races and epochs so much as "the different channels by which an interchange of ideas has been effected." In such a world, the prominence of the Mediterranean, and the Europe it watered and fed into world dominance, is not Providential fate, but geographic accident. Into this basin funneled the labors of three continents. Credit, then, where credit is due: "True cosmical views" are not "a work of a single people, but the fruits yielded by reciprocal communication, and by a great, if not general, intercourse between different nations."[62]

So he starts with the Mediterranean, where, as Plato said, we have settled "like ants or frogs around a swamp" (a nicely deflationary metaphor), where a broken volcanic coastline riddled with harbors and islands encouraged travel by sea and therefore the earliest attempts to relate land, ocean, and the earth's interior fires. Thus did formation of the planet itself influence the course of human events. Egyptians, brilliant as they were, were confined and isolated, while the less artistic Phoenicians had a powerful influence, spreading trade and commerce, and most of all the alphabet, wherever their ships touched land. Later it was the maritime Greeks who scattered abroad the seeds of cultivation. Yet how little we really know, Humboldt reflects; how much "ancient cultivation, even among European nations, has been lost without our being able to discover any trace of its existence." His history of the contemplation of the universe is missing many participants. But of course it includes the Macedonian emperor Alexander the Great, a kind of ancient Napoleon who surrounded himself with naturalists, geometers, historians, philosophers, artists, who by reaching deep into India and Africa brought three continents into communication, extending the sphere of human knowledge across the globe. "Great currents of migration, advancing like ocean currents between masses which are themselves unmoved, become objects of cosmical importance." None more profoundly so than the Arabs, who enlightened a darkened Europe and founded modern physical science, mathematics, and medicine. Progress, Humboldt emphasizes again and again, depends on *im*purity, mixture, interchange, of objects and products, ideas, languages, technologies. A nation that is geographically stranded, or one that closes its borders, will stagnate like an inland sea, its waters never amplifying the Cosmos.[63]

Thus was America, until the Columbian era, when the Cosmos finally rounded the globe. Of course the people Columbus named Indians were there first, one of the great currents of migration out of Asia, bearing with them the seeds of art, science, and religion, seeds that took root and grew into brilliant new civilizations; and much later the Norse had been there too, fishing for generations off Newfoundland. But neither Amerindian nor Norseman had sent out streams that flowed into the expanding Cosmos. Columbus, however, did, setting out with centuries of preparation at his back—encyclopedias, classical learning, advances in Africa and Asia, new instruments like the compass from China—and, most crucially of all, returning, opening the channel, uniting at last the Old World with the New. The men of those days knew the "glory of events" would survive down the ages, and were motivated not just by gold and God but by "the poetry of life," a charm we are losing: "Not only one hemisphere, but almost two thirds of the earth, were then a new and unexplored world, as unseen" as the dark side of the moon. The "free and grand exotic forms of nature" awoke everywhere "the impulse to direct observation," and from then on "a new and more vigorous activity of the mind and feelings, animated by bold aspirations and hopes which can scarcely be frustrated, has gradually penetrated through all grades of civil society," sparking the revolutionary movement for freedom of government and of religion. The discovery of the New World awoke Europe to the Renaissance of art and science.[64]

The emergence of Cosmos was completed when, at this same period, the telescope revealed the depths of space. Now were opened to view the heavens as well as the earth, initiating the epoch of science whose results Humboldt had sketched in his first volume. In this new era, the human mind would develop not under "the incitement of outward circumstances," but "under the influence of its own subjective force." The spiraling, ever-widening cycle of pleasure and knowledge had ignited, and "henceforward the work in the world of thought progresses uninterruptedly, each portion continually contributing its aid to the remainder. None of the earlier germs are stifled." From everywhere knowledge gathers force and direction: Copernicus, Newton, Kepler, Galileo, Franklin, Cuvier. We learn the speed of light, and seek forces deep in the fabric of matter and "in the delicate cells of organic tissues" that still escape our senses.[65]

The succeeding three volumes, which detailed the results of science, never had many readers. This is a shame, for Humboldt's thinking continued to develop in his last years as he wrestled with the implications of his historical, diachronic approach to science. As Margarita Bowen has detailed, some of the last pages he wrote were meditations on the way facts depend on concepts, on the necessary incompletion of knowledge, on what Humboldt called "the great and complex community that we call nature and world," or, as Bowen phrases it, "the concept of the global ecosystem." As she continues, Humboldt "saw this community as 'the combination into a *natural whole* of elements that have the capacity for evolution,' including in it 'the *intellectual life in the cosmos* . . . the world of thought and feeling.'"[66] Humboldt never finished *Cosmos*, though he finally ceased writing it. Thanks to the proliferation of scientific knowledge, the last volumes were obsolete even as they hit the press. Yet his notion that the global ecosystem must include the ecology of knowledge reached beyond his contemporaries to our own times, when so many of us live in virtual worlds while the real one, the one we didn't construct, tips into a cascade of climate change that we ourselves, in our long fit of absentmindedness, caused, that we cannot control, and whose consequences we can only dimly foresee.

But Humboldt's introduction to volume 3 indicates the direction his thoughts were tending: man's sublime desire to strive toward the infinite, to grasp all that is revealed to us, deludes him "into the idea that they had reached the goal, and discovered the principle which could explain all that is variable in the organic world, and all the phenomena revealed to us by sensuous perception." Over the next dozen pages he offers yet another pocket history, this time of the notion of philosophic reductionism from the ancient Greeks through Newton. "It is indeed a brilliant effort," he concludes, but impossible. Nature's variety cannot be reduced to mathematical principle, to the laws of physics. Causal relations are too complex; the phenomena we would explain are too diverse. The cosmographer feels himself "penetrated by a profound consciousness" that the fruits of observation and analysis "are far from having exhausted the number of impelling, producing, and formative forces."[67]

Humboldt was investigating dynamic, even chaotic, systems—weather, currents, biological change and metamorphosis, physical migrations and cultural exchanges—in which interactive properties emerge over time in ways that are irreducibly complex and unpredictable. Humboldt's science reached toward a paradigm shift that would not begin for another 150 years. Boas took from Humboldt the romance of the eachness of things, but what Humboldt romanced in *Cosmos* was the allness of things, *through* their eachness: the earth spinning and cooling in interstellar space, winds coursing across her face, stirring currents in her oceans, subterranean fires thrusting up continents and mountains to catch the clouds and turn them into rain and then the rain into forests, and rivers that became highways for peoples in canoes, then boats, then ships carried on ocean currents before the winds to distant continents where they find, to their astonishment, that everything is the same, but yet different. Into that vibration between difference and sameness Humboldt poured his poet's soul.

The Face of Planet America

Who has plotted the steps toward the discovery of beauty? You have got to be in a different state from common. Your greatest success will be simply to perceive that such things are, and you will have no communication to make to the Royal Society.

HENRY DAVID THOREAU

We have come full circle: standing in the scientific laboratory, holding the kinglet, seeing again that puff of feathers and that brilliant ruby crown, still baffled by its singular surprise but knowing, now, what to do. Boas had his kinglet moment among the Inuit of Baffin Island, and it led him to see that science, when it turned away from history, culture, the humanities, had lost something real. To reclaim it, he turned not away from science but toward Humboldt's Cosmos. There was, there still is, a way to romance the kinglet, to perceive, as Thoreau said, "that such things are."[1] Humboldt taught a generation of American poets and scientists how to cherish their own kinglet moments, and because he did, because he ignited in others his spiral of delight, our circle does not end here after all. The journey is just beginning.

The Apocalypse of Mind: Emerson and Poe

Ralph Waldo Emerson was one of Humboldt's earliest and most steadfast American readers. When Edward Everett arrived at Harvard late in 1819, trailing clouds of glory from Germany, he soon became Emerson's first intellectual hero and "for a time, his personal idol." It must have been Everett, fresh from meeting Humboldt, working to introduce him to America, who urged Emerson to add Humboldt to his reading list; in any event, in June 1821 something or someone prompted Emerson to jot down instructions to read "Humboldt's Work on America." Eighteen months later he mentioned Everett's article on Humboldt in a letter to a friend, and in another two years it was he who was advising a student to read Humboldt's "Travels." From then on into the 1870s Humboldt appears frequently in Emerson's writings, particularly in the early 1830s (when Emerson purchased a copy of William

McGillivray's biography while reading the *Political Essay* and the *Personal Narrative*), then, again, in the mid-1840s after Emerson had read *Cosmos*. Humboldt was a constant and life-long presence in Emerson's imaginative life, associated always with that first exhilarating influx of the new European ideas.[2]

Emerson thought of Humboldt as, above all, a Great Man, "the encyclopedia of science," "the Napoleon of travellers," one of the "colossi" of the age who suggests "that a certain vastness of learning, a *quasi*-omnipresence of the human soul in nature is possible." Over the decades, Emerson associated him with Aristotle and Archimedes; Copernicus and Newton; Columbus, Cook, and Scott; Napoleon and Goethe.[3] Why? There were his virtues, of course—his courage face to face with a jaguar, his tough body, his "unweariable" mind—but more important, Humboldt *sees* more: where other men see only "a dull dumb unprofitable world," Humboldt lights up that world, making the sea and the land "break forth into singing, and all the trees of the field will clap their hands." In the age of "association," he is a model of self-reliance who "gathers an University of science into himself." Was it Humboldt whom Emerson had in mind when he wrote that the American Scholar must be "an university of knowledges"? He seemed the very embodiment of the age itself: it was, Emerson twice declared, "the Age of Humboldt."[4]

In intellectual terms Emerson placed Humboldt second only to Goethe, joining them as "distributing eye[s]" who by ordering the world made it beautiful. In Emerson's own distributing eye, Humboldt was a leader of the exciting new Kantian science: what was the "charm" that every fact in nature—"every ore, every new plant, every new fact touching winds, clouds, ocean currents," the secrets of chemistry—possessed for Humboldt? "What but that much revolving of similar facts in his mind has shown him that always the mind contains in its transparent chambers the means of classifying the most refractory phenomena," of resolving them from chaos into "bright reason" and so giving man "property" in every district of the globe? All things on the earth and in the heavens "become to him merely a kind of mnemonics or inventory of his inward kingdom." Humboldt, that is, moves beyond mere understanding, which can only accumulate facts, to a higher reason, which fuses facts into insight, showing that man can set "the shining maze in order . . . carrying their law in his mind," seeing his idea "realized up yonder in giddy distances and frightful periods of duration." Throughout the 1830s, when Emerson was reading intensively in modern science, Humboldt became his model for the natural scientist who fused the chaos of natural particulars into a meaningful whole directed by a governing intelligence. A new day was dawning: "All this sky full of cobwebs are now forever swept clean away. Another race is born. . . . Humboldt and Herschel have arrived."[5]

The interesting thing here is that Emerson is leading Humboldt down a path Humboldt resisted. True, Humboldt held that the perceptual unity of the mind, our conviction that the world we see is somehow a single ordered whole, is itself a fact in the Cosmos. Yet Humboldt's own Cosmos outruns the human mind, forever opening new horizons as we advance, whereas Emerson's universe mirrors us, reflecting to us, and through us, the mind of the creator. That is, the laws Emerson reads in nature are not Humboldtian patterns and regularities but ideas of God embodied in our own creative and ordering intelligence. In his prospect, Emerson thinks it is we who will outrun the universe: "All this outward universe shall one day disappear, when its whole sense hath been comprehended and engraved forever in the eternal thoughts of the human mind." The universe will then fold up into an apocalypse of the mind. Or as Emily Dickinson put it:

> The brain—is wider than the Sky—
> For—put them side by side—
> The one the other will contain
> With ease—and You—beside—

In short, Emerson's Humboldt encloses the universe in a single mind, a brain wider than the sky. As he wrote after reading *Cosmos*, "I delight . . . in Humboldt, who can represent in their order and symmetry the vast and the minute of the system of nature, so that if this world were lost out of space, he could almost report it from his brain." The claim would have horrified Humboldt, but it chimed perfectly with Emerson's desire to reduce all nature to a single governing principle, a "code" of laws so concise it could be "written on the thumbnail, or the signet of a ring." Emerson never read the third volume of *Cosmos*, where Humboldt repudiated exactly this kind of idealist reductionism.[6]

Humboldt coined a number of new scientific terms—"magnetic storms," "volcanic paps," "geologic horizon"—and these delighted Emerson: in such "new objects & new names one is delighted with the plastic nature of man as much as in picture or sculpture." Emerson saw in Humboldt a poet whose memorable phrases "are the very mnemonics of Science" and whose vocabulary yields a "poetic pleasure." While drafting his breakthrough essay *Nature*, Emerson reminded himself to review his Humboldt notes for the chapter "Language." Although the final text of *Nature* does not mention Humboldt by name, Emerson's theory of language subsumes his understanding of

Humboldt as one of the "wise men" who "pierce this rotten diction and fasten words again to visible things," recovering man's primary creative relationship with God. Thus Emerson applauded Humboldt's introduction of the ancient Greek word *kosmos*, which Emerson himself had used in an 1835 lecture: "The ancient Greeks called the world κόσμος or *Beauty*; a name which in our present artificial state of society sounds fanciful and impertinent." Humboldt had just the year before worried about the impertinence, but in the event, he received Emerson's blessing: our language has "no word to convey the face of the world," wrote Emerson, none that "expresses that power which seems to work for beauty alone. The Greek Kosmos did: and therefore, with great propriety, Humboldt calls his book which recounts the last results of Natural Science, *Kosmos.*" For Emerson, language is a kind of resource, a tool of the mind. In giving us this new word, Humboldt is handing us an important new tool—a new object, a new part of the domain of knowledge—or as Emerson concluded in "Language," "a new weapon in the magazine of power."[7]

Thus it is not surprising that Emerson's most interesting allusion to Humboldt comes in "The Poet," where the man whose "glance is stratification" becomes a model of the "true poet." Not, that is, a mere "music-box," but "a beholder of ideas, and an utterer of the necessary and causal," who stands "out of our limitations, like a Chimborazo under the line, running up from the torrid base through all the climates of the globe, with belts of the herbage of every latitude on its high and mottled sides." Emerson's readers would have instantly understood the allusion: his description reproduced in words Humboldt's famous image of Chimborazo belted into climate zones, and Humboldt's ascent of Chimborazo was one of the most celebrated feats of the era (though as Emerson noted in 1834, the British in the Himalayas had finally broken his thirty-year record). Chimborazo was the icon of Humboldt, of his physical daring and mental comprehensiveness in the sublime volcanoes of the Andes. The poet as mountain: Humboldt towers above us, not disappearing into the mist but arraying before our eyes the planet itself, not scattered and fragmented across illimitable space but gathered and stacked into a logical causal sequence that was not hidden and occult but open and legible to everyone.[8]

Emerson captured Humboldt's openness in other ways as well. Unlike so many touring American intellectuals, he never met Humboldt, but in 1842 he breathlessly quizzed a Harvard student just returned from Berlin for the latest gossip: "Then you have the grand Humboldt the Napoleon of travellers an encyclopedia of science, a man who knows more of nature than any other one in it—What can you tell me of him?" A day or two before, he had meditated in his journal on the satisfactions of traveling to France or Germany, taking ship to "go man-hunting" and find "rapport" with the personal magnetism of the great. "Seeing Humboldt or Wordsworth . . . satisfies the conditions and we can express ourselves happily." He hunted less for heroes than for what he called "representative" men, those who "represented" or expressed something in each of us; and there was indeed a certain democratic, Everyman quality to his Humboldt—why, right there in the country church Emerson could see "the cousins of Napoleon, of Wellington, . . . of Humboldt," who needed only a little air and sunshine and society to call out "the right fire from these slumbering peasants." How call out that fire? How lift "the cover of another hive & see the cells and suck the honey"? Why, "he can study Humboldt until he can talk with Humboldt," and thus get "the clue he wants." We are all, potentially, Humboldts. Interestingly, Caroline Healey Dall associated Emerson with his hero, a man of the world who conducts business "not in the Exchange or in State St but up on the plain of the intellect": "He has also a wonderful knowledge of affairs—though his funds are in no bank. He is the millionaire of Cosmos."[9]

Emerson was one of the first in the United States to get his hands on a copy of *Cosmos*. He wrote excitedly to Caroline Sturgis in September 1845, while passing his copy to her, "Here is Humboldt's first part, if you have daylight enough in the Chapel to read the bad type. But the book is good, and promises better: it has the breadth of march that always distinguished the encyclopedic man."[10] In his journal he rhapsodized, "The wonderful Humboldt, with his extended centre & expanded wings, marches like an army, gathering all things as he goes. How he reaches from science to science, from law to law, tucking away moons & asteroids & solar systems, in the clauses & parenthesis of his encyclopaedical paragraphs!" In Emerson's essay "Wealth," Humboldt in effect makes every reader a millionaire of Cosmos: "The reader of Humboldt's 'Cosmos' follows the marches of a man whose eyes, ears, and mind are armed by all the sciences, arts, and implements which mankind have anywhere accumulated, and who is using these to add to the stock." For Emerson, *Cosmos* capped Humboldt's role as a scientific revolutionary, one of those who studied nature "under the light of ideas," showing that "modern science, with all its tongues, teaches unity."[11]

Emerson capped his own views on Humboldt in an address to the 1869 Boston centennial, organized by his friend Louis Agassiz. At first he begged off speaking publicly, but he gave in and delivered a charming memorial, the full text of which has been lost, though his journal preserves a draft of his remarks. Humboldt now appears as "one of those wonders of the world," like Aristotle and Newton, who appear now and then "as if to show us the possibilities of the Genus Homo, . . . whose eyes are natural telescopes & microscopes &

whose faculties are so symmetrically joined that they have perpetual presence of mind, & can read nature by bringing instantly their insight & their momentary observation together." The faculties of others "are a committee that slowly, one at a time, give their attention & opinion;—but his, all united by an electric chain,—so that a whole French Academy travelled on his shoes." Put him "on any sea or shore" and his instant recollection "of the past history of every other sea & shore illuminated this." "You could not lose him" concluded Emerson; "you could not detain him; you could not disappoint him." Place this "man of the world" anywhere and he would bring to that place his "armed eye," a "Lynceus who could see through the earth, & through the ocean." Emerson's last image looked back over thirty years to his essay "The Poet," in which he wrote, "As the eyes of Lynceus were said to see through the earth, so the poet turns the world to glass, and shows us all things in their right series and procession." The more Emerson read and thought on Humboldt, the more Humboldt became the representative man, not of science, but of the poetry of nature, the man who showed us how to look the universe into order and beauty—into Cosmos.[12]

The American writer whose appropriation of Humboldt most closely resembled Emerson's is, perhaps surprisingly, Edgar Allan Poe. Though Poe loved to skewer the pretensions of the Transcendentalists, those "frogpondians," scholars have noted his genuine respect for Transcendentalist philosophy. This may be most evident in Poe's last major work, *Eureka*. Harold Beaver observes that both Emerson and Poe "were cryptographers studiously deciphering the workings of nature" in a post-Newtonian universe in the throes of change. And both turned to Humboldt in their search for what Emerson in *Nature* called "an original relation" to that universe. Poe's *Eureka*, the culmination of his lifelong fascination with science, opened with the words, "With Very Profound Respect, This Work is Dedicated to Alexander von Humboldt." In its general sweep as well as in specific details, *Eureka* is a response to, and a retelling of, Humboldt's *Cosmos*.[13]

Eureka began as a two-hour lecture entitled "On the Cosmogony of the Universe," which Poe delivered to a "scant audience" in New York on 3 February 1848. As one of the few who attended recalled, "He appeared inspired. . . . his eyes seemed to glow like those of his own raven." Reviews were respectful: "The work has all the completeness and oneness of plot required in a poem, with all the detail and accuracy required in a scientific lecture," reported one newspaper; another praised the lecture as "a nobler effort than any other Mr Poe has yet given to the world." Poe wanted to print fifty thousand copies (enough to rival *Cosmos*), but was given only fourteen dollars and a print run of five hundred, which sold "very slowly." He had hoped that the lecture and book together would successfully launch his new literary journal, *Stylus*, but died in Baltimore while rallying Southern subscribers. *Eureka* never found a large audience. At least one contemporary reviewer suspected a hoax, although others in the twentieth century found in Poe's science an anticipation of Einstein, Planck, and Eddington—a reasonable claim given that, while Poe's intent is quite different, some of his sources, notably Michael Faraday, do indeed stand at the head of modern physics.[14]

Poe himself disavowed any pretensions to science. He called *Eureka* "a prose poem," "an Art-Product alone:—let us say a Romance," a "Book of Truths" offered to the dreamers "for the Beauty that abounds in its Truth." Poe's truth is this: the universe is "but the most sublime of poems," authored by God, who in authoring us as part of the universe made man its true creator. Humboldt, of course, says nothing quite like this, and it has been suggested that Poe did not actually read *Cosmos*, but only looked over some of the early reviews. Yet Poe evidently knew German well enough to have gleaned what he needed from the original, even well enough to have translated by himself a lengthy passage from Humboldt.[15] In any case, the opening paragraphs of *Eureka* lead up to an explicit acknowledgment of Humboldt, on whose work Poe repeatedly hangs his own meditations. The structure of *Eureka* mirrors that of *Cosmos*, but unlike Humboldt, Poe does not split the subjective, human element into a separate volume. Instead he wraps it back into his sketch of the physical universe—in essence, folding and integrating Humboldt's stereoscopic twin pillars, universe and mind, into a single unified framework.

The foundation of that framework—and the scope of Poe's ambition— lies in his opening proposition that "*In the Original Unity of the First Thing lies the Secondary Cause of All Things, with the Germ of their Inevitable Annihilation.*" To "illustrate" this proposition Poe proposes to take a Humboldtian "survey of the Universe," with one key difference. The "nearest approach" to what Poe has in mind is Humboldt's *Cosmos*, but Poe complains that Humboldt "presents the subject, however, *not* in its individuality but in its generality." Humboldt's theme is "the law of *each* portion" of the universe, as it is "related to the laws of *every other* portion"—not a bad description, actually. The problem is that Humboldt's approach necessarily involves him in such a "multiplicity" of points, so much detail, that it precludes "all *individuality* of impression." That "individuality" is what Poe is after—meaning, an impression of the universe as one, not many. Both Poe and Humboldt want to take the aesthetic view of the whole, and like Humboldt, Poe too projects himself onto the top of a volcano to take in that view. The scene is vast; it is diverse; but (and here is where their differences emerge) it cannot be comprehended in "its *oneness*" except by "a mental gyration on the heel" that blurs it all together. No one thinks to do this (certainly not Humboldt), but Poe offers a corrective: "We need so rapid a revolution of all things about the central point of sight that, while the minutiae vanish altogether, even the more conspicuous objects become blended into one." What would vanish is (again echoing Humboldt) "all exclusively terrestrial matters"—that is, everything that concerns man. "The Earth would be considered in its planetary relations alone. A man, in this view, becomes Mankind; Mankind a member of the cosmical family of intelligences." Poe replays Humboldt, but reverses the polarity: instead of starting with the multitude of individualities and weaving them together into a tapestry of patterns, Poe starts with the One, the originating Idea, a God-vision so elevated that the merely sublunary world in which we all live vanishes into the mist.[16]

Again like Humboldt, Poe begins with an outline of scientific method intended to justify his own alternative approach (though Poe's is satirical), then offers two possible directions of travel: from earth outward to deep space, or from deep space back to earth. Like Humboldt, Poe chooses the latter, with the caveat that the universe is *not* infinite. "Infinity" is not a reality but "the *thought of a thought*," and the truth is that the universe is self-enclosed, a sphere whose center is everywhere and circumference, nowhere. There is rather an infinity of beginnings, and the beginning is "the *Godhead*," who out of his volition made from "Spirit," or "Nihility," matter in its primitive simplicity and Oneness. God's volition causes matter to radiate outward, diffusing across space according to the law of repulsion, but, as Newton showed, every particle of matter is always attracted to every other, so the expansion is limited. The instant God withdraws his volition, particles of matter will collapse together, seeking their primal Oneness, and "the majestic remnants of the tribe of Stars flash, at length, into a common embrace." The universe will annihilate itself, from the Big Bang to what cosmologists like Stephen Hawking playfully call "the Big Crunch."[17]

Poe knows his theory is an intuitive leap, and he can't prove a bit of it—although he offers enough evidence from physics, astronomy, and cosmological theory to show he is surprisingly well-read. There are echoes, for example, of Faraday's theory of matter as a field of force, in which individual atoms were, in Emerson's phrase, interpenetrating "spherules" of force with no particulate existence at all. (Faraday couldn't prove his idea either, but just before his death James Clerk Maxwell did, and Einstein built his theory of relativity on them both.) But Humboldt would have argued that Poe looked not forward to Faraday and Einstein but backward, to the "scholastic rationalism" of his contemporaries whose "purely ideal science of nature" was corrupting the youth of Germany. As Humboldt wrote, more or less exactly at the same moment as Poe, the popular impulse "to reduce all the phenomena of the universe to one principle of explanation" based on "opposite polarities and the contrasts presented by + and −" dated back to Aristotle. Humboldt thought that building a universe on the forces of attraction and repulsion solved nothing, only reproducing "the same facts in a disguised form."[18]

Had Poe lived to read Humboldt's criticism, he would have found it irrelevant. After all, he claimed to be writing poetry, not science, and to be supplementing Humboldt's "eachness" with his own vision of "allness," the universe in its "individuality" or oneness. In his view, the balanced opposition of "Attraction" and "Repulsion" did not just form matter, and therefore the very matrix and dynamic history of the physical universe; their polar interaction restored mind as the universe's formative, creative force. "The former is the body; the latter the soul: the one is the material; the other the spiritual, principle of the Universe. *No other principles exist.*" The two principles, material and spiritual, "accompany each other, in the strictest fellowship, forever. Thus *The Body and the Soul walk hand in hand.*" Having established this, Poe, like Humboldt, offers an exhilarating roundup of the evolution of the solar system according to Laplace's nebular theory. In Poe's version, spinning and cooling suns shed planets like snakeskins, and pulses of solar energy fire the ascent of life to man himself and—who knows?—perhaps a race superior to man. The forces at work in the solar system can hardly be comprehended—imagine, says Poe, the force required to move the planet Jupiter, which could contain over a thousand Earths. "Yet this stupendous body is actually flying around the Sun at the rate of 29,000 miles an hour—that is to say, with a velocity 40 times greater than that of a cannonball! The thought of such a phaenomenon cannot well be said to *startle* the mind:—it palsies and appals it." It is in this section that Poe translates a passage from Humboldt on the astonishing fact that we find many groups of stars "*moving in opposite directions,*" which means they cannot be revolving around some central body.[19]

Poe professes that his vision does not need scientific proof, for "it is by far too beautiful, indeed, *not* to possess Truth as its essentiality." If beauty becomes the standard of proof, what is Poe's standard of beauty? An organic wholeness in which parts and whole are reciprocally creating each other in "an absolute *reciprocity of adaptation.*" Poe sees this as the mark of the divine: in human constructions "we see no reciprocity. The effect does not re-act upon the cause." However, "in Divine constructions the object is either design or object as we choose to regard it." We can take cause for effect, or vice versa; "we can never absolutely decide which is which." The more a work of art is organically interwoven, the more it approaches the Divine: in a work of fiction, the incidents of the plot should be so arranged "that we shall not be able to determine, of any one of them, whether it depends from any other or upholds it." While human plots will always be imperfect, "The plots of God are perfect. The Universe is a plot of God." The universe is "the most sublime of poems," authored by God, as is evident in its perfect "symmetry and consistency"—aesthetic standards woven into the fabric of the universe guaranteeing that the beautiful is the true. Poetry, then, is our best guide to truth.[20]

And the conditions of Poe's universe perfectly satisfy those for poetry. Its end is contained in its beginning, "out of the bosom of the thesis," creation leading inevitably to annihilation, and annihilation to a new creation in a perpetual cycling—universes without end. Thought realizes itself in matter, matter in returning to unity sinks into "Nothingness," "into that Material Nihility from which alone we can conceive it to have been evoked—to have been *created* by the Volition of God." Out of the old, a new is born, "swelling into existence, and then subsiding into nothingness, at every throb of the Heart Divine."[21]

But Poe is still not finished. That "Heart" is not out of us, it *is* us. "And now—this Heart Divine—what is it? *It is our own.*" Poe's tell-tale heart of the universe beats out the final truth, that God is in each of us, the star-stuff, matter and spirit, of which we are made, each of us—human, animate, inanimate—"infinite individualizations of Himself." All around are diverse myriads of "individual Intelligences" that will someday, like the view spinning on a mountaintop, "become blended—when the bright stars become blended—into One." "Bear in mind," Poe concludes, "that all is Life—Life—Life within Life—the less within the greater, and all within the *Spirit Divine*."[22] *Eureka* is Humboldt passed through the alembic of Poe: imagine him with raven-bright eyes, reading these words, an Ancient Mariner come back from his cosmic voyage to seize us by the arm and tell us the wonders he has witnessed. Poe responded to Humboldt with the intensity of a conversion experience, but he was dead within a few months. Had he lived longer, what more might Poe have done with his new vision?

The Face of Nature: Thoreau, Church, and Whitman

Emerson and Poe both seamlessly adopt Humboldt into their own systems of thought by folding physical, external nature back into mind—first into the mind of God, as the Creator of nature, then by extension into the mind of man, which is both part of the Creation and, as part and particle of God, its creator as well. For both, the crucial third term joining man and nature, soul and body, was God the Creator of each and all; and for both, the primary reality was not the physical face of nature but something deeper and higher, transcendent, hidden by the mask. Both seek to strike through the mask, to reach what Emerson called an "apocalypse of mind." By contrast, Humboldt did not use God to join humans and nature. Neither materialist nor God-centered, he did not doubt there was mind in nature—not, however, a single transcendent God-mind, but myriads of human and animal minds forged through natural processes across time-scales ranging from evolutionary to historical to personal. The "myriad individualities" that Poe wished to merge into a single cosmic consciousness, Humboldt wished to thread together like Pacific pearls, each one gleaming. Yet without God, how could the dualism between humans and nature, poetry and science, be overcome? Without God, what was left but a soulless materialism? Many Romantics, like Emerson and Poe, adapted Humboldt's own thinking to restore Humboldt's omission.

One of them was Coleridge. As my allusion to "The Rime of the Ancient Mariner" suggests, Coleridge was yet another who had met Humboldt, read his works, and wrestled with what Dometa Wiegand calls "a difficult phe-nomenological problem," how to relate data about the material world to mind and cognition. Wiegand names this "phenomenological ecology," a scientific perspective "seeking to retain empirical physical data but also con-sidering emotional and aesthetic effects of the world to be *data* about the world and its objects"—what Bowen calls Humboldt's "ecology of knowl-edge." Wiegand argues that the Ancient Mariner learns to recognize this eco-logical relationship between thought and nature, first when he realizes that "the slimy things" that crawl "upon the slimy sea" are alive, even as he is alive; next, when he begins to look at them, to see their different colors and shapes and motions, the way they leave tracks in the water: "This intake of sensory impression as data is the bridge between the material reality of the world and the interior psychological world." This alliance, this empathy, al-lows the Mariner to feel his connectedness to the physical world and through it, the realm of transcendence. Coleridge thus joins us on Humboldt's bridge. But did he take his reasoning from Humboldt? As Wiegand warns, influence cannot be described in such an oversimple way, for it flows "in all directions, interrupted, rethreaded, connected and severed in millions of inconceivable ways with as many unforeseeable consequences. All people are imbricated in a tangle of history." Yet in reevaluating Coleridge, she asserts the importance of remembering that his "*entire* life occurred in what might be aptly named the 'Age of Humboldt.'"[23]

So did the entire life of Henry David Thoreau. Late in *A Week on the Concord and Merrimack Rivers* he asked a profoundly Humboldtian question: "Is not Nature, rightly read, that of which she is commonly taken to be the symbol merely?" Thoreau was fascinated by the reports of explorers, and it was probably in Emerson's library that he first picked up McGillivray's biography of Humboldt. In one of his earliest essays, "A Walk to Wachusett" of 1842, Thoreau looks longingly at the mountains of his Concord horizon and imagines himself "with Humboldt, measur[ing] the more modern Andes and Teneriffe," step by step across the landscape. What follows is his own exploring expedition, albeit only to the top of a local landmark. Already Thoreau is casting himself as the traveler who stays home, seeing the familiar and local with the intensity of the visitor who in his passage comprehends the local in its planetary relations. In his passage to Walden Pond, a journey of barely two miles that took him over two years to complete, Thoreau entertained passersby between rambles in the woods and supply runs home. He also wrote his first book, a travel narrative of the river excursion with his best friend and elder brother John when John was still alive, and the Thoreau we know, the lonely man who walked the face of the planet, was not yet born.[24]

Many things happened to give birth to the Thoreau who became one of America's greatest writers, the one above all most sensitive to nature's rhythms and seasons. One was John's tragic death, for in his grief Thoreau turned to nature, to life forces that promised to sustain him past all human loss. Another was his sojourn at Walden Pond, where he learned to live with nature's days and seasons. A third was his 1846 excursion to the Maine wilderness, where he was shaken to the core by his encounter, on Mt. Ktaadn, with "a force not bound to be kind to man," a "vast, Titanic, inhuman Nature" that sought to drive him out. Thoreau's response was not to wince away from the wild, but to embrace it, to call for "*Contact! Contact!*" rather than transcendence. Thoreau began a search for a way of working with nature that would give him that intimacy of contact, while retaining the deep sense of meaning that Emerson had given him in their decade of friendship. The rush of newly available writings by Humboldt coincided with that search. Starting in 1849, Thoreau read *Cosmos*, then *Ansichten* (in both translations), transcribing passages into his notebooks, then the new translation of *Personal Narrative* and Hermann Klencke's biography, all in the years before he published *Walden*. Along with Humboldt, he read Darwin's narrative of his own South American voyage and the many reports of the Humboldtian exploring expeditions to the American West and beyond to polar seas.[25]

"A man receives only what he is ready to receive," wrote Thoreau late in life; "We hear and apprehend only what we already half know." The writings of Humboldt came to Thoreau just when he was ready to receive them, when they could clarify what he already half knew. The turning point in his life, from the fall of 1849 through the fall of 1850, coincided precisely with his first immersion in Humboldt. As Brad Dean observed, in this year Thoreau "dramatically reoriented" his life: "He adopted a daily routine of morning and evening study and writing separated by an extended afternoon excursion into the countryside around his hometown; earned a livelihood by periodically surveying his neighbor's woodlots; and began intensive, formal studies of natural history, particularly botany, and of aboriginal cultures, especially those of North America." He began dating his daily journal entries, which grew in length to essays often of several pages, and stopped scissoring out pages for use in his other writings, thus ensuring "the scientific integrity of the data he collected during his afternoon excursions." He became a scientific traveler in Concord, adopting the philosophy and working methods of Humboldt and his followers.[26]

It was a rigorous program, and Thoreau wrestled with his models as he differentiated his project from theirs: "But this habit of close observation— In Humboldt—Darwin & others. Is it to be kept up long—this science—Do not tread on the heels of your experience[.] Be impressed without making a minute of it. Poetry puts an interval between the impression & the expression—waits till the seed germinates naturally." Sometimes he worried that this "habit of close observation" was killing off his poetic spirit, and he struggled to compensate, to keep always in view his own higher purposes. After one long day spent boating on the river, he filled his journal with page after page of nature notes, then concluded, tellingly, "Every poet has trembled on the verge of science." Many entries read like this one: "July 25. Dodder prob the 21st ult. Blue-curls. Burdock prob yesterday. . . . Cerasus Virginiana Choke cherry just ripe. . . . Cynoglossum Morisoni Beggar's lice roadside between Sam Barretts mill & the next house E. In flower & fruiting probably 10 days." Not long ago such painstaking scientific detail was seen as a symptom of the decay of one of America's great literary talents into a manic collector of soul-destroying scientific data. But looked at through the lens of Humboldt's phenomenological ecology, one can see that Thoreau is following Humboldt's dictum: "Individuality of observation can alone lead to truthful representation of nature." At the beginning of his career Thoreau had warned, "Let us not underrate the value of a fact; it will one day flower in a truth." For them both, the truth of nature could be grasped only through careful observation of multitudes of details, which far from being dry and pointless, come to life by the light of the whole they compose, their glow visible to the educated imagination. As Thoreau said, "A true account of the actual is the rarest poetry." For years he kept one notebook for "facts" and another for "poetry," but by 1852 he was no longer sure which was which, "for the most interesting and beautiful facts are so much the more poetry and that is their success." Perhaps if his facts were "sufficiently vital & significant—perhaps transmuted more into the substance of the human mind—I should need but one book of poetry to contain them all."[27]

Walden became a volume in that book. In the midst of his new program of reading, walking, and writing, Thoreau took out the manuscript he had set aside after the commercial failure of *A Week*, and over the next three years he doubled its length, filling it with facts that had fused into poetry and turning it from a good book into a great one. Walden Pond, that bundle of incident and observation, looms in our imaginations because each detail, in its density of connections both to lived experience and to abstract systems of thought, illuminates some concrete moment in our own experience. Walden Pond was a pocket Cosmos, "my own sun and moon and stars, and a little world all to myself." Self-contained, self-reliant, yet netted up with the rest of the planet, as Thoreau remarks: "Thus it appears that the sweltering inhabitants of Charleston and New Orleans, of Madras and Bombay and Calcutta, drink at my well." From this pond, "a good port and a good foundation" for his business "with the Celestial Empire," Thoreau will sail "before the mast and on the deck of the world," "taking advantage of the results of all exploring expeditions, using new passages and all improvements in navigation; charts to be studied, the position of reefs and new lights and buoys to be ascertained, and ever, and ever, the logarithmic tables to be corrected" lest by miscalculation the vessel splits upon a rock; "universal science to be kept pace with, studying the lives of all great discoverers and navigators, great adventurers and merchants, from Hanno and the Phoenicians down to our day"—labor enough "as demands a universal knowledge."[28] Thoreau's local would always speak to the cosmic: *Walden*, like *Eureka*, was a response to Humboldt's *Cosmos*.

In his journals one can see Thoreau following the precepts of Humboldtian science: explore, collect, measure, connect. From the port of Walden he explores the universe in his backyard—"I have travelled a good deal in Concord," he joked—and he tells us we can do the same: "Nay, be a Columbus to whole new continents within you, opening new channels, not of trade, but of thought." "Why do precisely these objects which we behold make a world?" he wonders; "why should just these sights & sounds accompany our life? . . . I would fain explore the mysterious relation between myself & these things . . . make a chart of our life—know how its shores trend—that butterflies reappear & when—know why just this circle of creatures completes the world."[29] Like Humboldt, Thoreau answered these questions by collecting and by measuring: he created one of the best botanical collections in the region, and his attic study was filled with bird nests, arrowheads, cones and seeds, hatching turtles, and curious insects. He even became a professional collector for a time, sending fishes, tortoises, snakes, mice, and a live fox from Walden Pond to Louis Agassiz at Harvard. He was especially adept at keeping track of stream depths and lake levels, and as a professional surveyor, he prided himself on the accuracy of his measurements. He took his own advice literally and made a navigational chart of Walden Pond, recording "how its shores trend" and measuring its depth with a precision verified by modern methods.

And everything was about connection. When he had mapped the pond, Thoreau observed "that the line of greatest length intersected the line of greatest depth *exactly* at the point of greatest depth" (an observation he confirmed by measuring a second pond), and he went on to connect these natural facts with human meaning: "Draw lines through the length and breadth of the aggregate of a man's particular daily behaviors . . . and where they intersect will be the height or depth of his character." Of all objects in nature he asked, like Emerson, What is its relation to me? For as he reminded us, "Shall I not have intelligence with the earth? Am I not partly leaves and vegetable mould myself?" "Contact" meant both this intimacy of earthly connection, "the perpetual instilling and drenching of the reality which surrounds us," and the view from on high—again, not in the misty way of Poe but in Humboldtian particularity: one ascends a mountain "to see an infinite variety far & near in their relation to each other thus reduced to a single picture." To suppress the details, Humboldt said, so the masses can be better seen. Like Humboldt, Thoreau spent most of his life not on mountaintops but in valleys, looking up and out, and his conclusion was the same: "The universe is wider than our views of it." No single law can capture all nature, for our laws are confined to the limited universe that we can detect; "the harmony which results from a far greater number of seemingly conflicting, but really concurring, laws, which we have not detected, is still more wonderful."[30]

From this paraphrase of Humboldt, Thoreau moves to a characteristically Thoreauvian image: "The particular laws are as our points of view, as, to the traveller, a mountain outline varies with every step, and it has an infinite number of profiles, though absolutely but one form." The brain is not wider than the sky, after all; we will never contain the mountain, or the sky, or the planet. Thoreau's convictions here allow what he called "the wild" to penetrate even the "garden and cultivated field and crops" of his home in Middlesex County, for even here, in land settled and abused for centuries,

Thoreau found places "wild as a square rod on the moon," spots that "are meteoric, aerolitic." Is not the whole globe an aerolite? Are not the stones of this earth "planetary matter," part of the Cosmos? "Are not the stones in Hodge's wall as good as the aerolite at Mecca?"[31] The lesson of Cosmos is that the earth is, literally, heaven under our feet.

The Cosmos may outrun our views of it, but the central fact for Thoreau remains our viewpoint, the fact of our human-centered subjectivity. For all his deep involvement in science, this conviction kept Thoreau from becoming a scientist in the modern sense, for he disavowed its central premise: "There is no such thing as pure *objective* observation—Your observation—to be interesting i.e. to be significant must be *subjective*[.] The sum of what the writer of whatever class has to report is simply some human experience—whether he be poet or philosopher or man of science." Thoreau wrestled with this critique of science several years later, when his own most scientific work was reaching its maturity: "I think that the man of science makes this mistake, and the mass of mankind along with him: that you should coolly give your chief attention to the phenomenon which excites you as something independent on you, and not as it is related to you." Not the objects in themselves are Thoreau's concern; rather, "the point of interest is somewhere *between* me and them." In the characteristic Humboldtian project, the goal is not an object but a relationship, and the teller is part of the tale. Knowledge or "truth" comes into being or "flowers" precisely in the space *between* subjective and objective, in the relationship created by the act of viewing, an act that composes a perceptual whole.[32]

Humboldt insisted that this reciprocity makes the human mind, its emotions as well as its concepts, a necessary component of the Cosmos, but this was not the direction science was taking in the 1850s—as Thoreau well knew. His involvement in Agassiz's research program garnered Thoreau an invitation to join the AAAS, but instead of being flattered, Thoreau was indignant. How, he fretted, could he make himself understood to them? He would be a "laughing stock of the scientific community . . . in as much as they do not believe in a science which deals with the higher law." When he finally returned the form, he politely declined, giving as his particular specialty "The Manners and Customs of the Indians of the Algonquin Group previous to contact with the civilized man," and adding that the "character" of his science could be inferred by his special attraction to "such books of science as White's Selborne and Humboldt's 'Aspects of Nature.'"[33]

The allusion to Gilbert White seems to align Thoreau with old-fashioned natural history, yet the allusion to Humboldt shows that Thoreau's science was looking forward as well. In the years after *Walden* his detailed analysis of the composition and history of plant communities led him to a theory of forest succession still accepted today, and beyond that, to pioneer concepts that are now called "ecological," although the word would not be coined until after his death. While developing his understanding of the causes and consequences of plant distribution, he stumbled across puzzles that baffled science. For instance, he found toad spawn in shallow pools of rainwater on the top of Mt. Monadnock, far from lowland swamps. How did they get there? Could toads really have hopped up the mountain? "Agassiz might say that they originated on the top," but Thoreau had already crossed swords with Agassiz and won, leading to a certain skepticism about his special creationism. Or take the pontederia in Beck Stow's pool. Since there was no connecting stream, "How did they get there?" Indeed, how did any plant get anywhere?[34]

By 1860, Thoreau had read Darwin's *Origin of Species* and was working out the answer. Thoreau was one of the first Americans to read Darwin's revolutionary book, and as Brad Dean says, "perhaps the only American who understood the subtleties of Darwin's argument and embraced that argument enthusiastically, without expressing a single reservation." On the contrary, Thoreau was energized by the implications of Darwinian evolution: "The development theory implies a greater vital force in nature, because it is more flexible and accommodating, and equivalent to a sort of constant *new* creation." When he read Darwin's call for research on the poorly understood phenomenon of seed dispersal, Thoreau realized his years of hard data and theorizing put him on the cutting edge of the day's most avant-garde science. He exploded into a frenzy of work, and within months he had organized his preliminary results into "The Succession of Forest Trees," his one scientific essay, and was drafting not one but two new book manuscripts. But his latent tuberculosis, perhaps worsened by an autumn spent in the rain and snow collecting data, flared up, and Thoreau died with his new books unfinished. Now that Dean has edited and published them, we can begin to gauge the loss: he was shaping stunning and wholly original works, and his voice would surely have had an impact on the Darwin debates. Humboldt had taught Thoreau how to take full advantage of the Darwinian and ecological revolution. Of all Humboldt's American children, none was more deeply in dialogue with Humboldt, and none did more to bring his philosophy, practice, and passion into the mainstream of American thought and culture.[35]

Some years ago, faced with the constellation of differences between such writers as Emerson, Poe, Coleridge, and Agassiz, on the one hand, and on the other, Humboldt, Darwin, and Thoreau, I proposed that there are not one but two competing narratives in Romanticism. Both share the impulse to read nature "whole," but they approach that whole in radically different ways. The dominant mode, "rational holism," conceived nature "as a divine or transcendent unity fully comprehended only through thought." By contrast, "empirical holism" "was an emergent alternative which stressed that the whole could be understood only by studying the interconnections of its constituent and individual parts." I argued then in depth, as I have above in brief, that "by acting as a Humboldtian naturalist, Thoreau participated in and helped to advance an alternative tradition of romantic science and literature that looked toward ecological approaches to nature and that was suppressed, then forgotten."[36] Yet the questions here are larger: *how* did empirical holism emerge, how effectively did it work to "bridge" humans and nature, and what happened to suppress it? For as I hope to have shown, Humboldt's was hardly a simple legacy. Various elements of his thought were taken up, across six decades, by figures both mainstream and dissident, who deployed his ideas in often contradictory ways.

To see this one need only look at the paintings of Frederic Edwin Church. A student of Thomas Cole, whom Barbara Novak calls "the first fully equipped landscape painter in America," Church inherited from his teacher the belief in the high and strenuous moral calling of landscape painters, "those priests of the natural church." Novak finds in Church "the paradigm of the artist who becomes the public voice of a culture, summarizing its beliefs, embodying its ideas, and confirming its assumptions." That voice was profoundly Humboldtian. Church was, in Novak's works, "Humboldt's most fervent admirer," and *Cosmos* hit Church with the force of revelation. Humboldt's prophecy of "a new and hitherto unknown brilliancy" for landscape painting once artists headed for the tropics sent Church packing to South America, where he sought out and lived in Humboldt's house in Quito. Church followed Humboldt's footsteps across the Andes, sketching studies for the epic landscape paintings that would make his reputation as one of the century's greatest painters. Everyone saw the connection: as a London art critic wrote, Church was "the artistic Humboldt of the new world": "At length, here is the very painter Humboldt so longs for in his writings; the artist who, studying not in our little hot-houses, but in Nature's great hot-house bounded by the tropics, with labour and large-thoughted particularity parallel to his own, should add a new and more magnificent kingdom of Nature to Art, and to our distincter knowledge." Church's paintings were displayed as spectacles. His 1859 masterpiece *Heart of the Andes* was exhibited in a specially lit room planted with palm trees taken from the original site, behind an architectural frame set off by elaborate curtains (fig. 14). Viewers were charged admission (in the first month receipts totaled over $3,000) and given special viewing tubes to study the details of the immense painting. Some brought binocu-

FIGURE 14. Frederic Edwin Church (American, 1826–1900), *Heart of the Andes*, 1859; oil on canvas, 168 × 302.9 cm (66⅛ × 119¼ in.): The Metropolitan Museum of Art, Bequest of Margaret E. Dows, 1909 (09.95). Image © The Metropolitan Museum of Art.

lars. Church sent his painting on to London and thence to Berlin, to show the master himself, but sadly, Humboldt died before it reached him. In June 1859, Church wrote his friend, the travel-writer Bayard Taylor, of his sorrow upon hearing "the sad intelligence of Humboldt's death—I knew him only by his great works and noble character but the news touched me as if I had lost a friend—how much more must be your sorrow who could call him a friend."[37]

Viewers of Church's paintings are immediately struck by their high level of detail. His mountains are studies in geology, his clouds are meteorologically exact, his plants and birds are rendered with the fidelity of the scientific illustrator. Yet these details are governed by the impression of the sublime whole to which they variously contribute, a whole visible only through the details that compose it. Giving spectators special viewing tubes was more than a publicity stunt: many a Church painting invites the viewer's participation in reading or composing miniature views out of its richly interwoven natural facts, then in standing back and admiring how the small composes the large and the large the immense, in an ascending scale (fig. 15). Other Church paintings invite a different kind of participation, triangulating the viewer in a particular relationship with the scene through the use of light. In *The Andes of Ecuador* (fig. 16) the viewer faces the lowering sun, whose light dissolves the features of the landscape in a luminous golden haze often

FIGURE 15. Frederic Edwin Church, *Heart of the Andes*, detail. Image © The Metropolitan Museum of Art.

FIGURE 16. Frederic Edwin Church (American, 1826-1900), *The Andes of Ecuador,* 1855: Oil on canvas, 48″ × 75″: Reynolda House Museum of American Art, Winston-Salem, North Carolina.

compared with Turner. The effect is not an "objective" part of the landscape, but, in Humboldt's word, an "impression" dependent upon the viewer's position in relationship to the sun and the land. Thoreau was fascinated by such effects, particularly the glow of scarlet oak leaves in autumn against a declining sun, "an intense, burning red" that can be seen only while standing at a certain distance. As the sun moves, the color grows, "partly borrowed fire, gathering strength from the sun on its way to your eye. . . . You see a redder tree than exists." In 1855, Thoreau stopped by the Athenaeum gallery in Boston to see this very painting. A few years later he remarked on how the landscape painter expresses beauty through knowledge—knowledge like his own: "How much of beauty—of color, as well as form—on which our eyes daily rest goes unperceived by us! No one but a botanist is likely to distinguish nicely the different shades of green with which the open surface of the earth is clothed,—not even a landscape painter if he does not know the species of sedges and grasses which paint it."[38]

"Their form is their history," said Humboldt of the features of the face of nature. The face of a Church painting is also a certain history, not just the one usually told about his apprenticeship to Cole, but another, Humboldtian one. Humboldt thought of himself as a "geographer," which means literally an "earth-writer." The problem of representation is everywhere central to Humboldt's work, for his style was intensely visual. Just as Humboldt's textual sight line can range from cosmic generalities to local specifics, so do his modes of representation range from highly abstracted maps and data sets, through maps dense with topographical detail, to highly specified topographical illustrations, to paintings rich in both natural detail and spiritual meaning. Having studied art under some of Europe's best teachers, Humboldt used visual representation as both a tool of observation and synthesis—"the pencil is the best of eyes," said his protégé Louis Agassiz, himself a gifted draftsman—and a source of aesthetic pleasure. But he had no desire to become a studio artist. Instead he issued a call for artists to follow his footsteps, and for a generation many did. Humboldt thus mediates or bridges what Edward S. Casey calls "truth-about," scientific factual instruction, and "truth to," an aesthetic presentation with a life of its own. He accomplishes this through his insistence on depth, both spatial depth, as in the vertiginous chute of *Natural Bridges of Icononzo* (see frontispiece), and historical depth. In Humboldt, since every surface evidences its past, every place is intertwined with event. For Humboldt, the visual is always the historical.[39]

One can see this in Humboldt's maps. In his study of the interrelation between cartography and landscape painting, Casey outlines three stages in the development of American survey mapping, beginning with the adoption by Congress in 1785 of Thomas Jefferson's "Ordinance for Ascertaining the Mode of Disposing of Lands in the Western Territory." This ordinance resulted in the notorious "Rectangular Survey," which superimposed on America's western lands a grid of six-mile-square units, "the Promethean posture of the pure Geometer who lays abstract patterns down upon the bare earth." The survey immediately ran into trouble with the irregular windings of the Ohio River, and in 1850 the verticality of California's mountains forced Congress to pass a special act allowing surveyors to depart from rectangularity. In the second stage, "the nongeometric is allowed to invade the geometric—but only in highly selected ways." Landscape features could be recognized, but only if they intersected with survey lines. Irregularities between the lines were to be recorded textually in a surveyor's field notebook, where the "face of the country" could be noted as "level, rolling, broken, hilly, or mountainous." Thus "what was unfit for the survey map, the unruly face of the earth, was confined to the survey book." The third phase of survey mapping incorporated such detailed topographical features "as rightfully belonging to a map and not just to an informal journal." Casey's exemplar is John C. Frémont, who instructed that maps should explicitly represent the actual forms of landmasses. In the work of the Army Corps of Topographical Engineers, Frémont's employer, techniques such as hachuring and contour lining pull us "down into the valley of the shadow of places," resulting in a survey "as polymorphic as the earth itself in its profuse local geography." Such maps are not flat and empty "but convoluted and plenitudinous," not just land maps, "sites in sheer space," but landscapes, even placescapes, "places on the land and of the land."[40]

Frémont, as we've seen, learned his mapping techniques from Nicollet, who had learned them from Humboldt, who in his own imaginary views from space used hachuring lines to pull us into the depths of valleys (despite the conflict they created with legible place-names) and who filled in blank spaces with commentary on place derived from his native informants (fig. 17). Thoreau's survey and resulting map of Walden Pond is in this tradition (fig. 18). Thoreau is most obviously concerned with ascertaining the depth of the pond (his neighbors pretend that it is bottomless) as well as the depth of a person's character. Yet this very act of surveying, as represented in the map, is itself an event in *Walden*; that is, Thoreau's map links up both his narrative as it unfolds in time, and the given and highly specific landscape that holds his narrated events together. The making of the map marks the intersection of the fiction of *Walden* the book, and the history of Walden the pond, representing to Thoreau's readers how landscape intertwines space and time. Space-time operates on earth as well as in interstellar space: there the railroad track, built only the year before Thoreau moved to the pond, cuts across the cove; there the house he built, there the holes he drilled in the ice to sink his measuring line. Even the delicate shake of the hand-drawn outline of the pond's shores recalls the moment the engraver's needle plowed the surface of the plate from which the map was printed.[41]

Humboldt's pictorial engravings step still more closely into his placescapes, holding together the middle ground between the science of earth writing and the art of landscape painting. On the scientific side, they are topographically exact, emphasizing cartographic fact: Humboldt stresses that in sketching "the great scenes of this savage nature" he seeks less "picturesque effect" than "an exact representation of the shapes of the mountains, the vallies by which their sides are furrowed, and the tremendous cascades formed by the fall of their torrents." Such would seem to be surface concerns, recording the "look" or the "face" of nature, but as we have seen, for Humboldt surface always reveals depths. He wants to grasp the inner dynamics of the landscape, what Casey calls its "morphic patterns," how it evolved into being. What's notable here is that Humboldt's nature is not made by divine agency but makes itself. Sediment settles and rocks form, mountains lift and water furrows their flanks, plants group together by soil and habit and climate, volcanoes grow by successive addition, all according to laws that Humboldt

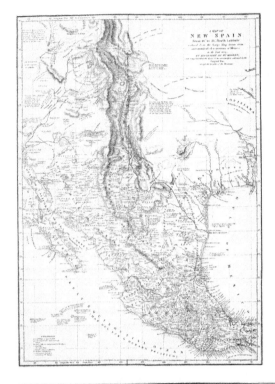

FIGURE 17. (a) Humboldt's map
of Mexico, in *Political Essay on the
Kingdom of New Spain.* Author's
collection; (b) detail.

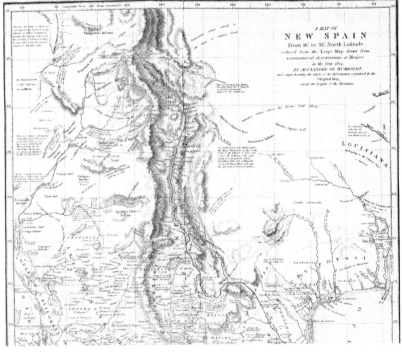

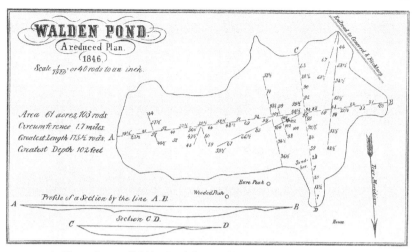

FIGURE 18. Thoreau's map of Walden Pond, from *Walden* (1854). Author's collection.

labors to formulate. So for example, *Chimborazo and Carguairazo* (fig. 19) is animated by an underlying causal sequence, revealed in the different morphic patterns of the two mountains: one still building, still layering to itself that lovely, perfect cone; the other, exploded and collapsed into a jumble of ridges bristling with points, needles, "and broken rocks falling into ruins." Whereas the catastrophe that gave Cotopaxi "the head of the Inca" could be dated only on an indigenous calendar, the catastrophe that shaped Carguairazo can be given a European date: the night of 19 July 1698, when mud slides laid waste the surrounding country and an earthquake swallowed thousands of inhabitants. The palimpsestic nature of Humboldt's engravings, interlayering human history with natural formation, is dynamic, causal, directional. Places are events, to the imagination that can travel into the depths of time; surfaces bespeak their past to whoever has the mind to listen.[42]

Humboldt's engravings, writes Ángela Pérez-Mejía, "greatly influenced the development of Latin American art and became the model for a new way of representing voyages in which science and art were equally privileged." Many of the cartographers who carried Humboldt's scientific methods into the American West were themselves artists, but it was the studio artists inspired by Humboldt who are remembered today. If Humboldt's engravings face on the one side toward cartographic and scientific truth, they face on the other the artistry of a Church. Like Church's paintings, Humboldt's engravings invite judgments of accuracy and one-to-one truth to reality. Yet in both cases such judgments can be misleading. Though they both based their finished images on sketches taken from nature, once in the studio, the resulting completed pictures offer not simply photographic "representations" of an already given reality,

FIGURE 19. *View of Chimborazo and Carguairazo*, plate 16, Alexander von Humboldt, *Vues des Cordillères* (Paris, 1810). Rare Books and Special Collections, University of South Carolina Libraries.

unknowable matter apart from mind, but *presentations* of an invented place, what Whitman called "the image-making faculty, coping with material creation, and rivaling, almost triumphing over it." On the plane of the picture, the observing eye "reimplaces" nature in an aesthetic act that resolves the old Cartesian/Kantian dualism between mind and nature. Humboldt shows us that nature is known by placing oneself in it, first experientially—slipping barefoot in mud, fumbling in a snowstorm, hearing the lugubrious cries of strange birds—then imaginatively, by reimplacing it in mind via the bite of the engraver's needle or the slick and dab of the brush. As Casey says, "In creating and viewing such a work of art, we disport ourselves—and thereby, paradoxically, find our place within it and within the world it presents to us." In the form of play that is art, the viewer is drawn in, invited to collaborate in the world it presents. The river in *Natural Bridges of Icononzo* spills at our feet, luring us up into the dizzying frame; the train of figures disappearing into the vastness of the plain of Tapia invite us to follow them up the snowy sides of Chimborazo; we hear the vast emptiness of the Great Square of Mexico echo with silence; the gaze of the *carguero* meets our eyes, asking us if it is we who will step into his empty chair and be carried into the mountain wilderness. To collaborate with these images is to learn, like Humboldt and Church, to read depth in surface.[43]

Yet there is a problem here. "Landscape is an instrument of cultural power," says W. J. T. Mitchell. It is not a noun but a verb, "a process by which social and subjective identities are formed." As "the 'dreamwork' of imperialism," landscape art discloses "both utopian fantasies of the perfected imperial prospect and fractured images of unresolved ambivalence and unsuppressed resistance." One can see this readily in the leading American landscape artists: Bierstadt and Moran went to the American West to become the painters celebrating Manifest Destiny. As Moran said, the artist had never before had a nobler opportunity to both minister to his love for the sublime and become a coworker with "the pioneer, the man of science, and the soldier, who cleared, surveyed and held this mighty continent and brought it under the mild sway of civilization." Church pursued not the east-west axis of imperial growth but the Humboldtian fantasy "of a pan-American unity from North to South, creating a pantheistic cosmos of the Sublime with the northeastern United States at its heart," in "an artistic analogy to the Monroe Doctrine" of 1823—which Jefferson had first broached in a letter to a sympathetic Humboldt. However beautiful we may judge these paintings to be, they were hardly innocent. During a pivotal period in American history, they worked hard to help Americans dream their millennial dream and forget the brutality that dream required.[44]

This work demanded the reframing of Humboldt's Enlightenment, European ideology into one more appropriate for the United States. First, it had to be translated into Christian theology. Humboldt's revolutionary vision of a planet-wide America was taken up into an American Christian millenarianism that dated back to the Puritans and had received new life during the American Revolution. Founding fathers such as Benjamin Rush, whom Humboldt had met in Philadelphia, were inspired by an evangelical vision of an America separated from a corrupt and dissipated England, triggering a social transformation that would spread across the globe; in the new millennium, governments would derive their powers from the people and be free "from the great disparities of wealth which characterized the old world." After the revolution this evangelical vision merged with a Protestant tradition that transformed "'untrammeled' earth into sacred ground," leading to an American ideology of a pristine and spiritually regenerative wilderness utterly evacuated of all corrupt and sinful human presence. If this meant "removing" Indians who had inhabited the "wilderness" for thousands of years, so be it. The idea of wilderness was conveyed to the American people less by actual wilderness, which was distant and by definition untouchable, than by representations of it. These sublimely heroic landscapes were not merely aesthetic statements but religious assertions of the presence of God in wild nature.[45]

Second, Humboldt's ideology had to be nationalized. Where the European had worried that colonial peoples' lack of monuments and historical memorials alienated them from the land and from each other, Americans discovered that landscape itself was "an effective substitute for a missing national tradition," a "repository of national pride." Cultivating heroic American landscapes became a patriotic act, and a democratizing one: they were available to all, not only in galleries but in theatrical panoramas, as Humboldt had mentioned. These were unrolled before mass audiences of thousands, making them literally "moving pictures," screened like movies with commentators and a musical background. Such landscapes often portrayed the birth of the nation. While a few gallery paintings, like Cole's *Course of Empire* series, were critical, most, like Church's *Mt. Ktaadn* (1853), were celebratory and idyllic. Nowhere in evidence is the stern inhuman wilderness that had so shaken Thoreau (ironically, on the very mountain Church makes into an icon of peaceful cultivation), nor the rampant deforestation that so horrified him. The deforestation had horrified Church too, but his painting shows a harmonious balance between wilderness and industrial development, an imperial gambit that, as Mary Louis Pratt observes, transforms the wilderness into an idyll of colonial civilization.[46]

Alexis de Tocqueville's comments are instructive here. The European thinks it is Europeans who really care about American wilderness. Americans themselves "scarcely think of it." They see the forests that surround them "only at the moment at which they fall by their strokes," and their taste for nature is only a passing fad. "Democratic peoples can amuse themselves well for a moment in considering nature; but they only become really animated at the sight of themselves." The spectacle that really fills their eyes is their own mighty power to transform the wilderness: "The American people sees itself advance across this wilderness, draining swamps, straightening rivers, peopling the solitude, and subduing nature." In this magnificent self-image, the American sees his own individual contributions to the destiny of the entire human race, a destiny presided over by "a general and constant plan according to which God guides the species." It struck Tocqueville that the American citizen looked in only two directions: at himself alone, or at his entire society, even the whole human race. There seemed no middle ground between "particular and very clear ideas, or very general and very vague notions." No wonder the publication of *Cosmos* erupted into a cult of Humboldt. If Tocqueville is right, even before Humboldt had published his last book or Church had picked up his first paintbrush, American democracy was uniquely prefigured to seize on Humboldt's aesthetic blending of the individual and the cosmic. In Humboldt's Cosmos they saw not his contested and cosmopolitan planet, not an idea that would liberate all the peoples of world, but Planet America: themselves.[47]

⚜

Barbara Novak notices a peculiar polarity in America, between "grand opera and the still small voice." On the one hand, loud, heroic epic landscapes; on the other, the small, quiet, introverted scenes of the American Luminists. Literature offers a parallel in Whitman and Dickinson. Both could be called "poets of the Cosmos," but Dickinson is closer to Emerson and Poe, for whom nature ultimately dissolves into the apocalypse of mind. In her quiet voice she runs a competition with God himself, and more than once it is she who comes out the victor. Here is one of her responses to the concept of Cosmos:

> I send Two Sunsets—
> Day and I—in competition ran—
> I finished Two—and several Stars—
> While He—was making One—
>
> His own was ampler—but as I
> Was saying to a friend—
> Mine—is more convenient
> To Carry in the Hand—[48]

So, of course, was Humboldt's: its immensity may stagger the imagination, but its five volumes fit snugly in a briefcase, and its first volume nestles the universe quite nicely in the palm of one hand.

Where Emily licked the paradox with a smile and enough energy left over to tend to her garden, Walt Whitman gave over his entire life to wrestling with it. "Begun in ripen'd youth and steadily pursued," he intones, "Never even for one brief hour abandoning my task, / I end it here in sickness, poverty, and old age." His task, of course, was *Leaves of Grass*:

> Haughty this song, its words and scope,
> To span vast realms of space and time,
> Evolution—the cumulative—growths and generations.

In short, to embrace—to *be*—the "kosmos." Literally so: Whitman introduced himself to the world in 1855 as "Walt Whitman, an American, one of the roughs, a kosmos." Eventually he revised the line to read, "Walt Whitman, a kosmos, of Manhattan the son." As David Reynolds has observed, the word "kosmos" was so important to Whitman "that it was the only one he retained in the different versions of his famous self-identification." Not everyone got the point. Charles Eliot Norton sniffed in his 1855 review: "That he was a kosmos, is a piece of news we were hardly prepared for. Precisely what a kosmos is, we trust Mr. Whitman will take an early occasion to inform the impatient public."[49]

Had he read his Humboldt, Norton would have known that, in Reynolds's words, "for Whitman as for Humboldt, 'cosmos' signified both the order of nature and the centrality of human beings." Or as Whitman put it in *Democratic Vistas*, his gloss on *Cosmos*: "the All, and the idea of All." For Whitman had indeed read his Humboldt. He was said to have written *Leaves* with a copy of *Cosmos* on his desk, and he took notes on Humboldt and preserved newspaper clippings about him. Whitman's self-identification carries the symptomatic trace of Humboldt's concept (Whitman even prefers to retain Humboldt's German/Greek "k"): the tight equivalence of homely detail and transcendent whole, "Walt Whitman" as simultaneously "a kosmos" and the son of Manhattan. It is as if one of Church's tiny figures turned to us and spoke, asserting his centrality to the painting. Norton may have found enlightenment in Whitman's 1860 poem "Kosmos":

Who includes diversity and is Nature,
Who is the amplitude of the earth, and the coarseness and sexuality of the earth, and
   the great charity of the earth, and the equilibrium also,
Who has not look'd forth from the windows the eyes for nothing, or whose brain
   held audience with messengers for nothing, . . .
Who, out of the theory of the earth and of his or her body understands by subtle
   analogies all other theories,
The theory of a city, a poem, and of the large politics of these States;
Who believes not only in our globe with its sun and moon, but in other globes with
   their suns and moons,
Who, constructing the house of himself or herself, not for a day but for all time, sees
   races, eras, dates, generations,
The past, the future, dwelling there, like space, inseparable together.

As Reynolds details, Whitman's poem suggests how deeply his sense of "kosmos" was indebted to Humboldt: he carries over Humboldt's theory of the earth and heavens, his earth-centered vision in a setting of distant planets with their own suns and moons, and his all-unifying mystical aesthetic. Reynolds finds his hunch confirmed in Whitman's notebooks, where he defined "kosmos" as "noun masculine or feminine, a person who[se] scope of mind, or whose range in a particular science, includes all, the whole known universe." In this definition, "kosmos" is virtually a synonym for Humboldt himself, in the language of his reviewers and of readers like Emerson.[50]

Whitman's use of the term follows a revealing trajectory. He first elaborates on "kosmos" as a term for the American poet, then, building on Emerson's suggestion in "The Poet" that "America is a poem in our eyes," as a name for America's destiny as the ultimate realization of the kosmos. The process begins in his 1855 "Preface," which Whitman opens with an allusion to Emerson—"The United States themselves are essentially the greatest poem"—adding the Humboldtian touch that here in the United States are "details magnificently moving in vast masses." In a verbal analogue of Humboldt's Chimborazo or of Church's paintings, Whitman claims that the American "incarnates" the land's "geography and natural life and rivers and lakes." Slowly he builds to his climax: "The American bards . . . shall be kosmos," including all they see, and countering all "out of the stronger wealth of himself." Whitman's cosmic poets will form an alliance with science, which is not their enemy but their "encouragement and support." Scientists are "the lawgivers of poets and their construction underlies the structure of every perfect poem. . . . always of their fatherstuff must be begotten the sinewy race of bards." He continues with an elegant paraphrase of Humboldt's aesthetics: "As the attributes of the poets of the kosmos concentre in the real body and soul and in the pleasure of things they possess the superiority of genuineness over all fiction and romance. . . . The poets of the kosmos advance through all . . . stratagems to first principles," and on and on, in Whitman's inimitable piling of phrases. Ultimately, the priests will die away. "Their work is done. . . . A superior breed shall take their place," namely, "the gangs of kosmos and prophets en masse. . . . Through the divinity of themselves shall the kosmos and the new breed of poets be inter- preters of men and women and of all events and things."[51]

Whitman's "gangs of kosmos" are superior because they are closer to na- ture. At the end of *Specimen Days* he warns that democracy cannot maintain itself "without the Nature-element forming a main part—to be its health- element and beauty-element—to really underlie the whole politics, sanity, religion, and art of the New World." What he means by this comes clear to- ward the end of *Democratic Vistas*:

> Nature, true Nature, and the true idea of Nature, long absent, must, above all, become fully restored, enlarged, and must furnish the pervading atmosphere to poems, and the test of all high literary and esthetic compositions. I do not mean the smooth walks, trimm'd hedges, poseys and nightingales of the En- glish poets, but the whole orb, with its geologic history, the kosmos, carrying fire and snow, that rolls through the illimitable areas, light as a feather, though weighing billions of tons.

America must be planetary in the Humboldtian sense, based in what Tho- reau called "the personality of such planetary matter . . . terrene, titanic . . . meteoric, aerolitic," recognizing that "when we are lifted out of the slime and film of our habitual life, we see the whole globe to be an aerolite." This is, for Whitman, the only ground—"the last, the highest, only permanent ground"—on which we must stand to criticize and cleanse all pretense, all that "violates or ignores . . . the central divine idea of All, suffusing universe, of eternal train of purpose, in the development, by however slow degrees of the physical, moral, and spiritual kosmos." As for America, in its amplitude it rivals "the operations of the physical kosmos," but as it develops it must come to realize the kosmos in both dimensions, body and spirit:

> As, within the purposes of the Kosmos, and vivifying all meteorology, and all the congeries of the mineral, vegetable and animal worlds—all the physical growth and development of man, and all the history of the race in politics, religions, wars, &c., there is a moral purpose, a visible or invisible intention, certainly underlying all . . .

Or, as he adds some lines later, "the All, and the idea of All," with a soul "buoyant, indestructible, sailing space forever, visiting every region, as a ship the sea."[52]

Whitman was not complacent. Humboldt knew his Cosmos to be an achievement, which meant it could be lost, broken, shipwrecked, corrupted; in the private despair of his last years he held on to the beauty of the physical Cosmos, in all its deep harmonies, to give himself hope that somehow humanity would weather the political storms that had betrayed his lifetime of hopes. Whitman too had weathered a few storms, and as he aged he came to a similar port. After supporting the Mexican War and Polk's expansionist policies, he found the extension of slavery into the new territories insupportable. As it became clear through the 1850s that the one entailed the other—that, as Horsman concludes, in Whitman's America "the world was to be transformed not by the strength of better ideas but by the power of a superior race"—Whitman recoiled from nationalistic imperialism to an ecstatic cosmic geography, turning, as Reynolds says, from the politics of manifest destiny to the politics of nature: wishful thinking, Reynolds concludes, and delusional too, although his poetry still stands "as testament to Whitman's struggle to plant poetic seed on volcanic political soil."[53]

If kosmos became Whitman's port in the storm, he did issue from it a challenge: "We sail a dangerous sea of seething currents, cross and undercurrents, vortices—all so dark, untried—and whither shall we turn?" Was America truly the nation that would carry humanity to the full flowering of Cosmos? Both Humboldt and Whitman hoped so, but feared, and yet still hoped. As Whitman concluded:

> We see, as in the universes of the material kosmos, after meteorological, vegetable, and animal cycles, man at last arises, born through them, to prove them, concentrate them, to turn upon them with wonder and love—to command them, adorn them, and carry them upward into superior realms—so, out of the series of the preceding social and political universes, now arise these States. We see that while many were supposing things established and com-

pleted, really the grandest things always remain; and discover that the work of
the New World is not ended, but only fairly begun.[54]

Delusional? Tocqueville would have found this "bombastic," another instance of American writers' need to "swell their imaginations" from their
"slight affairs" to vast conceptions and excessive depictions, forsaking the
great for the "gigantic." It is easy to accuse Whitman of being an apologist for
manifest destiny: did he not say that the United States would soon include
Canada and Cuba, that "the Pacific will be ours, and the Atlantic mainly
ours," that "the individuality of one nation," America, will "lead the world,"
that America's "Passage to India" will weld together all the lands of the globe?
Does it really help matters to assert that Whitman's passage to India was really "passage to more than India," to the "Kosmos"?

> Of you O prairies! of you gray rocks!
> O morning red! O clouds! O rain and snows!
> O day and night, passage to you!
> O sun and moon and all you stars! Sirius and Jupiter!
> Passage to you![55]

The answer has to be no. Ironically, the very social concerns Humboldt had
spent a lifetime agonizing over—deforestation and environmental degradation, the oppression of indigenous peoples, political disenfranchisement
of minorities and dissidents, the horrific injustice of slavery—are the ones
that disappear out of these epic, ecstatic "Humboldtian" landscapes. Even
as American audiences adored them, the wilderness they apotheosized was
being, as Tocqueville did not fail to note, laid bare by the axe. But the paintings, like Whitman's kosmic prophecies, made these contradictions bearable.
A totem wilderness, whether framed on a wall or outlined on a map, solved
nothing, really; but ah! The paintings lift us out of the troublesome earth, riddled with our intractable dilemmas, into cosmic space where all the harmonies are restored. Humboldt thought that lush New World wilderness paintings would compensate Europeans for their impoverished environment, that
New World poets would transport them to wilder landscapes, freer shores.
Could he have imagined that under the cover of their dreamwork, imperialism would ravage the very nature he hoped they would save?

Dwelling: Susan Cooper, Muir, Marsh

Humboldtian landscapes are both specific and cosmic, place-centered and
transcending all places. Are not our porch-stones, Thoreau asks, "as good
as any corner-stone in heaven?" Humboldt, the quintessential cosmopolitan

who traveled the globe, known everywhere in Western society, came to America to discover the Cosmos. Thoreau was already in America, but it took Humboldtian science to help him to realize it: in 1845, Thoreau devoured Frémont's fresh-off-the-press narrative of his exploring expedition to the Rockies and California until, as he reports, he shamefaced threw the book aside and asked "where it was then that *I* lived." The answer at that moment happened to be, at Walden Pond. Down Thoreau burrowed, "through the mud and slush of opinion, and prejudice, and tradition, and delusion, and appearance, that alluvion which covers the globe," through Paris and London and New York, "through church and state, through poetry and philosophy and religion, till we come to a hard bottom and rocks in place, which we can call *reality*, and say, This is, and no mistake." Thoreau would work and wedge his feet downward not *through* nature but *to* nature, to a "Realometer" that would gauge the flood of "shams and appearances," to absolute bedrock, and thereon build a new life. How different from Whitman's call for "Passage, immediate passage! The blood burns in my veins! Away O soul! hoist instantly the anchor!" Or from Emerson's call for a philosophy "of fluxions and mobility" in "this storm of many elements. . . . We are golden averages, volitant stabilities, compensated or periodic errours, houses founded on the sea." These three writers are all engaged in a dance of skepticism and faith. In a world of "cross and under-currents," where stars cross paths, planets roll, continents tilt, empires bubble and burst, where is it then that *we* live? Do we found our houses on Emerson's sea and like Whitman "sail forth—steer for deep water only," or do we cleave to Thoreau's "hard bottom and rocks in place" and there found, and build, and dwell?[56]

Dwelling, in ecocritic Greg Garrard's words, asks not for passage but for planting, cultivation, stewardship, a relation to the earth "of duty and responsibility" implying "the long-term imbrication of humans in a landscape of memory, ancestry and death, of ritual, life and work." Dwelling takes place on the middle ground between wild and civil, there on the border where, as Susan Fenimore Cooper observes, "the peculiar tendencies of the age are seen more clearly." To be a borderer is to enact a meeting and blending: to be Thoreau working his bean field, apprehending the natural world "through human labor practices" in a Georgic tradition of creative engagement with the land that, as Michael Ziser argues, traces back to Hesiod and, one could add, forward to Wendell Berry.[57] Or to be Susan Cooper, a third-generation pioneer cultivating a frontier no longer wild yet not entirely tame, living in a domestic space rich with surprise. Or refusing to cultivate, calling like Cooper for pocket parks, a bench on the village green beneath a few trees, or like Thoreau for forests where trees might "stand and decay for higher uses—a common possession forever, for instruction or recreation," or like John Muir fighting to save the wilderness of Yosemite. Or imagining, with George Perkins Marsh, the whole earth desolated by "human crime and human improvidence," "fast becoming an unfit home for its noblest inhabitant" unless it is rescued by a new ethic of global stewardship. Thoreau, Cooper, Muir, Marsh—all four borrowed from Humboldt and transformed his language and concepts into an American idiom of moral responsibility for what Cooper calls "our common home."[58]

Susan Fenimore Cooper grew up in the wilderness settlement founded by her grandfather, William Cooper, awakened, as she watched it transform into a bustling county seat, to the fragility of the face of nature. Her first book, *Rural Hours*, appeared in 1850 and sold so well that by 1855 it had gone through seven editions. As Lawrence Buell shows, Cooper's work, "the first major work of American literary bioregionalism," both anticipated and influenced Thoreau's *Walden* and established the context in which Thoreau's work was read. *Rural Hours*, long forgotten, is now respected as a classic of American environmental writing, in which, as Buell says, "seeing things new, seeing new things, expanding the notion of community so that it becomes situated within the ecological community," has deepened the sense of place by reframing the familiar. Cooper does this in a characteristically Humboldtian way, as when she admires the common flower mayapple, and notes that it is also found in Central Asia: "One likes to trace these links, connecting lands and races, so far apart, reminding us, as they do, that the earth is the common home of all." And she interprets Humboldt in the characteristically American way, by absorbing his picture of the universe into traditional natural theology. As Rochelle Johnson writes, Susan Cooper's ideas "are framed by her belief in a cosmos in which a Christian God provides not only for humankind but also for all other life forms. In this cosmos, everything exists for a purpose determined by God," and God's providential design centers on humans. "It is consistent with God's will, then, that humans convert the wilderness into a land that is shaped and cultivated."[59]

Yet Cooper was increasingly fearful of the consequences of that human cultivation. "Sixty years have worked a wonderful change" in her neighborhood: "the forest has fallen upon the lowlands, and there is not a valley above us which has not been opened. Another half century may find the country bleak and bare." She is sorry to see fine old forest colonnades "falling rapidly before the axeman," and groves that might have stood for centuries "with increasing beauty . . . wantonly destroyed." Game birds are "rapidly diminishing" due to "reckless extermination"; wild animals are "not as yet . . . entirely exterminated," but many "will entirely disappear from our woods and hills, in the course of the next century." Populations of summer birds are down by more than half, "a sad change" that could be reversed if laws against shooting them were enforced. Nor is this problem merely aesthetic: the "wholesale slaughter of the smaller birds" inevitably results in the explosion of damaging insect populations.[60]

All these cumulative losses lead Cooper to a pointed critique of the capitalist system that allows a beautiful grove of virgin pines to be cut down at the whim of their owner, merely because they are private property. The character of the country would be lost "if ever cupidity, and the haste to grow rich," were to destroy the forest "and leave these hills to posterity, bald and bare, as those of many older lands." Yet trees, as she pleads in a passage strongly anticipating Thoreau, "have other values" than "their market price in dollars and cents . . . they are connected in many ways with the civilization of a country; they have their importance in an intellectual and in a moral sense." As Humboldt insisted, deep ties interlink the character and intellect of a people with their natural environment. To scar the face of nature is to scar the people it nourishes. But Cooper worries that we may already be damaged beyond repair: "Shall we sell the ashes of our fathers that a little more coin may jingle in our own pockets?"[61]

Her solution was knowledge. If people knew more about the flora, fauna, and geology of their land, they would approach it, as Johnson observes, "more humbly and gratefully and with less greed." As she continues, Cooper's writings are an extended argument "for a sustainable balance" between culture and nature, approached through an ecological awareness that actions have unanticipated consequences: slaughter too many birds to decorate your hats, and you may lose your crops to insects. Nature writings like Cooper's were on the front lines of an emerging battle, and engaging her audiences emotionally would, she hoped, lead to intellectual enlightenment. The ascending spiral from pleasure to knowledge would lead to a kind of ecological wisdom: nature has values beyond the jingling of coins. All elements of the Cosmos have their purpose, for if the whole is constituted by its particulars, then every one is needed, none can be spared or destroyed without cost. Cooper hails the new fashion for accuracy in descriptions of nature, which once were "vague and general," but now are "more definite and accurate." The taste for landscape painting and a natural style in gardening no doubt play a role, but the causes must be deeper: "The head called for more of truth, the heart for more of life." Writers and painters learned "to look at nature by the light of the sun, and not by the glimmerings of the poet's lamp. And a great step this was, not only in art, but in moral and intellectual progress." Cooper argues that Humboldt's three forms of nature description—literature, painting, and gardening—were not escapist entertainments but the engines of a new cultural revolution.[62]

That revolution would be realized not by passive spectatorship but by active participation. Cooper compares the varied impressions produced by an excursion through tame and easy country to enjoy an isolated view—a cascade, a small lake, a ruin—with a gradual ascent up a mountain. The ascent itself adds to the enjoyment:

> Every turn we reach in the climbing path, every rood we gain in elevation, opens some fresh object of admiration, or throws what we have already seen into a new light. . . . Even the minute beauties which we note one by one along the ascending pathway, the mountain flower, the solitary bird, the rare plant, all contribute their share of pleasure; the very obstacles in the track, the ravine, the precipice, the torrent, produce their own impression, and add to the exultation with which we reach at length the mountain-top, bringing with us a harvest of glowing sensations gathered by the way, all forming delightful accessories to the greater and more exalted prospect awaiting us at the goal.

The true passage to Cosmos is not found, but forged, step by step, moment by moment. A landscape merely seen is a pleasure, but like jingling coins, a moral, emotional, and intellectual dead end. Only "the climbing path," the one that demands active participation in constructing the landscape, leads to exaltation. Cooper's own ascent through the thickets of science, religion, and philosophy at a time when old beliefs were shattered and new ones still forming leads her to what Tina Gianquitto calls "a mode of representation—that of the picturesque aesthetic—that can accommodate her competing desires" for scientifically accurate knowledge of nature, for an aesthetic elevation at the prospect of the whole, for moral empowerment. As she observes, "The picturesque eye organizes those varied elements into a unified panorama and allows Cooper to submit nature to Humboldt's 'process of thought'; it reveals a cohesive narrative of nature and nation and shows the cultural, historical, and moral progress of a civilized community." The accretion of details, however apparently trifling in themselves, leads to an apprehension of the whole, or in Gianquitto's words, "a panorama that reveals God and consequently produces a 'moral . . . energy that could be translated into action.' For Cooper, this moral energy extends outward from the environment, to the home through the village, and into the nation." Through the agency of God, Humboldt's Cosmos is domesticated and, as with Whitman, put in service to the American prospect.[63]

Though her thoughts center in the garden, Cooper does distinguish a special quality of "grace" in the wild, as when she writes of wildflowers that "there is a unity, a fitness, in the individual character of each plant to be traced most closely, not only in form, or leaf, and stem, but also in the position it chooses, and all the various accessories of its brief existence. It is this that gives to the field and wood flowers a charm beyond those of the garden." Like Humboldt, she worries that gardeners may "pervert . . . the very nature of a plant," as when they strip leaves from a rosebush to make the flowers more conspicuous. Fortunately, Cooper can enjoy the uncommon grace of the common wild rose—a pleasure enhanced by the knowledge that "M. de Humboldt mentions . . . in his travels in South America he never saw one, even in the higher and cooler regions, where other brambles and plants of a temperate climate were common." Yet Cooper does believe that "the hand of man generally improves a landscape. The earth has been given to him, and his presence in Eden is natural; he gives life and spirit to the garden." And vice versa: "Gardening is a civilizing and improving occupation. . . . Persuade a careless, indolent man to take an interest in his garden, and his reformation has begun." For proof of the "civilizing effects of large gardens" she points to those in France, Italy, and Germany, where grounds are open to the public and no one thinks to abuse or injure the plants.[64]

Cooper's most interesting use of Humboldt comes in what Buell calls "the most comprehensive short treatise on the history of environmental consciousness in world cultures from ancient times to the present"—although Buell does not mention that much of Cooper's "treatise" is, in fact, a long paraphrase of Humboldt's essay in *Cosmos* on the history of the feeling for nature, so the claim for Cooper's originality must be qualified. Cooper does acknowledge her source, quoting several times from Humboldt, though an interesting tension develops when her narrative reaches the modern era. Cooper credits "the more deeply felt appreciation of the works of the creation" not as Humboldt does to the diffusion of civilization—the arts and sciences, commerce and exploration, new instruments of vision—but to the union of this diffusion with Christianity, which connects man to earth with "close and endearing ties." Humboldt had credited the rise of Christianity with "a beneficial influence on the condition of the lower classes by inculcating the social freedom of mankind," and with expanding "the views of men in their communion with nature," which was seen as part of "the greatness and goodness of the Creator." However, Humboldt does not privilege Christian belief, while for Cooper Christianity is intensely liberating and comforting precisely in confining one to a system "which leaves nothing to chance, nothing to that most gloomy and most impossible of chimeras, fate, but refers all to Providence," to God's omniscient wisdom. Man, if "the lord of the earth and all its creatures," yet is still "subordinate" to God, adoring him with "an overpowering, heartfelt, individual humility." Cooper does not argue with Humboldt's secularism; she quietly overwrites his voice with her own.[65]

As with Church and Whitman, from nature as scripture Cooper turns to nature as nation. She observes that America is uniquely in need of the "refreshments" and "solace" derived from "country life." In a nation vexed with "the fever of commercial speculations, the agitation of political passions," the dizzying progress of science and confusing controversies of philosophy, and above all the spirit of ambition "so wearing upon the individual," Americans more than any others need "the shade of the trees and the play of healthful breezes to refresh our heated brow." Because of these influences, the poles have shifted: it is now the country rather than the town that ennobles, elevates, and civilizes. Moreover, nature gives Americans what Humboldt said they lacked: as Cooper chimes in, "They have no Past . . . there are no grave monuments of former generations standing in the solemn silence of a thousand warning years along their streets." Hence, the works of the Deity, "The everlasting hills—the ancient woods—these are his monuments—these tell him of the past, and not a seed drops from his hand but prophesies of the future." As for those other monuments, the ones left by the Indians, only traces remain. Springs in the woods recollect their absence, for "we feel assured that by every fountain among these hills, the Indian brave, on the hunt or the war-path, must have knelt ten thousand times, to slake his thirst." Place-names on the map also recollect absences. Cooper notes without comment Humboldt's observation "that Matanzas, *massacre*, and Vittoria, *victory*, are frequently scattered over the Spanish colonies." Fortunately, continues Cooper, "the red man" had already named the natural features of our country, and she urges their names be preserved. Beautiful or harsh, "they have still a claim" of historical and linguistic interest. Behind her stand Humboldt and Nicollet, struggling with foreign sound systems and patiently inscribing Indian names onto their maps. Now, a generation later, since "a name is all we leave them, let us at least preserve that monument to their memory."[66]

Cooper sees that the ancient forests are falling, the wild creatures have fled. The "red man," "lord of the land" for thousands of years, "no longer treads the soil; he exists here only in uncertain memories, and in forgotten graves." Geological memories survive—mountains still bear "the record of earth's stormy history"—but the face of nature, that palimpsest of the land-scape's natural and cultural history, is being erased, and with it, all that marks America as God's westering messenger, carrying the future on her back. Cooper herself was being erased, too, along with an entire tradition of writing that valued careful and alert descriptions of living nature. Rochelle Johnson has built on her earlier work to understand why, and has come to a disturbing conclusion: the very elements of culture that Humboldt thought would lead humans toward a deeper understanding of nature—literature, landscape art, gardens—led instead away from it. The regeneration Humboldt thought they would inspire led not outward to nature but inward to the mind: "The metaphors for nature came to replace what they originally only described." The result was, paradoxically, not to awaken popular curiosity about "nature without metaphor," as sought by Cooper and Thoreau, but to turn nature into metaphor, deadening it and alienating Americans from it. By enclosing nature into mind, a brain wider than the sky, Americans lost sight of the ways living and chaotic nature exceeds our capacity for reason or perception. In essence, the nineteenth century replaced the wild rivers of Humboldt, Thoreau, and Cooper with the tamed and utilitarian "canals" of ideologically entrenched and dominant conventions.[67]

Cooper responded by pointing her readers back to the wounded landscape. In 1893, when she was eighty, she published "A Lament for the Birds," in which she recorded that where once every village tree had borne two, three, even four or five birds' nests in its branches, "Today you may perhaps discover one or two nests in a dozen trees." Rachel Carson was off by seventy years: it had long since become a silent spring. But without monuments, without memory, what human alive would ever know it?[68]

John Muir's favorite monuments were glaciers, those rivers of ice whose "cross and under-currents" were measured not in minutes but in centuries. Now that glaciers, too, are vanishing, soon Muir's writings will be like Indian place-names, relics to their memory. It was Humboldt who helped Muir find those glaciers and who helped him frame the powerful idea, central to Muir's philosophy, "that going to the mountains is going home" and thus, as Cooper had suggested, that preserving wild nature is a necessity of modern civilized life.[69]

America's preeminent "Green Man" grew up not there but in Scotland, running with the village gangs, tormenting cats, and delighting in dog fights with what he called "the natural savagery of boys." When John was eleven, his stern Calvinist father moved the family to the American frontier of Wisconsin, where John worked seventeen-hour days to clear the land, making the wilderness say corn, potatoes, and wheat instead of trees, and witnessing the wholesale competitive slaughter of songbirds by the bloody bagful. Here Muir turned to the beauty of nature as "compensation for hardship and pain." He also turned to reading, especially the books of travelers. In his autobiography Muir recalled his mother saying, "Weel, John, maybe you will travel like Park and Humboldt some day." To which his father cried out, "Oh Anne! dinna put sic notions in the laddie's heed." As Steven Holmes observes in his sensitive biography of the young Muir, the writings of Humboldt "were too abstract and scientifically sophisticated for Muir to fully grasp at this point," and the teenaged Muir found the adventures of Mungo Park, a fellow Scotsman, much more appealing. Muir's real encounter with Humboldt came years later when, to escape the Civil War draft, he wandered and botanized in the Canadian wilderness. This was "his first solitary immersion in nature," and as he reflected later, "I entered at once into harmonious relations with Nature. . . . I felt a plain, simple relationship to the Cosmos." Eventually he settled into a job at a mill working as a mechanic and an inventor. In his loneliness he started a correspondence with Jeanne Carr, the wife of his favorite college professor at Madison. In his first letter to Carr, dated 13 September 1865, he confessed his frustrations: he dreamed of college, of inventing machinery, of studying medicine—but to each dream the answer came, "You will die ere you can do anything else." Above all other dreams was one: "How intensely I desire to be a Humboldt."[70]

Holmes finds Muir's invocation of Humboldt at this point in his life "particularly significant and complex." Humboldt's philosophy and field methods "clearly echoed Muir's scientific education and . . . would come increasingly to characterize his thought and activity," but he was never as "intellectually sophisticated" as Humboldt. Thus specific intellectual influences are hard to trace; more important "was the figure of Humboldt the person." Like Emerson, Muir regarded Humboldt "as a sort of 'representative man,' the embodiment of a human life expressing itself fully in the disciplined yet passionate study of the natural world." Humboldt provided an emotional inspiration and aesthetic sensibility that Mungo Park's British imperialism could not, and moreover, Muir found in Humboldt "a role model of a mature and socially acceptable identity for an adult male." Thus Humboldt became for Muir "a sort of imagined adjunct mentor," and though it would be years before Muir could fulfill his Humboldtian dream, already Humboldt's cosmopolitanism, "his embodiment of 'the science which is of no country,'" helped Muir transmute his focus on "'the dingles and dells of Scotland'" into a broader, more cosmopolitan interest in the natural world from the Arctic to the Amazon. Muir began to imagine one particular way "to be a Humboldt": a long walking tour from Wisconsin through the southern United States to South America.[71]

In March 1867, an accident at the factory blinded Muir's right eye. Though he eventually regained his sight, the trauma ended his fascination with machines, and as part of his recovery he undertook a long botanical ramble during which the dream of a thousand-mile walk took shape. Muir felt it was less "a conscious decision" than the result of "elemental forces" that emerged from within or converged from without to carry him along to the wilderness. He set out on 1 September 1867, bound for Florida and points south. But something happened to him along the way. The farther south he walked, the more alien the land and the people felt. As Holmes writes, this "both energized and alienated Muir," charging him with conflicting emotions of awe and repulsion.[72]

The crisis came in Florida when he fell ill with malaria, "a macabrely appropriate capstone to what was supposed to be a triumphant scientific march to South America." Like Thoreau on the slopes of Mt. Ktaadn, Muir discovered a nature that was not bound to be kind to him—that could, and nearly did, kill him. How, then, could man be the lord of the universe? Muir's inherited Calvinism fused with the voice of Humboldt to fundamentally reorient his thinking. In a climactic passage in *A Thousand Mile Walk to the Gulf,* Muir put it this way:

> Now, it never seems to occur to these far-seeing teachers that Nature's object in making animals and plants might possibly be first of all the happiness of each one of them, not the creation of all for the happiness of one. Why should man value himself as more than a small part of the one great unity of creation? And what creature of all that the Lord has taken the pains to make is not essential to the completeness of that unit—the cosmos? The universe would be incomplete without man; but it would also be incomplete without the smallest transmicroscopic creature that dwells beyond our conceitful eyes and knowledge.

As Muir continues, "This star, our own good earth, made many a successful journey around the heavens ere man was made," and "whole kingdoms of creatures" rose and disappeared without us. Our own disappearance would raise no extraordinary commotion. The universe, independent of man, needs us not; but the Cosmos is "incomplete" without us—and without every link in the great interwebbed tangle of life.[73]

Muir reluctantly gave up his plans to follow Humboldt to the headwaters of the Orinoco. Instead, he shipped to California in hopes of regaining his health—interestingly, a journey that reprised the reputation of California as the arcadia whose environment would cure all ills, a reputation based on Humboldt's medical theories as carried to California earlier in the century by doctors trained in Benjamin Rush's Philadelphia. If he could not fulfill his dream of seeing the Andes, then Muir would see the Sierras instead, joining the long line of Humboldtians who substituted North for South America. In California his ideas grew increasingly Humboldtian, as he moved from his concern with God as the external force shaping nature, a kind of cosmic landscape gardener, to what Holmes calls "the landscape itself as an active force, imbued with a divine light and life energy and flowing according to laws hidden from humans but ultimately beneficial to all beings." As for the role played by humans, Holmes states that Muir increasingly saw them as "literally composed of the same stuff as the natural world. . . . . Accordingly, the human emotional response to natural beauty is itself an elemental process," as Humboldt had insisted by arguing that the products of his "spiritual labor" belong as essentially to the Cosmos as the phenomena of nature. And Muir was turning away from botany to Humboldt's own specialty, geology. By 1871 he "was totally committed to and engaged by glacial work." It was in the rugged California mountains that Muir learned how to see what Humboldt called the "face of nature." As Muir wrote:

> The most telling thing learned in these mountain excursions is the influence of cleavage joints on the features sculptured from the general mass of the range. Evidently the denudation has been enormous, while the inevitable outcome is subtle balanced beauty. Comprehended in general views, the features of the wildest landscape seem to be as harmoniously related as the features of a human face. Indeed, they look human and radiate spiritual beauty, divine thought, however covered and concealed by rock and snow.[74]

In his study of the "Humboldt current" in America, Aaron Sachs identifies Muir as one of its primary carriers, particularly the Muir of the Gulf states and of Alaska. Sachs find Muir at his most Humboldtian in his 1881 cruise on the *Corwin*, where he encountered a wilderness that "could bewilder as easily as it could inspire," and peoples who lived with that wilderness and brought it to life "through their skills as guides, builders, makers of clothes, storytellers, educators": Aleut and Inuit, Tlingit and Chukchi. The scientific racism Muir expressed in California, where he found the Indians repulsive, recedes, at least for a time. And, curiously, Sachs finds that in Muir's copies of Humboldt's books, "next to passages about trees and plants and natural scenery in general, Muir spent the most time on Humboldt's musings about Indians." Yet unlike Boas, Muir never embraced Humboldt's radical cultural pluralism, and in later years, he "push[ed] Indians out of the picture": "By the time Muir founded the Sierra Club in 1892, turning full-time to political activism on behalf of wilderness areas, he no longer seemed to be thinking about unity, or diversity, or the concept of cosmos; almost all the radicalism had seeped out of his agenda." Humboldt's friend George Catlin had envisioned a series of national parks "where the world could see for ages to come, the native Indian in his classic attire, galloping his wild horse, with sinewy bow, and shield and lance, amid the fleeting herds of elks and buffaloes. . . . A *nation's Park*, containing man and beast, in all the wild and freshness of their nature's beauty!" But Muir's legacy was quite different, an ideal of wilderness emptied of the human presence. It was Muir who would prevail, for unlike Catlin, he had the institutional power and government support to carry out his vision. Sachs concludes that "Muir's rich social and ecological visions had been reduced to 'the cause of saving samples of God's best mountain handiwork.'" Where the sublime landscapes of Church and Bierstadt had been peopled with vignettes of rural folk or bands of roving Indians, national parks that were the property of all had to evacuate the few, no matter that they called the wilderness home. Eden had room only for the innocent, and after a century spent denying Indians their humanity, white Americans could see in them only their own guilt. As modernization continued to demand conquest, Humboldt's vision of the Cosmos was fragmenting and withering.[75]

The last Humboldtian exploring expedition set sail in 1899, one hundred years after Humboldt and Bonpland sailed from the port of La Coruña: the Harriman expedition to Alaska. Muir was recruited to it by C. Hart Merriam, whom Sachs calls "the last great Humboldtian scientist bred in America," along with John Burroughs, the bird artist Louis Agassiz Fuertes, photographer Edward S. Curtis, and a full interdisciplinary team of ornithologists, botanists, zoologists, geologists, and an Indian advocate. Most had been recruited at the interestingly named Cosmos Club in Washington, D.C., founded by John Wesley Powell "to bind the scientific men of Washington by a social tie and thus promote that solidarity which is important to their proper work and influence." The multiauthored expedition report reads like a version of Humboldt's *Political Essay*, informed by a century's progress in science and technology but more naïve in its utter lack of skepticism about race and empire. It covers a similar array of topics: the narrative of the journey, geology and geography, landforms, plants and animals, native peoples, history of exploration, economic evaluations. Yet it now takes a compendium of specialists to produce such a volume, which reads like an issue of *National Geographic*, a gentleman's holiday in a secure and comfortable ship amidst an exotic landscape whose challenges are merely aesthetic: Humboldt as ecotourist.[76]

Holmes, however, stresses a different side of Muir than does Sachs, a different Humboldtian legacy, one that associates nature not with distant and exotic lands but with the familiar landscapes of home in a way more reminiscent of Susan Cooper. Muir wrote of his first moment of revelation in the California mountains, "We are now in the mountains and they are in us, kindling enthusiasm, making every nerve quiver, filling every pore and cell of us." He is "home," in other words, not when he is in the mountains but when the mountains are in him, in that mysterious Humboldtian communion of nature with mind and soul. The result is a verb, a process that might be called "enhomement," as when Muir writes of a camp in the Cascades: "Never was mortal wanderer more blessedly homed." As Holmes explains, "He *says* it is the place itself that has homed *him*, but in terms of the narrative it is Muir who speaks this house into existence. . . . It was both the environment *and Muir's interpretation of it* that 'homed' him—gave him a home, and allowed him to feel at home in it."[77]

Here is a clue to what is perhaps Humboldt's greatest legacy: the wanderer, ever sailing on turbulent seas, ever in passage to somewhere else, mobile even in Berlin as he pitched and struck his tent at his king's command, who yet finds a way to be at home everywhere because his mind and heart can enter every sky, be altered by every land, from dusty Cumaná to mosquito-ridden Esmeralda to precarious Quito, each step, each port a note or a chord in the gathering harmony of the Cosmos. Humboldt the wanderer knew alienation, but he also knew that alienation was the condition of modernity, of that outward reach to interweave the blue planet, all its places and beings and stars, its hopes and wars and betrayals, into the Cosmos. As Muir discovered on the Gulf of Mexico, as Thoreau discovered on Ktaadn, that very alienation was itself the necessary next step in the Cosmos, in the ability to bespeak not just one place, but the earth itself, as home.

If this is true, then in some ways George Perkins Marsh is the most profoundly Humboldtian of all Humboldt's American children. A polymathic scholar from Vermont (and the cousin of James Marsh, who popularized Coleridge in the United States), Marsh spoke twenty languages, served as a congressman from 1843 to 1849, became the American minister to Turkey from 1849 to 1854 and to Italy from 1861 until his death in 1882. He was, in short, the quintessential American cosmopolitan who seemed to know everyone in America and nearly everyone in Europe, including many of Humboldt's associates. As a congressman Marsh spoke out against slavery, the Mexican War, and U.S. expansionism, and helped to found and shape the Smithsonian Institution. From his base in Turkey he aided Louis Kossuth and other refugees from the failed 1848 revolutions, and traveled throughout the Mediterranean and Middle East. Back in the United States he successfully urged the introduction of camels to the American Southwest (a proposal originally floated by Humboldt in a process amusingly recorded by Möllhausen's sketches). In his day Marsh was best known for his studies of Scandinavian languages and folklore, his promotion of a democratic history of the common people, and his many publications in linguistics. In our own day, Marsh is known for one thing: publishing *Man and Nature* in 1864. Marsh's biographer, David Lowenthal, calls it, next to Darwin's *Origin of Species*, "the most influential text of its time to link culture with nature, science with society, landscape with history." It sold over a thousand copies in a few months, and in a decade was regarded as "a classic of international repute," with an early impact on American timber culture and a rapid spread in Europe. After a period of neglect, Lewis Mumford resurrected it, in 1931 calling it "the fountainhead of the conservation movement," a position from which it has not since wavered.[78]

Marsh was steeped in the writings of Humboldt, and he claimed his landmark book was conceived in the philosophical and imaginative tradition of "the new school of geographers" led by Humboldt and Ritter. His message was based on the planetary leap from parochial environment to the earth itself as home. This leap was not metaphoric but terrifyingly literal:

> The earth is fast becoming an unfit home for its noblest inhabitant, and another era of equal human crime and human improvidence, and of like duration with that through which traces of that crime and that improvidence extend, would reduce it to such a condition of impoverished productiveness, of shattered surface, of climatic excess, as to threaten the depravation, barbarism, and perhaps even extinction of the species.

The process was accelerating. Marsh had already guessed the shape of the future in 1847, when in an address to a county agricultural society he applied Humboldt's warning of the "calamity" caused by deforestation in Mexico and Venezuela to the once-wooded hillsides of Vermont. In a single generation the changes were so striking that it hardly took scientific training to notice them: "any observing person" could see them, if he or she understood *how* to see, "and every middle-aged man who revisits his birthplace after a few years of absence, looks upon another landscape." Like Thoreau, Cooper, and Muir, Marsh too was a witness to the ecological transformation sweeping the continent, but unlike them, he took a very Humboldtian step: he traveled overseas, where he could see for himself evidence of over two thousand years of global ecological disruption. What Marsh witnessed in the Mediterranean and the Middle East was sobering. Human action, primarily deforestation, had over the centuries brought the "face of the earth" in parts of Europe, Africa, and Asia "to a desolation almost as complete as that of the moon." This had happened in plain sight, in the brief span of human history. Lands once covered with "luxuriant woods, verdant pastures, and fertile meadows" were now so blasted, eroded, and barren that no recovery was possible. Was this past to be America's future?[79]

Marsh declared his alliance with Humboldt in his 1860 essay "The Study of Nature." Here he runs through a long series of characteristically Humboldtian ideas, starting with the influence of nature on human history, following the emergence of a liberating and egalitarian global economy by the spread of arts and science along new networks of transport across continents and oceans, and ending with a universal embrace of diverse humanity as one family bonded by shared interests. We live in a new era, says Marsh, unknown to the Greeks, in which modern science has made "the sublime conceptions of extended space, of prolonged duration, of rapid motion, of multiplied numbers, and of earthly grandeur and beauty and power" part of our "mental constitution." Far from killing poetry, science had enriched it with "a grander and even more poetical truth." In the new planetary prospect of science, the very breeze on one's cheek bespeaks "images, now of tropical luxuriance, and now of polar desolation; speaks in turn of the palm-tree and the iceberg, of equatorial days shared equally by darkness and light, and of climes where the entire year knows but one summer day and one winter night." Marsh celebrates landscape painting, but reminds his readers that even the greatest painting pales beside the works of inanimate nature. If the innate childhood sympathy with nature be not aborted, then the adult will see beauty and instruction everywhere, in Old Worlds and New, for despite analogies of form and climate, "neither tree, nor shrub, nor flower, nor grass, nor bird, nor quadruped, nor fish, nor creeping thing, is specifically identical in both. All this diversity of form he recognizes as varied manifestations of the same universal laws."[80]

Readers of Humboldt will recognize every one of Marsh's points as central Humboldtian themes, and Marsh makes sure to credit his source and mentor: "The popularization and general diffusion of science, of which Humboldt was the great apostle, has been of infinite service to the cause of art, because it has brought exact knowledge of at least the external forms and characteristics of visible objects within the reach of a large class of observers, to whom that knowledge would otherwise have been unattainable." Troubled, like Francis Lieber, by growing misunderstanding of Humboldt, Marsh defends his hero, "this greatest of the priesthood of nature," from allegations that his large views of science were due to his lack of mathematical training. In his defense, Marsh asserts values that, as Lowenthal observes, were descriptive of Marsh himself: "A learning which embraced the whole past history and present phase of every branch of physical research, and which was moreover graced with the elegances of all literature and dignified with the comprehensive wisdom of all philosophy, cannot but be a reproach to narrower natures, which see and appreciate truth, not in the mutual interdependence of apparently unrelated knowledges, but only in the naked proportions of number and magnitude." It is Humboldt whom Marsh credits for the growing bond between science and the arts, especially landscape painting. To John Ruskin's criticism of the shallowness of New World landscape painting, Marsh returns the Humboldtian reply that the hand of man need not be present in a landscape to awaken admiration in any soul in sympathy with creative nature.[81]

There is, however, one significant point on which Marsh departs from Humboldt. "The life of man is a perpetual struggle with external nature," he asserts. Only "by rebellion against her commands and the final subjection of her forces" can man achieve his higher ends, for "he is the rightful lord, and Nature the lawful, though unwilling slave." Marsh issues a declaration of war: "Wherever he fails to make himself her master, he can but be her slave. In this warfare there is no drawn battle." Like Susan Cooper, Marsh reframes Humboldt through Christian doctrine, and like hers, his point is that man, as a free moral agent placed by God as the lord of creation, has not only fearful destructive power but the potential to be a powerful regenerative force. Marsh repeats the Miltonic doctrine of the "Fortunate Fall": our expulsion from Eden was, if not exactly a blessing, "at least a means and condition of all blessings." We will win our emancipation, our return to Eden, by reclaiming through the knowledge and power of science all that we lost, and more.[82]

Man and Nature is dominated by these twin themes of destruction and reclamation. Part of Marsh's power is his homely appeal: while science slowly accumulates evidence, any sensible person can see what's really happening: "But we are, even now, breaking up the floor and wainscoting and doors and window frames of our dwelling, for fuel to warm our bodies and seethe our pottage, and the world cannot afford to wait till the slow and sure progress of exact science has taught it a better economy." We must look not with the slow eye of science but with the quick eye of the historian, who has learned to read a landscape's past from its present appearance, to see how, as Humboldt had said, "its form is its history." The Greeks and Romans, taken up with their human drama, were insensible to nature's beauty, and as a result more than half their empires, even provinces they celebrated for their fertility, are now "either deserted by civilized man and surrendered to hopeless desolation, or at least greatly reduced in both productiveness and population." Who but an environmental historian would know that the barren, stony land of Greece was once covered with forests and deep topsoil? That brooks were once rivers, and dry lands were once veined with freshets? That harbors closed by shoals and sediments were once deep waters and busy commercial ports? A territory "larger than all Europe" which once sustained a population equal to our own has been effectively withdrawn from human use. Only recently has the population of Europe been "half awakened" to the extent of the debt they must pay, a debt imposed on them by the "prodigality and the thriftlessness" of their ancestors.[83]

Through the telescope of time Marsh could see what few had seen before. Humboldt's new and improved geography shows that all elements of nature act reciprocally on each other, for nature embraces not only the globe itself "but the living things which vegetate or move upon it, the varied influences they exert upon each other, the reciprocal action and reaction between them and the earth they inhabit." As Marsh continues, Humboldt's geography teaches us that "every plant, every animal, is a geographical agency." Emerson had claimed that all human operations "taken together are so insignificant, a little chipping, baking, patching, and washing," that they do not change the face of nature. On the contrary, argues Marsh: "But man is everywhere a disturbing agent. Wherever he plants his foot, the harmonies of nature are turned to discords. . . . the face of the earth is either laid bare or covered with a new and reluctant growth of vegetable forms, and with alien tribes of animal life," in an agricultural revolution which, however vast, is trivial compared with the unanticipated forces it has released.[84]

We were sent, Marsh writes, to transform the earth and instead we have gone to war with it: felled the forests that bind the soil to the earth's rocky skeleton, broken up the mountain reservoirs, torn open the plant cover of plains and prairies, destroyed the fringe of plants that stabilizes coastal sands, "ruthlessly warred" on all the tribes of wild animals. This "almost indiscriminate warfare" must stop; nature herself fights back against "the ravages committed by man." "She avenges herself upon the intruder" by setting loose energies meant to assist us, but now turned against us: deluges of rain that scour the deforested mountains down to barren rock, tear apart the hills, and spread the eroded soil into swampy plains. Yet humans can become creative agents instead, if only we would knowingly turn the power of our science and technology to rebuilding instead of overthrowing.[85]

Marsh can identify the staggering scale of human destruction through his deep sensitivity to ecological relations. Little things can have huge effects: a Parisian manufacturer invents silk hats, and suddenly halfway around the world beaver populations rebound and the geography of a continent is altered. Farmers slaughter birds under the mistaken belief that they eat grain and seeds, unleashing hordes of noxious insects, their real food. Since birds are vulnerable to severe weather and to habitat loss, human settlement unintentionally destroys even those tribes not targeted by the general war-fare. (One thinks of the population crashes underway today among migra-tory songbirds, as global warming resets seasonal clocks in their summer breeding grounds, their winter feeding grounds, and at every point in their passages between; when did you last see an indigo bunting or a Baltimore oriole? Don't count on ever seeing another.) Marsh piles on examples, gath-ered from all corners of the globe, to overwhelm his readers with one point: "The equation of animal and vegetable life is too complicated a problem for human intelligence to solve, and we can never know how wide a circle of dis-turbance we produce in the harmonies of nature when we throw the smallest pebble into the ocean of organic life."[86]

Like Thoreau, Marsh calls for preservation of forest lands to be "at once a museum for the instruction of the student, a garden for the recreation of the lover of nature, and an asylum where indigenous tree, and humble plant that loves the shade, and fish and fowl and four-footed beast, may dwell and per-petuate their kind." And there was an economic benefit too: "Economically managed," an untouched forest could soon yield a regular income, and the collateral advantages would be tremendous. Were the Adirondack forests to fall, New York would soon look as barren as the French Alps. Even the char-acter of the nation would benefit if the incessant fluctuations of the American landscape were stabilized by restoring a just proportion of woods to fields. America would then become "a well-ordered and stable commonwealth, and . . . a people of progress."[87]

Wilderness advocates have been hard on the technocratic Marsh, yet Muir admired his work (and used his arguments to secure watershed protection for Yosemite soil and forests), while Marsh shared much with Thoreau, whom he read and admired, and with Susan Cooper, who issued similar pleas for ecological awareness. It's tempting to read the nineteenth century through the wisdom of our own times, but as Prescott warned, we should be wary of judging the past "by the lights of our own age. We must carry ourselves back to theirs," taking their own point of view, thus extending to them the same justice we would ask of our future. The historian David Lowenthal similarly calls for an approach to the past that respects its integrity on its own terms: "The past is a foreign country . . . they do things differently there," he re-minds us.[88]

This relativistic way of thinking about history is a legacy of the nineteenth century, built on the cornerstones laid by the Humboldt brothers. While Wilhelm was one of the architects of historicism, Alexander regarded sci-ence, too, as a form of history, hence a form of art, of *poiesis* or making. To him the point of telling a history was not to universalize but to particularize, to romance the "accidental individualities" that gave charm and character to a landscape, a people, an era. The face of the present is, like Muir's mountain, cleaved and jointed by its unique history, each facet telling a different story and each story, as Muir said, "essential to the completeness of that unit—the cosmos." Humboldt's telling was taken up by a range of poets—Emerson and Poe, Thoreau, Church, and Whitman, Susan Cooper, Muir, and Marsh— among multitudes of others who no longer have a voice in the twenty-first century. They each spoke Humboldt, but in their own way, using him to perform their own necessary cultural work. In this geography of the past, already steeply oversimplified, each voice is essential, each a contributing agent, and in their chorus they bespeak something new: a discourse, a language, that declares that humans and nature are imbricated into one vast, tangled, and infinitely various and ever-changing whole. Humboldt may have been silenced, but not before he taught the builders and dwellers of Cosmos to speak.

Epilogue: Recalling Cosmos

It is true that the progress of science may die away, but then its essence will have been extracted. This cessation itself will give us time to see that Cosmos, that esthetic view of science which Humboldt prematurely conceived.

CHARLES SANDERS PEIRCE, *1863*

Time was catching up with Humboldt. As he said to Benjamin Constant, "I am quite aware that principles are imperishable, but unfortunately I am not a principle." To the end he drove himself in what he called "a race to death," desperate to realize his great vision of Cosmos, to put his principles before the world—and to make enough money to hold off his creditors just a little longer. His American visitors marveled at his animation and deep courtesy, his pleasant voice and lively eyes, his thin white hair (always tousled in photographs), his immaculate plain black suit and extraordinary capacity for work. Yet to friends Humboldt was writing that he was starting to fray: to Arago he admitted he was losing "the sureness of muscular movements" and felt unsafe reaching for books from a ladder, descending a staircase, or climbing into a carriage. To his old friend Baron von Bunsen he mourned the "loss of the phosphorous of thought, or a loss of weight in the brain, as the new school would say." He still tried to answer the thousands of letters he received every year, too proud (and too poor) to hire a secretary, until, just a few weeks before his death, he published a public notice begging his correspondents "in both hemispheres" to leave him in peace to finish his work. He had become a celebrity, a monument: as John Lloyd Stephens said, "The other monuments of the city would remain; but he might pass away." When Bayard Taylor informed him that young Washington Irving had turned seventy, Humboldt sighed, "Ah! . . . I have lived so long that I have lost the consciousness of time. I belong to the age of Jefferson and Gallatin, and I heard of Washington's death while travelling in South America." Francis Lieber remarked that "God allowed him days so long that he passed into history before he passed away from among us."[1]

In February 1857, Humboldt suffered a stroke that paralyzed him briefly, terrifying his friends. Varnhagen confided in his diary, "Humboldt's loss would be irreparable. He is a counterpoise to so much that is mean and contemptible, which, after his death, would boldly seek the light and the glory in its own depravity." But it was Varnhagen who passed first, in October 1858, leaving Humboldt to write to his niece that she alone could comprehend how "entirely isolated" he now felt. He withdrew from court affairs and pushed on with *Cosmos*, taking time in February 1859 to celebrate the birthday of George Washington with a band of American patriots. The occasion must have inspired him. One who was there reported home that, with "the stars and stripes encircling his manly brow, he said: 'I am half an American.'" Days later, in the midst of the fifth volume of *Cosmos* and the granite rocks of Russia, he paused, and instead of writing on, reviewed the finished pages and added his footnotes. On April 19, the manuscript went to the printer and Humboldt took to his bed, "exhausted and feverish." His mind was clear, but his body was failing. Too weak to speak at the end, he died peacefully at his Berlin home, on 6 May 1859, with Wilhelm's family at his side.[2]

The funeral was the largest and grandest Berlin had ever seen for a private citizen. Humboldt's family, followed by state and royal officials, diplomats, six hundred students, and hundreds more followed his hearse through the streets of Berlin, draped with black and crowded with mourners. The procession might have been even longer, except that the clergy stayed away, all but the one officiating minister and seven freethinking disciples of Schleiermacher. Among the students marched those from the United States with whom he had been celebrating just weeks before—"a cluster of mourners from afar . . . the only foreign nation thus represented." Humboldt was buried at Tegel, beside his brother Wilhelm and Caroline his sister-in-law.[3]

A wave of memorial orations and obituaries flowed across America. Many of Humboldt's friends used the occasion to assess his work and importance, which at this close view were seen to be staggering, though there was some disagreement whether he should be claimed by all the world, or only the world of science. It was not, as both Asa Gray and Edward Everett pointed out, that Humboldt had reached preeminence in any single field, but that no one else could bring all the sciences "into one field of view, and compare them as one whole, through their relations and dependencies." As Everett added, Humboldt himself was "an intellectual 'Cosmos' akin to the scientific 'Cosmos' of his own formation." William R. Alger literalized the metaphor: "The isothermal lines and vegetable zones belted his brain as they lie around the globe itself. The magnetic, meteorological, and astronomic observatories

established throughout the world sent in the reports to him as if they were
the nervous outposts of his personal intelligence." If Humboldt's brain contained the Cosmos, his body became the globe itself, each correspondent in
his all-connecting network a node on his central nervous system. At one pole,
Agassiz claimed Humboldt for the world of science, stressing his exclusion
of "every thing that relates to the turmoil of human society, and to the ambitions of individual men." Humboldt's travels and researches "completely
changed the basis of physical science," in a scientific "revolution" as profound
as the social revolution in France. Not a scientist in Europe (and Agassiz knew
them all) but had received favors from Humboldt and was deeply in his debt,
including himself. By contrast, Lieber opened the lens all the way: "There
is a void without Humboldt," he mourned; he couldn't imagine the world
without him. While others had emphasized Humboldt's "imperial" survey of
the sciences alone and insisted that he had "excluded" the turmoil of human
society, Lieber disagreed, stressing that Humboldt's "comprehensive view"
connected nature with man and the movements of society, "with language,
economy, and exchange, institutions and architecture."[4]

Agassiz may have tried to contain Humboldt within the world of science,
but the world at large agreed with Lieber. The American cult of Humboldt
climaxed in 1869, when a tsunami of celebrations of Humboldt's hundredth
birthday moved across the continent, from coast to coast, North and South,
in great cities and small towns. In New York City, ships in the harbor were
bright with bunting and buildings were hung with American and German
flags and portraits of Humboldt; City Hall was bedecked with Humboldt
banners and a crowd of twenty-five thousand—"two-thirds of them Germans"—gathered in Central Park to hear politicians, diplomats, professors,
and dignitaries give speeches and witness the unveiling of a bust of Humboldt;
that evening, torchlight parades, concerts, and banquets were held across the
city. President Ulysses S. Grant attended the celebration in Pittsburgh, Pennsylvania; Boston, Philadelphia, Newark, Paterson, and Jersey City joined in
with their own celebrations; so did Baltimore, Maryland; Washington, D.C.;
Richmond, Virginia; and Charleston, South Carolina. Upstate New York celebrated festivals in Albany, Buffalo, Poughkeepsie, and Syracuse (where the
parade stretched over a mile); so did the Midwest, from St. Paul to Milwaukee, Chicago to Indianapolis, to Dubuque, Iowa (where the parade stretched
for two miles). Celebrations were held in Louisville, Kentucky; and Mem-
phis, Tennessee; in Cincinnati, Cleveland, Toledo, and Dayton, Ohio; New
Albany and Evansville, Indiana; Springfield, Quincy, and Aurora, Illinois.
Peoria unveiled a Humboldt monument, Pittsburgh laid the cornerstone for
a Humboldt fountain and telegraphed "electronic greetings" to Humboldt's

friends Agassiz in Boston and Liebig in Munich. St. Louis held three separate celebrations—a concert, a parade, and evening speeches—and in San
Francisco the celebrations went on for three days. The *New York Times* marveled at the "exceptional" honors paid across the nation, as across the globe,
to Humboldt, who "has no special claim upon the citizens of this Republic,
and yet it is safe to say that in no part of the world . . . will his name be pronounced with respect to-day by so many lips as in this Republic, which did
not exist when he was born." Humboldt's hundredth birthday, 14 September
1869, was a grand national holiday, proclaimed and celebrated in virtually
every corner of the United States.[5]

Take little Dubuque, Iowa, where the local German society joined with
the Iowa Institute of Science and the Arts to celebrate the institute's opening
day on the Humboldt centennial. Rain postponed the parade until afternoon,
when it pranced through town in all its glory, led by a bust of Humboldt, a
band of German singers, various German societies, and a car of "young ladies"
representing all thirty-eight states. Following them were cars of butchers,
miners, and coal, the Typographical Union, cigar and soda and ice makers,
the telegraph, the postal service and the railway, lumber and sewing machines,
flour, barrel makers, stoves, and tailors, cured meats, and on and on. High
culture was represented by ornithology, geology, zoology, botany, conchol-
ogy, and two bears; low culture by a "Music and Burlesque" car which "created much amusement." "Humboldt teaches us what nature will produce,"
read the banner on the soda makers' car; "This was a man," proclaimed the
tinners; "Glory to Humboldt, the beloved of all centuries," declared the cabinet makers, and for good measure, "Honor to the man who worked for us
all," "The greatest man of this century." The heartland parade of American
science, agriculture, industry, and civic pride stretched for two miles and took
an hour to pass. Then everyone assembled at the town square, ten thousand
people under banners that read "Humboldt—the friend and patron of merit
and talent," and "Humboldt, the pathfinder in Natural Sciences, the man
of the people." After a long series of speeches by local worthies extolling the
"cosmopolitanism" of science and the "commonwealth" of a globe united by
truth, the procession reassembled and marched to the doors of the new institute, where a new bust of Humboldt was unveiled—"may the spirit of Humboldt . . . preside over our Institute," petitioned its new president—and the
doors were thrown open to the crowds. An address by a former president of
the AAAS followed that evening, and the event raised $1,314 for the fledgling
institute.[6]

In contrast with this populist outpouring was the much more formal celebration held in Charleston, South Carolina, by the town's several German

societies. Three long hours of speeches in a hot and crowded hall were concluded by placing a laurel crown on a statuette of Humboldt while a band
played "Was ist der Deutschen Vaterland." This was the last gasp of the
Charleston circle of naturalists, once second only to Boston but now crippled
by the Civil War, who had played so central a role in the debate over polygenesis. Old John Bachman stood first, but too weak to read his speech, he asked
that it be read by another. It was a loving, warm portrait of Humboldt as a
personal friend from his youth through the maturity of his career, full of anecdote and reminiscence. Bachman made no mention of controversies past
or present, other than a bitter reference to "Sherman's Vandal Army" who in
burning Columbia, South Carolina's capital, had destroyed all but one of his
letters from Humboldt, precious memorials of their long friendship.[7]

Quite different was the address by John McCrady, professor of mathematics at the College of Charleston, who had been trained by Louis Agassiz to
become the South's first and thus far only specialist in marine invertebrates.
McCrady hated Boston, loathed abolitionists, and spent his life extolling the
South as a higher civilization, the next stage in human development. After
fighting for the Confederacy he returned to find that he, too, had lost his
books and manuscripts to Sherman's army. In 1869 McCrady was still fighting, now to rebuild Southern science, and he used the occasion to define
the exact nature and utility of Humboldt's science for his cause. Even more
urgently than his teacher, Agassiz, McCrady needed to "separate" that science from the rest of Humboldt—as he said, he would exclude "the barren
obverse" of Humboldt's life, surrounded as it was by the "foibles of humanity." Clearly he had in mind Humboldt's famous abolitionism. One of the
Dubuque speakers had exulted that the triumph of Humboldt must silence
all friends of the Confederacy; conversely, in order to speak about Humboldt
this friend of the Confederacy needed to strip away his science from the dross
of society. What was left, though, was "pure gold": Cosmos, "the great work of
our age." In a careful and thoughtful analysis, McCrady specified Humboldt's
greatness as his ability to become "experientially acquainted with the inner
working life of each science," allowing him to do what no one else could, and
what, in the increasing specialization and fragmentation of knowledge, desperately needed to be done: tackle the great neglected problem of the *unity* of
the Cosmos. This was the essential first step, enabling all professional scientists from then on to understand how their own fields of research related to
all others. It was in this sense that Humboldt was, quite literally, the founder
of modern science. Where Bachman had seen science, society, and personal
history fused into one "personal narrative" in the classic Humboldtian fash-

ion, McCrady, representing the next generation, salvaged the heroic "gold" of science from the poisonous mud of social conflict and civil war.[8]

The Dubuque festival had showcased the friendship between Germany and America, putting on civic display their common goals and values, all identified with Humboldt: representative democracy, the productive labor of the working man, the aspiration of the common people for higher knowledge, the cosmopolitan cultivation of the arts and sciences. But as the Charleston festival hints, fissures were beginning to open. Where old Bachman offered a fable of the bashful commoner's son hobnobbing happily with Europe's elite scientists, young McCrady mounted a technical argument showing the need for discipline, expertise, and rigorous professional training like that he had received in Agassiz's laboratory. The nature of science in America was changing, actively reshaping Humboldt's legacy to serve new ends. This process was opening yet another fissure, too: Bachman, a Lutheran minister, had felt no conflict between science and religious faith, but he had already found himself forced to defend his hero from charges of atheism. Science, for centuries the handmaiden to religion, was assuming a new and radically dangerous character, and by 1869 the fault lines were visible everywhere. Even in Dubuque, the scientists' addresses, which celebrated the liberation of the human mind from all "fanaticism and superstition," jarred uneasily against the invocations of the presiding clergy, for whom nature spoke "the thoughts of God." As one speaker exulted, "Humboldt's Cosmos . . . is a new book of Genesis, a history of creation." If *Cosmos* was the new Bible, what of the old? Just where was God in Humboldt's Cosmos?[9]

In *Cosmos*, Humboldt had pulled his punches. This irritated Harriet Martineau, who observed that where his English translators used the word "creation," Humboldt's German used phrases such as "the frame of things, the universe, the collective phenomena of nature, or the like." An educated reader could infer what Humboldt meant, but his unwillingness to speak out plainly on the question of religion was, she thought, a moral failure. Edward Everett thought it was simply a function of science: it was "a strange assertion . . . that the 'Cosmos' is a system of philosophical atheism, slightly veiled from motives of prudence." Humboldt clearly intended to treat only of physical nature, simply leaving out spiritual and moral questions as beyond the purview of science. Or as the Free Religion advocate Francis Ellingwood Abbot declared in Toledo, if Humboldt is atheistic, then so is the census report. Nevertheless, the orthodox were alarmed that a work describing God's creation should be so very cryptic about the creator. As William Rounseville Alger bluntly asked, "Is Humboldt in hell?"[10]

Perhaps he was: his letters to Varnhagen had shocked the pious by revealing a Humboldt who, in private, pulled no punches in his contempt for the clergy, "those black coats ... who know how to enslave mankind anew." There was a reason the church had boycotted Humboldt's funeral. He had, at best, shown an utter lack of seriousness about Christian doctrine. To Varnhagen he mentioned writing long ago that all religions had "three distinct parts—First, a code of morals, very pure and nearly the same in all—next, a geological dream—and thirdly, a myth or historical novellette; which last becomes the most important of all." When Varnhagen lent him his copy of David Friedrich Strauss's *Life of Jesus, Critically Examined*, a watershed work of German Higher Criticism that punctured Christian myth, Humboldt confessed he was "delighted" with the book (and returned it with a copy of Byron's *Don Juan*!) He made merry with one earnest old man who, Humboldt wrote, tried to convince him that the souls of "bed-bugs and musquitoes, are included in the scheme of salvation, and destined to go to heaven. So they threaten me up there too," joked Humboldt, "where I shall find the animal souls, well known to me from the Orinoco, chanting a hymn of praise." After publication of these letters, several Humboldt sarcasms on the church circulated widely. In one, an ultra-right-wing religious conservative tried to corner Humboldt in public by asking him, "Your Excellency frequently goes to church, 'now-a-days,' do you not?" To which Humboldt replied evenly, "Your 'now-a-days' is very kind of you. You allude, undoubtedly, to my adopting the only road which, at present, could lead to my promotion." In another—a one-liner Humboldt used on several of his American visitors—he would show off his pet chameleon, pointing out how it was able to direct one eye upward and the other downward: "Our parsons only were able to do the same, with one eye directed to heaven and the other to the good things of this world."[11]

Alger's conclusion was that, if Humboldt was in hell, "he is there in pretty good company," with "Socrates, Aristotle, Archimedes, Cicero, Bruno, Newton, Locke, Priestley, Lavoisier, Franklin, Goethe," none of them orthodox either. In worshiping the Cosmos rather than a personal God, "he was more religious in his unbelief than most men are in their belief." But for many, the question was less whether Humboldt was in hell, than what kind of hell his views could create on earth. In this respect the Boston celebrations are particularly revealing.

Boston's official celebration of the Humboldt centennial was sponsored by the Boston Society of Natural History, which gathered together not just scientists but all the luminaries of the region. As the organizing committee said in its letter appealing for financial support, "The physicist, the naturalist, the philologist, the geographer, the student of history, of literature and of the arts, are all his debtors, and we ask the cooperation of all." The main event was an afternoon address by Louis Agassiz, with the proceeds from ticket sales establishing a Humboldt scholarship to support "young and needy persons" studying at his Museum of Comparative Zoology. Agassiz was preceded and followed by a lush selection of German music: a Bach organ prelude, excerpts from Mozart's *Magic Flute*, sections of Beethoven's Seventh Symphony, and a song by Humboldt's old friend Felix Mendelssohn. Agassiz's two-hour lecture was a personable ramble brimful of wisdom and anecdotes, including memories of the day the great Humboldt, hearing of his distress, had dropped out of the clouds to rescue the starving Agassiz and lift him up into the great halls of science. Over a long dinner at a nice restaurant, Humboldt instructed Agassiz in *how* to be a scientist: "How to work, what to do, and what to avoid; how to live; how to distribute my time; what methods of study to pursue." They sat together during Cuvier's famous lectures refuting Geoffroy Saint-Hilaire's evolutionary theory, Humboldt whispering jibes against Cuvier into Agassiz's ear. While Agassiz spoke he stood next to two portraits of Humboldt, one painted by a Mexican artist in 1803 and the other by the Boston artist Moses Wight in Berlin in 1852. The Boston Music Hall was packed, with delegates from learned and scientific societies around New England, representatives from Harvard, Yale, Brown, Bowdoin, and Dartmouth, the mayor of Boston and the governor of Massachusetts.[12]

That evening the paintings were moved to Horticultural Hall, where they joined other Humboldt mementos—a palm frond that had rested on his coffin, his writing pen, several letters—as backdrop for the reception. Agassiz relaxed while other speakers took the stand: first Thomas Wentworth Higginson, Transcendentalist, abolitionist, colonel of a black regiment in the Civil War, and "preceptor" to Emily Dickinson, to celebrate Humboldt's scholarship and love for the arts and literature; then the Unitarian minister Frederic Henry Hedge, one of the founders of Transcendentalism, to proclaim that Humboldt's only true peer was Aristotle; then Ralph Waldo Emerson with his brief address, completing the troika of surviving old-line Transcendentalists. Professor E. J. Young recounted his 1854 visit with Humboldt; then a copy of the Wight portrait was unveiled and presented to the BSNH. Charles T. Jackson, vice president, accepted it graciously, pointing to the importance of Humboldt as a model for young scientific travelers. Mayor Shurtleff read a resolution from the city of Boston honoring Humboldt and adjourned the company to supper, after which Oliver Wendell Holmes regaled the assembled company with his poem "Bonaparte, Aug. 15th, 1769.—Humboldt, Sept. 14th, 1769":

> For God's new truth he claimed the kingly robe
> That priestly shoulders counted all their own,
> Unrolled the gospel of the storied globe,
> And led young Science to her empty throne.

"Hero of knowledge, be our tribute thine!" concluded Holmes, completing his apotheosis of the "peaceful conqueror" over the despot Napoleon. Higginson then read a poem by Julia Ward Howe, poet, abolitionist, and author of "The Battle Hymn of the Republic":

> Shine out, O West! illumined by his traces,
> Ere the cramped world took notice of thy state;
> He gave the record of thy virgin graces,
> And in prophetic vision saw thy fate.

Finally, several letters were read from invitees who could not attend, notably Ambassador Theodore S. Faye, who narrated the Humboldt River anecdote, and the poet John Greenleaf Whittier, who reminded the audience that Humboldt's "generous and hearty interest in the cause of freedom in the United States can never be forgotten by those of us who, in dark and evil days, were cheered by his approval and sympathy." By evening's end the scholarship fund stood at $9,035.71, plus any profit raised from the sale of the proceedings. All in all the day was a most impressive display of New England's "universal culture."[13]

The vexed question of religion recurred several times, striking the one discordant note. The freethinkers Higginson and Emerson (who was himself a hero to British Dissenters) did not mention religion at all, though Holmes stepped onto thin ice with his vision of Humboldt leading Science to the kingly throne vacated by the priests. But the thought that Humboldt had displaced religion disturbed the Reverend Frederick Henry Hedge, who objected that Humboldt could not have pursued his inquiries without a belief in God—indeed, though Humboldt may not have been a believer, the marvelous Humboldt *himself* "is the most convincing proof of the truth he missed." Agassiz worried the question for several pages, noting that "the modern school of Atheists" were claiming him as their leader, but Humboldt was "no scoffer" (really?) and he "had too logical a mind to assume that an harmoniously combined whole could be the result of accidental occurrences." Most tellingly, Ambassador Faye's dispatch from Berlin struck an ominous note: he had "observed indications in Europe that some parties will attempt to convert the celebration of his centennial anniversary into a demonstration in favor of materialism and atheism."[14]

Indeed. Even as Faye's warning was ringing through Horticultural Hall, the very cabal he feared was holding forth at Boston's Music Hall, filling it for a second time that day with Humboldt celebrants. But this event, the *German Humboldt Festival*, was quite different in tone. Its leader was Karl Heinzen, the German Forty-Eighter notorious today for establishing the theoretical justification for "selective homicide as a spark to general revolt"—that is, terrorism. Humboldt had admonished his audiences that the oppressed would, by the laws of nature, revolt to seek justice. Heinzen merely rephrased Humboldt's argument: if reactionaries did not accede to the calls for justice, then only murder and violent revolution would bring freedom. Humboldt was Heinzen's hero and, as he said, the hero of Germans everywhere, but most especially those in the United States, "which the great thinker united to Germany by an intellectual tie." But Americans, announced Heinzen, had gotten Humboldt exactly wrong. Since they had presumed to claim him as the second discoverer of themselves, it was high time for Germans to turn the tables and become "*the second discoverer of Humboldt*" for America. Heinzen's Germans had been invited to the Agassiz affair, but had turned the invitation down because only at a separate festival could Germans freely proclaim the truth about Humboldt, the truth that Americans didn't want to hear: that their hero, the great liberator and enlightener, "was an atheist and a materialist." Every literate German knew this, and if it horrified the Americans, then that just showed how shallow their knowledge of Humboldt was. Humboldt's nature was self-creating and self-organizing, and the goal of his science was to dissolve the dreams of myth and replace them with the rule of reason and scientific law. Just steps away from the Boston Music Hall was the Boston Court Hall, where by state law, Humboldt would have been imprisoned for a term of two years—still, even in 1869, the legal punishment for religious heresy. Had Humboldt actually come to Boston, the good city fathers, now so full of his praise, would have thrown him in jail.[15]

More was to come from Adolph Douai, a German American socialist and abolitionist who protested that not only did Humboldt not believe in God, Agassiz "had no right to participate in a memorial meeting in honor of the German scientist." Agassiz was a racist who, as anyone could read in his 1850 *Christian Examiner* article, "had asserted his belief in Negro inferiority and helped to provide a rationale for slavery." That the racist and proslavery apologist Agassiz was allowed to memorialize Humboldt, who had stood all his life against slavery and for racial equality, was a travesty. As Heinzen warned, those Americans who preen themselves on holding Humboldt in such high esteem should consider what he really thought of them. While they may have, at long last, abolished physical slavery, there was also "a species of intellectual slavery, much more degrading and pernicious than bodily slavery, since it is the mother of all others." And this was exactly the condition of America: intellectual slavery to consumerism and greed. Nor was it just the Germans who used Humboldt as a club to beat American smug complacency. That same day, Robert Ingersoll told an audience in Peoria, Illinois, that Humboldt had proclaimed that a new day was dawning. Science was "freeing the soul—breaking the mental manacles—getting the brain out of bondage—giving courage to thought—filling the world with mercy, justice, and joy." This was the true legacy of Humboldt, to whom "millions" were doing homage at that very moment. Humboldt stood for truth, reason, and love of nature, against superstition, the "stupidities" of theological metaphysics, and the ancient fear of God. "The world is his monument." Upon "the eternal granite of her hills he inscribed his name," and with it, the most sublime of truths: "THE UNIVERSE IS GOVERNED BY LAW."[16]

Dr. Samuel Osgood had written back from Berlin to tell America that in Germany, Humboldt's name was synonymous with "'modern times,' and Young Germany is wholly modern." The young, the radicals, the freethinkers, the socialists, were taking Humboldt as their patron saint. In 1869, Humboldt pointed to the future, not to the past: to universal peace and healing after decades of war and revolution, to the reign of reason, to the liberation of the human mind, to the dawn of a new age. In the United States, Humboldt was an icon, a celebrity of rock-star proportions on whom Americans projected their hopes and ideals. His name inspired a nationwide fervor of admiration and emulation. It wasn't just a household word, it was a point of reference, a rallying cry for cosmopolitan civic pride, for modern culture, for high science and the working man and a utopian future. Everywhere—coast to coast, north and south—he was known and beloved, so important that different cultural groups fought to claim his legacy for their own.[17]

How times have changed: today, scholars spend entire careers steeped in nineteenth-century America without learning that Humboldt ever lived. Stop a thousand educated Americans on the street and ask them who Humboldt was. Nine hundred and fifty will ask blankly, "Who?" (The other fifty will have come from somewhere else.) The name that graced a century, that defined an age, that anchored the map of America, has vanished.

What happened?

❧

A silencing this total, an erasure this complete, requires several levels of explanation. "Ignorance is often not merely the absence of knowledge but an outcome of cultural and political struggle," observes Londa Schiebinger, and there were many layers to this struggle in the United States. First, there was the very function of monuments to the past in postbellum America. The Civil War shattered the old world, and building it anew required burying what remained. One sees this in the memorials of Emerson: his biographers cast this philosopher, so deeply engaged with science, as unscientific because he was not systematic in the Comtean, positivist sense. Transcendentalism was dismissed as endearing but dotty and quite irrelevant to modern science and a new generation. One sees an identical move in the standard biography of Humboldt, published in English in 1873. In its conclusion, Alfred Dove shovels a great deal of dirt on Humboldt's coffin when he writes, "Science had in the mean time undergone many important changes . . . the aspect of science in 1834, of which 'Cosmos' was the record, had already passed away." Acts of veneration, via statues, fountains, parks, and massive life-and-letters biographies, were an efficient way to, in effect, bury the venerated, to make sure the dead stayed dead, less to remember than to forget. This is especially vivid in the Boston centennials: to the Germans, Humboldt was alive, vital, profoundly and urgently relevant. To Agassiz and company, he had passed into history and his relevance was fading fast.[18]

Then there is the historical fact that Humboldt was not a professor and never sought recruits, so he did not leave behind a school or group of students to perpetuate his name, as did, for instance, Agassiz. Worse, as several of his memorialists pointed out, his name was not attached to one single, commanding discovery. He was not a scientist so much as a "metascientist," reaching, as Charles Sanders Peirce surmised, toward a future that would rethink the very grounds of knowledge. Newton's name will live so long as gravity remains of interest; Darwin's name still provokes fights; but Humboldt's? Even the "Humboldt Current" has been renamed the Peru Current. It might have been tied to the foundation of ecology, which could have carried considerable cachet, except that the field was not named until Humboldt's German disciple Ernst Haeckel coined the word for Humboldt's connective science, which thus carried his own name, not his master's, into posterity. So Humboldt's name slips into obscurity even here: when twentieth-century environmentalists sought their intellectual forebears, they found Thoreau and Muir and Marsh, but they couldn't find Humboldt, from whom all three had drawn their inspiration. Agassiz was aware that though Humboldt was ubiquitous, Americans were already losing sight of him, and he thought he understood why: "Every school-boy is familiar with his methods now, but he does not know that Humboldt is his teacher." His mind is hidden from us "by the very abundance and productiveness it has caused." In a score of fields, scientists aided by Humboldt built on his work to make names for themselves that live on when his has been forgotten. As Kellner observes, Humboldt's is "the anonymity of a great teacher." Or as Whitman wrote, "The proof of a poet is that his country absorbs him as affectionately as he has absorbed it." Perhaps the United States absorbed Humboldt so completely that he has disappeared from view, though he is everywhere one looks.[19]

There is another way that Agassiz's memorial addresses point directly to himself, and through him to the irresistible development of American science, as culprits in the suppression of Humboldt. Agassiz had good reason to wish to bury Humboldt. For one thing, his mentor had been an enthusiastic proponent of evolution, and while Agassiz was expending every penny of his intellectual capital howling down Darwin, he likely heard Humboldt at his ear whispering jibes against—himself. As Douai had charged, Agassiz had taken his mentor's ideas in a very different direction. Humboldt knew this, but, characteristically, refused to cross an old friend. At one point in a letter to Agassiz, finding himself on the brink of argument, Humboldt breaks off with an apology: "I pause here, for I feel that I must annoy you, and I care for you too much to run that risk." The distance between them grew, for it was not by promulgating Humboldt's egalitarian, cosmopolitan, and materialist ideas that Agassiz made himself the visible center of American science, but by adapting them (like Guyot, his colleague now at Princeton) to a racist, nationalist, and idealist ideology much more responsive to the needs of the powerful.

The racism and nationalism have been sufficiently sketched herein, but it is worth spending a moment on the idealism. Paradoxically, it was Agassiz, with his old-fashioned religiocentric idealism and fanatic opposition to Darwin, who carried American science forward. Agassiz played a key role in crafting the ideology of scientific objectivity, according to which the "facts," not the scientist, proclaimed the laws of nature. As Agassiz reiterated relentlessly, exactly what the "facts" proclaimed were the providential thoughts of God. By thus separating the object of science from the human observer, then mandating that the object "speaks" or "proclaims" itself through the channel of the (passive) scientist, Agassiz could claim to be propounding the truths, not of Agassiz, but of God. The ideological load this could carry is most obvious in the article that Douai referred to, where Agassiz regretted that the thought of God, as revealed by the scientist with the courage to face the facts, was the Negro's tragic inferiority. His students, who nearly all became Darwinians (the exception was McCrady, the Charleston zoologist), easily dispensed with his theological underpinnings while retaining the rigorous, laboratory-based scientific structure erected on them: from then on, the facts of nature spoke through the transparent agency of the scientist, making science the one un-impeachable standard of truth, holding forth all the authority it had inher-ited from religion. In this process Humboldt was necessarily displaced, for he had foregrounded the voice and subjectivity of the scientist, always making his choices and actions and feelings present in the text as he tried to read order into the bewildering complexity of nature, tried to record, translate, and mediate its many voices. This "arresting self-reflexiveness," as Suzanne Zeller observes, was the saving grace of Humboldt's science: "The cosmic laws which he aspired to grasp lay always in the search itself."[20]

Yet the positivist ideology of scientific objectivity overshadowed such con-cerns, and remains so dominant today as to make rational alternatives almost unthinkable. According to it, Humboldt's voice is anachronistic and obsolete, *Cosmos* a mere "curiosity." He is too literary to be scientific, even as he is too scientific to be read as literature. Yet Humboldtian science was far too useful to be jettisoned. The solution was to strip away, in the fashion of McCrady, his offending subjectivity, his quaint concern with the intellectual and emotional role of the observer and the moral status of the witness, and make use of the very practical and fruitful product that remained. Historians of science today build careers exploring these methods and results under the rubric of "Hum-boldtian science." One can watch the process taking shape in Agassiz's labo-ratory, the training ground for the new science whence came the next great wave of productive engagement with nature, the next big thing. With a little push from Agassiz's former students, the noisy old Humboldt simply faded away. His efforts to develop the fledgling science he called "cosmography" had earned him the reputation of a scientist above all, and spawned a score of sci-entific specialties, but as science continued to develop and professionalize, it bypassed Humboldt's holistic cosmography for the very reductionism he had criticized. What happened was a bit like the drama in Humboldt's engraving of Mexico City, but in reverse: now it was the New World that had conquered the Old, taking it apart stone by stone and with those very stones erecting its own cathedrals and palaces. Science is a palimpsest too, its form revealing its history.

In 1887, Boas entered the field as a dissenter, reclaiming Humboldt's cosmography as the way to the future, past the dead-end fragmentation of knowledge. McCrady had thought the "gold" of Humboldt's science was the integrative framework it provided, and it is possible today to see Humboldt pointing not to the past but to the future. His integrative model should ap-peal to the same audience recently riveted by Edward O. Wilson's landmark volume *Consilience*, for Humboldt, too, offers a framework that embraces not just the positive results of the wide array of physical and natural sciences, but the subjectivity of the scientific observer, and with it, the history of scientific thought and its cultural connections all the way out to economics, politics, literature, and the arts. As the BSNH organizing committee had pleaded, we need the cooperation of all these branches of "universal culture." "Consilience" was, ironically, coined by yet another Humboldt follower, the British philosopher of science William Whewell, to name the Humboldtian concept of the "leaping together" or convergence of truths into a new and larger truth. Whewell's metaphor for consilience was, just like Humboldt's in *Cosmos*, a river system: the landscape of knowledge is interlaced with streams, tributaries, and channels connecting fields and disciplines like watersheds. Upriver and down, no field "reduces" to any other; each retains its own integrity and identity even as each contributes its stream to the great hydrologic system of global knowledge. How different from Wilson's twentieth-century appropriation of this profoundly Humboldtian concept: Wilson insists that all life be "reduced" to the laws of physics and chemistry, even art and poetry, which no longer have their own emergent integrity as distinct forms of knowledge but are merely epiphenomenal to the commanding laws of science. Few humanists can accept such a view, and as a result, the "two cultures" of literature and science continue on their separate paths, unabated. [21]

Whewell's consilience was an expression of the age of Humboldt; Wilson's *Consilience* was an expression of the age of reductionism. What the twenty-first century needs, now that biodiversity as well as cultural diversity are everywhere in crisis, is a neo-Humboldtian concept of Cosmos. The "two cultures" is at a dead end. What Rochelle Johnson calls the "canalization" of thought has cost us terribly, for the humanities and the sciences, as they have propped each other up in their mutual half blindness, have between them allowed environmental crises to escalate until today serious and thoughtful people foresee, in a single lifetime, the end of life on this planet as we know it—as the human species has *ever* known it. As we face the unthinkable, Humboldt's Cosmos gives us a new concept to think with, a way to reorient and transform disciplines and divisions that threaten not just to leach the poetry out of our technologically driven lives, as we worried not so long ago, but to cut away the very ground under our feet. Humboldt's Cosmos is about regrounding imagination, about reimagining science, about reminding us that humans are natural and nature is human. This division that built the modern world no longer makes any sense; we must begin to think of humans and natures as braided and roped and woven together in a resilient new Casiquiare exchange. What the spectacular success of scientific specialization has hidden is not just the integrative framework for netting together the many forms of knowledge and of beauty, but the very need for such a framework. We have no language for speaking of humans and natures as a single, law-governed, dynamically interactive and reflective whole. So we face the future, this peril we have precipitated, shorn of tools and of the organs of perception that tools provide. Cosmos was the fabric that Humboldt spent a lifetime weaving, the one fabrication that, he thought, would make all our technologies and visions and hopes and poems speak to each other. Like groundwater, it's still there, under our feet; in a dry time what we need are deep, taprooted ideas that can draw it back into our cultures and literatures.

The elements for a "cosmographical," integrative perspective continued to be a part of twentieth-century thinking, as various intellectuals, like Boas, re-discovered Humboldt and carried his philosophy and methods into our own time, particularly in geography: Lewis Mumford, Carl Sauer, Murray Book-chin, Yi-Fu Tuan, among others. Yet they remain dissident voices. Humboldt had challenged the ideology of his time so profoundly that finally his challenge could not be assimilated into the modern project. As Mumford observes, the movement formulated by Rousseau made major contributions to science, for it was continued by Goethe and Humboldt and a whole generation of nineteenth-century naturalists headed by Darwin and Wallace. "But in the long run it was ineffective because it could not be attached to the Power Complex without forfeiting its own principles and ideals. Unlike the situation in Defoe's fable, 'Robinson Crusoe,' the ship abandoned by the romantic castaways had not been wrecked, but became increasingly seaworthy and was headed for more distant ports." Helmut de Terra's reflections follow a similar line: Humboldt's "attractive practical philosophy might well have made educated people better conscious of their common purpose, rather than of their differences. It might have prevented the excesses of colonialism and eased class struggles." But Eu-rope rejected Humboldt's portrait of the Cosmos, for "a more materialistic ori-entation . . . was destined to replace Humboldt's visions with other values."[22]

It could also be argued, as Sandra Nichols does, that Humboldt's writing simply went out of style. The (self-admitted) faults of his prose—long, con-voluted sentences, frequent digressions, a lack of structural unity—were not aided by his Victorian translators, who gave English readers nearly the only Humboldt we have. Yet his letters snap and sparkle with wit, and he was uni-versally admired as one of the best talkers of the day, charming the glitterati in salons from Caracas and Quito to Washington, D.C., to Paris and London. His lectures packed huge halls to overflowing, and Nichols is surely right in guessing that in our day, Humboldt would have been a media star, like Carl Sagan, Stephen Hawking, or E. O. Wilson himself. With a bit of help, his books might well have continued into the twentieth century as classics, shelved next to Thoreau and Muir. But writers do not survive in the literary canon by merit alone. They are taken up or abandoned through particular decisions made by publishers, reviewers, advertisers, critics, and readers. Many of the giants of American literature, Thoreau, Poe, Melville, Whitman, Douglass, Stowe, had to be recovered by a handful of critics, sometimes only one, who energetically and successfully lifted them from obscurity by making the case for them as major writers. As Theresa Kelley and Paula Feldman have argued in championing female Romantic poets like Humboldt's friend and translator Helen Maria Williams, only zeal, advocacy, and political power can break the silence imposed by the canonization of select male Romantic poets. Had Humboldt found such a champion, he could have won new and updated translations that would have kept his books in circulation. [23]

Given the temper of the times, no doubt Humboldt would have been canonized as a "nature writer," narrowing his circle of readers but still keeping him alive, as happened, for some decades, to Thoreau. Currently, some eco-critics are making the case that all texts, not just so-called nature writings, need to be read in terms of the relationship they bear to the natural world, but the key to Humboldt, as to Thoreau, is reading his "nature" writings in relation to the *social* world. This points to another problem: Humboldt's baggy hybrid works were dismissed precisely because they fit in no one genre, least of all the reigning genres of "literature," fiction and poetry. Harriet Martineau, for one, is profoundly annoyed with Humboldt for making *Cosmos* "a hybrid production between poetry and science," words she does not mean as a compliment but as a diagnosis. Yet his very hybrid, hypertext quality could make him attractive today, in a postmodern environment that welcomes experimental forms and increasingly embraces the digital humanities. Database, says Whitman scholar Ed Folsom, is the new genre of the twenty-first century—what happens, he wonders, when we make such "rhizomorphous" work into a database, "put it online, allow for the webbed roots to zig and zag with everything the database incorporates?" [24]

There was also the schism over religion, which in the United States was, as we have seen, tied in complex ways to both nationalism and to scientific culture. The proceedings and reports of the 1869 Humboldt centennials show a country still fighting the battles of the late 1700s: was the United States a cosmopolitan nation based on reason, enlightenment, and liberation from superstition, as Jefferson had believed? Or was it still a Puritan stronghold resting on God's word and providential design? A wide spectrum of positions had opened up around Humboldt, from conservatives upset with his failure to acknowledge the creator, to moderates like Agassiz and Bachman who appropriated him as a Christian in spirit if not in doctrine, to freethinking Transcendentalists like Emerson and Higginson, to radical materialist atheists like

Heinzen and Ingersoll. But this debate was swamped by the epic controversy over the Darwin revolution, and Darwin's name replaced Humboldt's as the flash point for religious controversy. By the time of the Scopes trial in 1925, the same language once used to hail Humboldt was applied to Darwin instead, and the debate has continued to replay the same positions for nearly a hundred years. Thus the religious debate over Humboldt's true character was absorbed by the Darwin debate over the true character of humanity. Nothing in Humboldt's *Cosmos* was nearly so exciting as arguing whether one's opponent was a monkey's grandson.

These were mortal wounds, but the coup de grâce was struck by an emerging American anti-Germanism. This was a part of a larger movement: where Humboldt had imagined "America" as a transnational entity that would revolutionize the globe, America the nation was closing in on itself, proclaiming its political borders the limit of its national identity and English as its one language. Already in 1869 voices like Heinzen's were being identified as dangerous and alien subversions of patriotic civic Christian nationalism. Faye's warning of a German conspiracy was a straw showing the way the wind was blowing. Sandra Nichols, among others, has documented what came next: in 1870–71, the Franco-Prussian War that forged Germany into an empire turned the United States into Germany's enemy, even as German immigrant numbers, already high, skyrocketed. Germans were flooding the country, settling into their own communities, holding onto their own traditions, celebrating their ethnic pride in a stream of festivals, reading their own German-language newspapers (by 1890, eight hundred existed in the United States alone), insisting on public school instruction in German. Forty-Eighters like Heinzen were especially troublesome, for these highly educated political refugees brought their "liberal, agnostic, and intellectual" arguments for reform into American politics: they were active in labor causes and became the "backbone" of the trade-union movement. When word spread that five of the six anarchists condemned to death for fomenting Chicago's violent 1886 Haymarket Riot were German, popular views of German immigrants and their subversive ideas deteriorated. One German custom particularly enraged Puritanic Americans: the holding of festive "beer gardens" on Sunday afternoons, with dancing, music, bowling, and a great deal of liquid refreshment. Reactionary groups rose in protest, passing blue laws to protect the Sabbath and forming temperance groups and English-only movements.

Friction increased as Kaiser Wilhelm pushed his gunboat diplomacy into South America and the Caribbean, and Germans became "Huns." Americans began to suspect the loyalty of German Americans, many of whom reciprocated with a defensive nationalism of their own. In 1914, when America learned that Germany had invaded Belgium, the country erupted into full-blown anti-German hysteria. German Americans were harassed, mobbed, and lynched. German language and authors were banned from public schools, book-burnings across the country encouraged Americans to cleanse their libraries of everything German, statues of Goethe and Schiller were vandalized, sauerkraut became "liberty cabbage" and hamburgers, "liberty steak." Cities, parks, and schools were renamed, the names of German families were anglicized. The impact, writes Nichols, was "profound," and though the hysteria subsided after the war, any last desire to hold on to a distinctive German heritage was annihilated by Hitler and the Holocaust. In the United States, the memory of German culture was driven underground, Nichols writes, "and with it, the memory of Alexander von Humboldt." Or as Aaron Sachs remarks, "there could be no Prussian heroes after 1917." Humboldt had once been celebrated all across the Americas, nowhere more so than in the country proud to claim him as "half an American." That Humboldt's name remains so vital in Latin America, where German culture was not purged from national memory, suggests that Nichols is right. He might have survived the Darwin debates over religion, the shift in literary taste, the evolution of modern scientific ideology, even the appropriation and dilution of his ideas, but Humboldt could not survive U.S. America's extraordinary capacity for xenophobia combined with the militarism of the German state.[25]

Need one remark on the terrible irony that the legacy of the man who lived to fight the imperialistic war machine was destroyed by that very machine? The erasure of Humboldt was, of course, merely collateral damage, like the looting and burning of Tegel that destroyed the family library and his drawings and paintings. The irony cuts still deeper into history: the fervent tone of the 1869 Humboldt celebrations was due to America's urgent need to heal after the terrible Civil War. Humboldt, that icon of unity, the hero of knowledge and peace, offered what Americans most needed, a universal culture of learning and an ideology of healing, a way to look beyond America's war to its place in the globe, its regenerated future in, and as, in Whitman's phrase, the kosmos—Whitman, of course, the Civil War nurse who spent the rest of his life trying to nurse America back to wholeness. Lincoln's Emancipation Proclamation had at last removed the great barrier that stopped Humboldt from endorsing the United States, which had finally fulfilled its promise to him. This lies behind the many declarations that in 1869, the United States, of all the countries in the world, had a uniquely important reason to honor Humboldt.

At the beginning of the Civil War, Frederic Church had painted Cotopaxi in full eruption, a violent and brooding image full of foreboding. Before the

FIGURE 20. Frederic Edwin Church (American, 1826–1900), *Rainy Season in the Tropics*, 1866; Oil on canvas 142.9 × 214 cm (56 ¼ × 84 ¼ in.): Fine Arts Museums of San Francisco, Museum purchase, Mildred Anna Williams Collection, 1970.9.

lurid volcano opens a chasm that splits the land in two. Many have suggested this painting's allegorical import, especially when it is paired with *Rainy Season in the Tropics* (fig. 20), painted just after the war ended. An allegory, without doubt, of God healing the planet with a rainbow. Yet I wonder. When politics got to be too much for Humboldt, when everything seemed to be slipping backward, he looked for redemption and hope to the deep harmonies of the Cosmos. Thoreau, too, turned from the pollution of Webster's Fugitive Slave Law to the purity of the wild water lily. "What confirmation of our hopes is in the fragrance of this flower!" he exclaimed. So long as nature can compound this fragrance, "I shall believe her still young and full of vigor, her integrity and genius unimpaired, and that there is virtue even in man, too, who is fitted to perceive and love it." This seems to have made sense in Thoreau's and Humboldt's century, but does it today? When we in the twenty-first century look to the Cosmos, we see a nature not "unimpaired" but melting away into rising seas and mass extinctions. If Humboldt's nature was harmonious, ours is chaotic, unpredictable, even terrifying—the best climate scientists in the world were caught off guard by the accelerating loss of the Greenland icecap and the sudden opening of polar seas. What have we done? Losing Humboldt cost us the Cosmos. Must it now cost us the planet as well?[26]

So in the spirit of hope I end not with nature but with art. There again in Church's rainbow is Humboldt's bridge, arcing over the chasm, a symbol of God's peace and also of Newton's science, for it was Newton who unwove the mystery of the rainbow by showing its scientific cause in the prisms of raindrops. Unwove? Hardly, answers Church. Carefully constructed according to the latest optical science, etched onto the canvas with a compass, Church's rainbow is yet stunningly beautiful. In the original painting it leaps off the canvas and lights up the gallery. A perceptual trick, an optical illusion? Rainbows are phenomena of perception. Like Thoreau's scarlet oak leaves, they come into being only as part of a perceptual triangle, sun through rain to eye. Or here, light through pigment to eye. Or, mind through body to the real, the phenomenal, the surface which bespeaks its depths.

The viewer rests at midlevel, not too high, just high enough to relate all the elements, rain and light and chasm, the village in the valley beyond, the rain-slicked road below, the passing travelers, Andeans leading pack mules, that red poncho a bolt of concentered energy. They have paused to retie saddlebags, and their voices echo across the chasm as the beat of the rain clears. A painting of passage: travelers crossing a pass, rain passing, time passing. One blink and it will be gone, the light faded, the travelers vanished around the corner. The viewer is in passage too, gazing through the curtain of rain to the rock face beyond, but it is the road that pulls us in, down into the valley of human community, looking ahead to the city where we might dwell.

Shall I admit that when I first saw this painting, I fought back tears? Why? For joy in its sheer beauty and in the skill of the artist, joy that such things be, rain and paint and high mountains, joy in all the memories it awoke of my own passages, long dull crossings snapped into brilliance by a bolt of light, chance beauty after harrowing storms. Humboldt would tell me that my tears were not trivial or embarrassing but part of the painting, part of the Cosmos, as is every kinglet moment. I could respond, But I know better. This is a Christian allegory; this is an ideological tool; this is an aesthetic machine. But this one time, I choose not to, for I am free to say instead, just this once, this is Cosmos, in all its enchantment and fragility.[27] The facts before me give me this unfettered freedom of feeling: wind scudding the clouds apart, ebbing rain glanced by a sunbeam, an optically exact rainbow arching over a geologically precise mountainside faced by botanically correct palms, a road too slick to trust on the back of a mule, a mule twitching his ears forward and back in thought, a rain-hazed village below, journey's end, as this moment, too, passes. Only now I am no longer alone but embraced by a community, the communion of Cosmos, present to me, one subjectivity standing in a gallery, one raindrop prism of consciousness.

In a recent essay, Wai Chee Dimock has suggested to professors and scholars like myself that it is our humanist obligation, through our professional rituals, to give voice to the dead, to lift from them "the gag order that comes with mortality." Their failures and their fallibilities are our own, and by embracing these failures, "we embrace the unfinished task of the dead," allowing them to come back as noisy ghosts, "garrulous beyond the grave." Our very weaknesses oblige us to give them a hearing and so acknowledge that authorship is too large to be borne by any one person, that it must be borne in "the deep field of time," "by the longitudes and latitudes of mankind and the full duration and extension of its recorded history." Few of our dead have been so completely silenced as that most garrulous of ghosts, Alexander von Humboldt, and today none deserves to be heard more than he. Perhaps, now, the frequencies of his texts, "received and amplified across time," will resonate with the wavelengths of today.[28] Humboldt saw the shape of a new world emerging, though like Margaret Fuller he thought that world was stillborn, abandoning him to mourn its defeat and write hopefully toward some future renaissance. It is our obligation to let him come back, to make noise in our aesthetics and politics and ethics and science, and hearing him, both to acknowledge his failures and to acknowledge in them our own. As Dimock says, the sequel—if there is to be one—is up to us.

Notes

Preface

1. Quoted in Stephen Jay Gould, "Church, Humboldt, and Darwin: The Tension and Harmony of Art and Science," in *Frederic Edwin Church*, ed. Franklin Kelly et al. (Washington, DC: Smithsonian Institution Press, 1989), 104.

2. Jorge Carrera Andrade, "Humboldt," *Selected Poems of Jorge Carrera Andrade*, trans. H. R. Hays (Albany: State University of New York Press, 1972), 230–31.

3. Nicolaas Rupke marvels that postwar Anglo-American Humboldt scholars operate "in virtually complete and blissful ignorance of the many hundreds of German publications on Humboldt, the brothers Humboldt, Humboldt and Goethe, Humboldt and classicism or idealism, Humboldt and Marxism and so on"; see *Alexander von Humboldt: A Metabiography* (Frankfurt am Main: Peter Lang, 2005), 191. For an introduction to German Humboldt scholarship, see Wolfgang-Hagen Hein, ed., *Alexander von Humboldt: Life and Work* (Ingelheim am Rheim: C. H. Boehringer Sohn, 1987). Curses on his Victorian translators belitter Anglo-American Humboldt scholarship. Jason Wilson's brisk but abridged translation of Humboldt's *Personal Narrative of a Journey to the Equinoctial Regions of the New Continent* (New York: Penguin, 1995) does not solve the problem. The most popular American edition of Humboldt's *Cosmos*, translated by Elise Otté, is available in an inexpensive reprint edition (2 vols., New York: Harper and Brothers, 1850, 1858; Baltimore: Johns Hopkins University Press, 1997). Spurred by the upsurge of interest in Humboldt, the University of Chicago Press is preparing new, full-text translations of Humboldt's *Essay on the Geography of Plants* and *Personal Narrative* including the *Political Essay on the Island of Cuba*.

4. The most complete biography in English is the two-volume *Life of Alexander von Humboldt* compiled by Karl Bruhns (London: Longmans, Green, 1873), an old-fashioned life-and-letters format stuffed with facts but unreliable and outdated generations ago. The best modern biography in English remains Douglas Botting's popular and well-illustrated *Humboldt and the Cosmos* (1973). An unpretentious and breezy update that relies on the same sources is *Humboldt's Cosmos* by Gerard Helferich (2004), which retells the narrative of Humboldt's South American voyage that was also well told by Victor Wolfgang von Hagen in *South America Called Them* (1945). Helmut de Terra's *Humboldt* (1955) benefits from some original research but is overwritten; Lotte Kellner's 1963 biography is dry and includes some errors. The most complete and authoritative account of Humboldt's life and work is that by Hanno Beck (1959–61), but

it has never been translated into English. Nicolaas Rupke gave up an attempt to write a new biography and produced instead a fascinating "metabiography" (2005). Although not a full biography, some of the sharpest writing on Humboldt's life by a U.S. American is in the opening chapters of Aaron Sachs's *The Humboldt Current* (2006). The upshot is that for all his world importance, not one single, authoritative, modern, critical biography of Humboldt exists in the English language.

Prologue

1. Alexander von Humboldt, *Researches concerning the Institutions and Monuments of the Ancient Inhabitants of America*, trans. Helen Maria Williams, 2 vols. (London, 1814; facs. ed., Amsterdam: Plenum Publishing, 1972), 1:45, 48; hereafter cited as *Researches*. This figure, now a major work in the canon of Aztec art, has been identified as Chalchiuhtlicue, "the goddess of groundwater." In late May to early June the Aztecs performed a rite to her at a shrine in the middle of Lake Tetzcoco, part of the most important festival for Tlaloc, the deity of rain, meant to summon the rainy season. See plate 17 and p. 261 in *The Aztec Empire*, ed. Felipe Solís (New York: Guggenheim, 2004).

2. *Researches* 1:52. The American who took Humboldt's invitation the farthest in this direction was William Hickling Prescott, who built on the work of Humboldt, among others, to recreate the civilizations of Mexico and Peru in his two classic histories, *The Conquest of Mexico* and *The Conquest of Peru*.

3. Ottmar Ette, *Literature on the Move*, trans. Katharina Vester (Amsterdam: Rodopi, 2003), 115.

4. Ángela Pérez-Mejía, *A Geography of Hard Times: Narratives about Travel to South America, 1780–1849*, trans. Dick Cluster (Albany: State University of New York Press, 2004), 42.

5. *Researches* 1:22–23. D. A. Brading suggests that Humboldt believed nature rendered the Indian peoples so savage that civilization had to be introduced by "missionaries" from Asia, but this interpretation leaves Humboldt's actual question unanswered: not where did this *civilization* come from, but where did the American *peoples* come from? See D. A. Brading, *The First America: The Spanish Monarchy, Creole Patriots, and the Liberal State, 1492–1867* (Cambridge: Cambridge University Press, 1991), 524–25.

6. *Researches* 1:14–15, 6, 31, 27. For Humboldt's role in unseating the "three stages" argument, see Fritz L. Kramer, "Eduard Hahn and the End of the 'Three Stages of Man,'" *Geographical Review* 57.1 (January 1967): 73–89.

7. *Researches* 1:4, 40–41, 42.

8. Peter Kropotkin, *Memoirs of a Revolutionist* (Montreal: Black Rose Books, 1989), 211–12.

9. Ramachandra Guha, *Environmentalism: A Global History* (New York: Longman, 2000), 26. Richard Grove has demonstrated Humboldt's attention to theories explaining environmental destruction on island colonies such as Mauritius and the Eastern Caribbean, and his central role in applying these theories on a continental scale. See Richard Grove, *Green Imperialism: Colonial Expansion, Tropical Island Edens and the Origins of Environmentalism, 1600–1860* (Cambridge: Cambridge University Press, 1995). Humboldt offers a favorable view of deforestation in *Essay on the Geography of Plants* (in *Foundations of Biogeography: Classic Papers with Commentaries*, ed. Mark V. Lomolino et al. [Chicago: University of Chicago Press, 2004]), 52; his key (and much more critical) discussions of deforestation are in the generally overlooked *Political Essay on the Kingdom of New Spain* (trans. John Black [London, 1811; facsimile ed., New York: AMS Press, 1966]), 2:24–27, 109–30; and more widely cited *Personal Narrative* (trans.

Helen Maria Williams [London: Longman, Hurst, Rees, Orme, and Brown, 1814–29]), 4:63–64, 134–54.

10. Aaron Sachs, "The Ultimate 'Other': Post-Colonialism and Alexander von Humboldt's Ecological Relationship with Nature," *History and Theory,* theme issue 42 (December 2003): 111–35.

11. Ursula K. Heise, "Science and Ecocriticism," *American Book Review* 18.5 (July–August 1997), 4+; posted at http://www.asle.umn.edu/archive/intro/heise.html, accessed 2/20/06.

12. The word "ecology" was coined in 1866 by the so-called "German Darwin" Ernst Haeckel, whose thinking was profoundly influenced by Humboldt and his followers. See Donald Worster, *Nature's Economy: A History of Ecological Ideas* (Cambridge: Cambridge University Press, 1977), 191–93. Worster calls Humboldt "a pioneer in biological ecology" (135). For specific links between Humboldt and modern ecology, see Malcolm Nicolson, "Humboldtian Plant Geography after Humboldt: The Link to Ecology," *British Journal of the History of Science* 29 (1996): 289–310.

Chapter One

1. John Locke, *Concerning Civil Government,* ¶49 (in *The English Philosophers from Bacon to Mill* [New York: Modern Library, 1967]), 422; Hans-Jürgen Grabbe, "Weary of Germany— Weary of America: Perceptions of the United States in Nineteenth-Century Germany," *Transatlantic Images and Perceptions: Germany and America since 1776,* ed. David E. Barclay and Elisabeth Glaser-Schmidt (Cambridge: German Historical Society and Cambridge University Press, 1997), 65–67. Locke is referring to America as he imagines it once was: a land without money, without commerce, without labor, and hence without economic value. Alexander von Humboldt, *Cosmos: A Sketch of a Physical Description of the Universe,* trans. E. C. Otté, 2 vols. (1858, 1850; Baltimore: Johns Hopkins University Press, 1997), 2:259, 264, 263. Hereafter cited as *Cosmos.*

2. Hermann Klencke, "Alexander von Humboldt: A Biographical Monument," in Hermann Klencke, *Lives of the Brothers Humboldt, Alexander and William,* trans. Juliette Bauer (London: Ingram, Cooke, 1852), 83, 145; Alfred Stillé, *Humboldt's Life and Character: An Address before the Linnaean Association of Pennsylvania College* (Philadelphia: Linnaean Association, 1859), 26. See also Karl Bruhns, ed. *Life of Alexander von Humboldt Compiled in Commemoration of the Centenary of His Birth by J. Löwenberg, Robert Ave-Lallemant, and Alfred Dove,* trans. Jane and Caroline Lassell, 2 vols. (London: Longmans, Green,, 1873), 2:390: "How often, too, is Humboldt laid claim to by America, where he is continually designated as a second Columbus." Hereafter cited as Bruhns.

3. Mary Louise Pratt, *Imperial Eyes: Travel Writing and Transculturation* (London: Routledge, 1992), 38–39. For Humboldt and Columbus, see Anthony Pagden, *European Encounters with the New World: From Renaissance to Romanticism* (New Haven: Yale University Press, 1993), 104–15; see also María DeGuzmán, *Spain's Long Shadow: The Black Legend, Off-Whiteness, and Anglo-American Empire* (Minneapolis: University of Minnesota, 2005).

4. The phrase "Napoleon of Science" is originally from Henry T. Tuckerman, "Alexander von Humboldt," *Godey's Lady's Book* 41 (1850): 133–38. Holmes's poem is reproduced in Louis Agassiz, *Address Delivered on the Centennial Anniversary of the Birth of Alexander von Humboldt* (Boston: Boston Society of Natural History, 1869), 86–88. The Napoleon anecdote is ubiquitous in the Humboldt literature. Allegedly, this was the only exchange of words between Napoleon and Humboldt. For the Frederick the Great anecdote see Victor Wolfgang von Hagen, *South America Called Them* (New York: Knopf, 1945), 88.

5. J. H. Elliott, *The Old World and the New, 1492–1650* (Cambridge: Cambridge University Press, 1970, 1992), 8, 14.

6. Alexander von Humboldt, *Views of Nature: or Contemplations on the Sublime Phenomena of Creation; with Scientific Illustrations*, trans. E. C. Otté and Henry G. Bohn (London: Henry G. Bohn, 1850), 156. Hereafter cited as *Views*.

7. The phrase is originally from Perry Miller, *Nature's Nation* (Cambridge: Harvard University Press, 1967).

8. John Lynch, *Simón Bolívar: A Life* (New Haven: Yale University Press, 2006), 23. This map, by Martin Waldseemüller, was the first to depict a Western hemisphere and a separate Pacific Ocean. Vespucci, unlike Columbus, recognized (South) America as a continental-sized landmass, an insight honored by the German mapmaker. The only copy of this map known to have survived was purchased by the Library of Congress for $10 million: see the Library of Congress website at http://www.loc.gov/rr/geogmap/waldexh.html.

9. Edward Everett, "Alexander von Humboldt," *Orations and Speeches on Various Occasions* (Boston: Little, Brown, 1879), 4:170; Botting, *Humboldt and the Cosmos* (New York: Harper and Row, 1973), 277. Steamship service began in 1839. When John Lloyd Stephens paid his respects to Humboldt, the instant he mentioned his involvement with the steamship company, Humboldt lit up with praise for the American government for its "statesmanlike" action in establishing the line, which must open new relations between Germany and the United States ("An Hour with Alexander von Humboldt," *Littell's Living Age* 15 [October–December 1847], 152). The definitive source for Humboldt names is Ulrich-Dieter Oppitz, "De Name der Brüder Humboldt in aller Welt," *Alexander von Humboldt, Werk und Weltgeltun* (Munich: R. Piper, 1969), 277–429. Henry A. Pochman notes that of all German authors, Humboldt had the eighth-highest number of translations from 1810 to 1899, and in over sixty authors listed, Humboldt is the only scientist; see *German Culture in America: Philosophical and Literary Influences, 1600–1900* (Madison: University of Wisconsin Press, 1957), 346–47.

10. Pratt, *Imperial Eyes*, 111–12, 119.

11. Ibid., 120, 131, 127; Edward W. Said, *Culture and Imperialism* (New York: Random House, 1993), xxv. It should be noted that Humboldt did deposit some geological specimens in Madrid. The controversy over Pratt's assertions has simmered for some years among U.S. Humboldt scholars. As Karl S. Zimmerer asserts, rendering Humboldt as an advance scout for European capitalism "misrepresent[s] the nature of his multifaceted and open-ended engagement with these topics"; see "Humboldt's Nodes and Modes of Interdisciplinary Environmental Science in the Andean World," *Geographical Review* 96.3 (July 2006), 355. For an example of this misrepresentation, see Susanne Zantop, "The German Columbus," in *Colonial Fantasies: Conquest, Family, and Nation in Precolonial Germany, 1770–1870* (Durham: Duke University Press, 1997), 166–67. That the questions Pratt raises are important and legitimate makes it all the more necessary to move past reductive and oversimplified polemics. Readers interested in pursuing this controversy further should consult Aaron Sachs, "The Ultimate 'Other': Post-Colonialism and Alexander von Humboldt's Ecological Relationship with Nature," *History and Theory*, theme issue 42 (December 2003): 111–35; Nigel Leask, "Alexander von Humboldt and the Romantic Imagination of America: The Impossibility of Personal Narrative" in *Curiosity and the Aesthetics of Travel Writing, 1770–1840* (Oxford: Oxford University Press, 2002), 243–98; Andrew Sluyter, "Humboldt's Mexican Texts and Landscapes," *Geographical Review* 96.3 (July 2006): 361–81; and Matti Bunzl and H. Glenn Penny, "Introduction: Rethinking German Anthropology, Colonialism, and Race," *Worldly Provincialism: German Anthropology in the Age of Empire* (Ann Arbor: University of Michigan Press, 2003).

12. Bruhns 1:377.

13. Sankar Muthu, *Enlightenment against Empire* (Princeton: Princeton University Press, 2003), 1, 7, 9–10, 259, 4–5.

14. See Aaron Sachs, *The Humboldt Current: Nineteenth-Century Exploration and the Roots of American Environmentalism* (New York: Viking Penguin, 2006). Matti Bunzl and H. Glenn Penny are struck by "the historical continuity of a basic intellectual position that valorized the particularity of each national and ethnic entity" in Germany. "What is even more remarkable is that this German attitude figured in persistent opposition to Western European ideas" ("Introduction," 11).

15. Nicolaas Rupke, *Alexander von Humboldt: A Metabiography* (Frankfurt am Main: Peter Lang, 2005), 18.

16. Quoted in Bruhns 1:391. Another son, by her first marriage, seems to have caused her much grief and soon disappears from the available sources. Julius Löwenberg does not fail to note the "remarkable" coincidence that the maiden name of the mother of "the Columbus of the nineteenth century" was von Colomb—the same as the first discoverer of the New World (Bruhns 1:10–11). As for the "von," in an 1805 letter to Marc-Auguste Pictet, who was preparing a biography for an English translation of Humboldt's works, Humboldt wrote: "In mentioning me, I should much prefer that you named me simply M. Humboldt. . . . It sounds more English, since the constant repetition of the *de* is very unpleasant to the ear." He requests Pictet to honor the family by mentioning him on *one* occasion only as "Frederick Alexander Baron von Humboldt": "This is a matter connected with certain *principles* with which you do not altogether sympathize, but which have been maintained by my brother and myself throughout life, leading us never to make use of any title, except in the most extraordinary cases, therefore never on the title-page of a book" (quoted in Bruhns 1:7–8). Nevertheless, the title pages of his books consistently use "von" or "de Humboldt."

17. Ronald Grimsley, "Jean-Jacques Rousseau," *Encyclopedia of Philosophy*, 8 vols. (New York: Macmillan, 1967), 7:221; Richard H. Grove, *Green Imperialism: Colonial Expansion, Tropical Island Edens and the Origins of Environmentalism, 1600–1860* (Cambridge: Cambridge University Press, 1995), 228; Jean-Jacques Rousseau, *Émile* (1762; London: Dent, 1974), 84; Bruhns 1:31 (undated letter to Karl Freieslaben).

18. Bruhns 1:14, 12, 391–92.

19. Bruhns 1:43; Lotte Kellner, *Alexander von Humboldt* (London: Oxford University Press, 1963), 8–9.

20. Bruhns 1:48, 26–27, 55–56.

21. *Cosmos* 2:20, 80. Grove makes the point that soon German naturalists would be preferred over other nationalities, as the best trained and most ideologically neutral. Ottmar Ette remarks that "Forster can probably be regarded as the actual creator of the artistic travel description in the German language" ("Transatlantic Perceptions: A Contrastive Reading of the Travels of Alexander von Humboldt and Fray Servando Teresa de Mier," *Dispositio* 17.42–43 [1992]: 167).

22. *Cosmos* 2:80. On Diderot, see Muthu 47–69. Forster's account of the trip he took with Humboldt, *Ansichten vom Niederrhein*, became a classic of German literature. The Temple of Liberty passage is cited in Nigel Leask, "Salons, Alps and Cordilleras: Helen Maria Williams, Alexander von Humboldt, and the Discourse of Romantic Travel," *Women, Writing, and the Public Sphere, 1700–1830*, ed. Elizabeth Eger et al. (Cambridge: Cambridge University Press, 2001), 244 (from Hanno Beck).

23. Aaron Sachs counts eleven: German, French, English, Spanish, Latin, Greek, Russian, Italian, Danish, Hebrew, and Sanskrit (*Humboldt Current*, 14). Humboldt studied in Hamburg

with Christoph Daniel Ebeling, whom Hans-Jürgen Grabbe calls "the most distinguished German scholar of the United States," and who used American politics to refute Europe's feudal order (Grabbe 65–67); in a letter Humboldt reported borrowing books on philology, history, and travel from Ebeling's "extensive library" (Bruhns 1:98).

24. Bruhns 1:63–64 (letter to Wegener of March 1789); Bruhns 1:102 (letter to Wegener of 23 September 1790); Bruhns 1:139–40 (letter to Freieslaben of 20 January 1794).

25. Joan Steigerwald, "Figuring Nature/Figuring the (Fe)Male: The Frontispiece to Humboldt's Ideas Towards a Geography of Plants," in *Figuring It Out: Science, Gender, and Visual Culture* (Hanover, NH: Dartmouth College Press/University Press of New England, 2006): 54–82, 77–78. Nicolaas Rupke calls the issue of Humboldt's sexuality "the most fractious of all." For an account of the controversy, including the ways Henriette Herz has historically been used to prove both Humboldt's philo-Semitism and his heterosexuality, see Rupke, *Alexander von Humboldt*, 196–202; see also Robert Aldrich, "Humboldt and His Friends," in *Colonialism and Homosexuality* (New York: Routledge, 2003), 24–29. Abundant documentation exists that Humboldt had, throughout his lifetime, several intimate and intense relationships with men; what is at issue is whether these relationships were homosexual or merely part of the German Romantic cult of friendship. It is possible, as many people assume, that travel legitimated his desire for sustained intimacy with men, and at least one critic (Michael Shortland, in an unpublished lecture quoted in Steigerwald 78) has suggested that travel allowed Humboldt to express both the masculine virtue of heroism and the feminine virtue of endurance. There is little suggestion in the literature that Humboldt and Bonpland were partners, although it is widely rumored that he and Carlos Montúfar were, a hypothesis offered to explain why, when living in Quito, Humboldt rejected the companionship of the brilliant young botanist Francisco José de Caldas, a student of the famous José Celestino Mutis, and chose instead the less obviously qualified Montúfar to accompany them for the rest of the trip. Steigerwald concludes, "There remains a continued reluctance to name Humboldt's sexual orientation, or to examine its relationship to his work," although that Humboldt challenged normative, naturalized conceptions of gender, in both his life and his work, is not in doubt (77–78). The question of Humboldt and gender needs to be more fully addressed.

26. Bruhns 1:134–35; Botting 25; Bruhns 1:152.

27. Bruhns 1:147–48.

28. Bruhns 1:205–8; Alexander von Humboldt also struck up a close friendship with Karl August. For Humboldt and Goethe, see Wolfgang-Hagen Hein, "Humboldt and Goethe," in *Alexander von Humboldt: Life and Work*, ed. Wolfgang-Hagen Hein (Ingelheim am Rheim: C. H. Boehringer Sohn, 1987): 46–55. Humboldt's electrical experiments, which bore on his understanding of the distinction between life and lifeless matter, are worth a chapter in themselves: for helpful discussions see Robert J. Richards, *The Romantic Conception of Life: Science and Philosophy in the Age of Goethe* (Chicago: University of Chicago Press, 2002), 316–21; and Michael Dettelbach, "Alexander von Humboldt between Enlightenment and Romanticism," *Northeastern Naturalist* 8.1 (2001): 9–20. Humboldt performed numerous experiments on his own body, and some of his descriptions are quite gruesome (Bruhns 1:150–51). His research was original and is credited with beginning the science of electrophysiology, but it never occurred to Humboldt to remove the organic tissue and just use the metals. When Volta did exactly this, he discovered the battery. Humboldt never forgave himself for not making the connection; if he had, as Gerard Helferich says, we would be measuring electricity not in volts but in humboldts (*Humboldt's Cosmos* [New York: Penguin, 2004], 106).

29. Bruhns 1:185, 187–88; *Cosmos* 2:82. Humboldt did not see Schiller's letter until it was published many years later, and he was hurt to learn that at the very moment when Schiller "saw me daily and overwhelmed me with tenderness," he could write such criticism behind his back. See Humboldt, *Letters of Alexander von Humboldt to Varnhagen von Ense from 1827 to 1858* (New York: Rudd and Carleton, 1860), 298. On the Goethean aspect of Humboldt's science, see Anne Buttimer, "Beyond Humboldtian Science and Goethe's Way of Science: Challenges of Alexander von Humboldt's Geography," *Erdkunde* 55.2 (2001): 105–20.

30. Johann Gottfried Herder, *Outlines of a Philosophy of the History of Man*, trans. T. Churchill, 4 vols. (1784–91; repr., New York: Bergman, [1800]), 14, 32, 228; see also Kate Rigby, *Topographies of the Sacred: The Poetics of Place in European Romanticism* (Charlottesville: University of Virginia Press, 2004), 72–75.

31. Bruhns 1:178. Michael Dettelbach observes that historical assessments of Humboldt are pulled between "the two poles of empiricism and idealism, Enlightenment and Romanticism," and argues that Humboldt unites these two poles in a way that had a strong currency for several decades; see "Humboldt between Enlightenment."

32. Richards 62; Roger Scruton, *Kant: A Very Short Introduction* (Oxford: Oxford University Press, 2001), 27.

33. Agassiz, *Address*, 75, 97. I explore romantic holism at length in *Seeing New Worlds: Henry David Thoreau and Nineteenth-Century Natural Science* (Madison: University of Wisconsin Press, 1995), chapter 2, where I coined the phrases "rational holism" and "empirical holism" to distinguish between top-down and bottom-up forms of reasoning that characterize contrasting styles of Romanticism. My discussion of Kant is indebted to Roger Scruton; Anne Macpherson, "The Human Geography of Alexander von Humboldt" (Ph.D. diss.: University of California, Berkeley, 1971); and Margarita Bowen, *Empiricism and Geographical Thought from Francis Bacon to Alexander von Humboldt* (Cambridge: Cambridge University Press, 1981). The subject of Humboldt's relationship to Kant, and hence to post-Kantian thought from Schelling to Coleridge to Emerson and beyond, is a large one rich with potential that needs to be addressed by a student of the history of philosophy interested in literary, scientific, and political implications. Kant's understudied political philosophy is of particular novelty and importance here. An excellent place to start is Robert J. Richards's *Romantic Conception*, which places Humboldt in the broader context of post-Kantian romantic science and philosophy, with particular attention to Schelling and German Naturphilosophie.

34. Bruhns 1:158.

35. Bruhns 1:238–39.

36. Alexander von Humboldt, *Personal Narrative of Travels to the Equinoctial Regions of the New Continent, during the Years 1799–1804*, 3rd ed., trans. Helen Maria Williams (London: Longman, Hurst, Rees, Orme, and Brown, 1822), 1:13; Hereafter cited as *PN*. Bruhns 1:246.

37. Quoted in Botting 65; Helmut de Terra, *Humboldt: The Life and Times of Alexander von Humboldt, 1769–1859* (New York: Knopf, 1955), 86–87; *PN* 1:iv; xix; *PN*, Williams's preface, 1:vii.

38. *PN* 1:iii. For additional sources on Humboldt and geography, start with Kent Mathewson, "Alexander von Humboldt's Image and Influence in North American Geography, 1804–2004," *Geographical Review* 96.3 (July 2006): 416–38. This entire special issue of the *Geographical Review* should be consulted.

39. Bowen 206, 208; Immanuel Kant, *Metaphysical Foundations of Natural Science*, trans. James Ellington (1786; Indianapolis: Bobbs-Merrill, 1970), 3–4; Samuel Taylor Coleridge, *The Friend, The Collected Works of Samuel Taylor Coleridge*, ed. Kathleen Coburn et al., 14 vols. to

date (London: Routledge, 1969–), 4.1:469; Macpherson 109. Kant never completed a work on the subject, but his lecture notes were gathered and published in 1802 as *Physical Geography*. Bowen finds them lacking in originality and dependent on secondary sources. See also Anne Godlewska, *Geography Unbound: French Geographic Science from Cassini to Humboldt* (Chicago: University of Chicago Press, 1999), esp. 111–27.

40. "I value my independence more and more every day, and for this reason I have scrupulously avoided accepting the smallest pecuniary assistance from any government" (Bruhns 1:287; letter to Willdenow, 21 February 1801); Harry Liebersohn, *The Travelers' World: Europe to the Pacific* (Cambridge: Harvard University Press, 2006), 80–81; *PN* 1:41.

41. *PN* 3:6; 1:34–40. In *Don Juan*, canto 4, stanza 112, Byron suggests some things Humboldt might rather have been measuring:

> Humboldt, "the first of travellers," but not
> The last, if late accounts be accurate,
> Invented, by some name I have forgot,
> As well as the sublime discovery's date,
> An airy instrument, with which he sought
> To ascertain the atmospheric state,
> By measuring "the intensity of blue:"
> Oh, Lady Daphne! let me measure you!

42. *PN* 1:xiv; 7:286–87; 3:350–51; 6:79; 7:285–86.

43. *PN* 2:45; 3:161; Herbert Wilhelmy, "Alexander von Humboldt in the Light of His American Journey," *Universitas: A German Review of the Arts and Sciences*, Quarterly English Language Edition 12.2 (1969): 47. The identification of the servant, whom Humboldt does not name, is made by Ingo Schwarz in *Alexander von Humboldt und die Vereinigten Staaten von Amerika: Briefwechsel* (Berlin: Akademie Verlag, 2004), 11, 106–7. A "zambo" might be of Indian and black and/or mulatto ancestry, as Humboldt detailed in *Political Essay on the Kingdom of New Spain* (trans. John Black [London, 1811; facsimile ed., New York: AMS Press, 1966]), 1:244–46; hereafter cited as *PE*.

44. *PN* 1:xxxviii–xxxix; Godlewska, *Geography*, 249; *PN* 1:xliii. That the suppression of biographical incident was a compositional choice Humboldt made while shaping his narrative for the public becomes clear when one compares the *Personal Narrative* with the journals, which are not nearly so shy about registering biographies and personal incident.

45. Scruton, *Kant*, 56–61. For Humboldt and objectivity, see my "The Birth of the Two Cultures," in *Alexander von Humboldt*, ed. Raymond Erickson et al. (New York: City University of New York), 247–58, http://web.gc.cuny.edu/bildnercenter/publications/humboldt.pdf; George Levine, *Dying to Know: Scientific Epistemology and Narrative in Victorian England* (Chicago: University of Chicago Press, 2002); on Humboldt's subjectivity, see John Ochoa, *The Uses of Failure in Mexican Literature and Identity* (Austin: University of Texas Press, 2004), 85–109.

46. *PN* 1:xviii. For more on Humboldt and hypertext, see Frank Baron and Detlev Doherr, "Exploring the Americas in a Humboldt Digital Library," *Geographical Review* 96.3 (July 2006): 439–51.

47. *PN*, Williams's introduction, 1:ix.

48. Bruno Latour, *Pandora's Hope: Essays on the Reality of Science Studies* (Cambridge: Harvard University Press, 1999), 88, 311; Deborah Kennedy, *Helen Maria Williams and the Age of Revolution* (Lewisburg: Bucknell University Press, 2002), 184–87. Nigel Leask sees Williams's role as so great

as to constitute virtual coauthorship, giving Humboldt's famous work a collaborative, female sponsorship: see "Salons, Alps and Cordilleras," 217–35; and "Alexander von Humboldt," 288.

49. *PN* 4:419. Oliver Lubrich offers the stimulating idea that this "breakdown" of Humboldt's text is deliberate, part of his experimentation with a genre that was itself breaking down: "This formal shattering performs the main message of the text. . . . Humboldt is, by no means, the author of 'totality,' which he is so often seen to be. His writing deals, rather, with the impossibility of grasping foreign reality and presenting it in the traditional forms of (metropolitan) literature. Humboldt's aesthetic is an aesthetic of lost certainty, his poetic is a poetic of de-authorized form." See "Alexander von Humboldt: Revolutionizing Travel Literature," *Monatshefte* 96.3 (2004): 380.

50. *PN* 7:472–73. For the supposed destruction of *Personal Narrative*'s final volume, see Nigel Leask, "Alexander von Humboldt," 297–98. Bowen suspects that "conservative intervention" suppressed the volume, as this was the period that saw the beginning of the "embargo" on Humboldt's social writings, and shortly afterward he terminated two other works of social commentary (249–51). By contrast, Macpherson argues that the *Personal Narrative* really was complete, with the "Essay on Slavery" as the intended conclusion ("Human Geography," 467–68).

51. Victor Wolfgang von Hagen, *South America Called Them* (New York: Knopf, 1945), 143. In *Personal Narrative* Humboldt includes a paragraph honoring Lavič, noting with sadness that, his health broken in prison, "he has sunk into the grave, without having seen the light of those days of independence, which his friend, don Joseph España, had predicted at the moment of his execution" (*PN* 6:75–76).

52. *PN* 1:xlix; Benedict Anderson, *Imagined Communities* (London: Verso, 1991).

53. Bruhns 1:286; *PN*, Williams's preface, 1:viii; Alexander von Humboldt, *Views*, x.; Muthu 162. As Kant wrote, "Even the good Mendelssohn must have counted on [hope for better times] when he exerted himself so zealously for the enlightenment and welfare of the nation to which he belonged. For he could not reasonably hope to bring this about all by himself, without others after him continuing along the same path." Quoted in Muthu 166.

Chapter Two

1. *PN* 1:233. For a postmodern view of Humboldt from this nomadic, cosmopolitan perspective, see Ottmar Ette, "The Scientist as Weltbürger: Alexander von Humboldt and the Beginning of Cosmopolitics," *Northeastern Naturalist* 8.1 (2001): 157–82; *HiN: Alexander von Humboldt im Netz* 11.2 (2001), http://www.uni-potsdam.de/u/romanistik/humboldt/hin/ette-cosmopolitics. htm.

2. Stephen Jeffries documents a debate at the Botanic Gardens in Melbourne, Australia that hinged on the public's lack of knowledge of Humboldt's plant geography, since those texts were still unavailable in English. Their resulting inability to understand the scientific goals of the director, Ferdinand von Mueller, "the Humboldt of Australia," led to his firing in 1873. See "Alexander von Humboldt and Ferdinand von Mueller's Argument for the Scientific Botanic Garden," *Historical Records of Australian Science* 11.3 (June 1997): 301–10; Andreas Daum, "Alexander von Humboldt, die Natur als 'Kosmos' und die Suche nach Einheit," *Berichte zur Wissenschaftsgeschichte* 23 (2000): 243–68, esp. 250–53.

3. *PN* 1:40–71. Gerhard Kortum and Ingo Schwarz note that the first volume of *Personal Narrative* is "a classic document in the history of oceanography"; see their article "Alexander von Humboldt and Matthew Fontaine Maury—Two Pioneers of Marine Science," *Historisch-Meereskundliches Jahrbuch* (*History of Oceanography Yearbook*), Band 10 (Stralsund: Deutsches

Meeresmuseum, 2003–4), 160, as well as Kortum's "Humboldt und das Meer: Eine Ozeanographiegeschichtliche Bestandsaufnahme," *Northeastern Naturalist* 8.1 (2001): 91–108. Fifty years after this voyage, Maury would work with Humboldt, building on Humboldt's work on ocean currents, to create the new science of oceanography.

4. *PN* 2:1; 1:119.

5. *PN* 1:99; Alfred W. Crosby, *Ecological Imperialism: The Biological Expansion of Europe, 900–1900* (Cambridge: Cambridge University Press, 1986), 82–99; *PN* 1:277–81; 7:129. According to Crosby, the Guanches migrated to the Canary Islands from Africa starting in 2000 BC. Europeans arrived on the Canaries as early as the 1290s, but not until 1402 did the campaign to conquer them begin. As he observes, "The Guanches deserve more attention than they have received," for they were the first people to be driven into extinction by modern European imperialism, which was born not in 1492 with Columbus but with the French expedition of 1402. See *Ecological Imperialism* 79–81, 99.

6. *Cosmos* 2:77. Saint Pierre was himself a philosopher and naturalist who "developed an apparently holistic and environmentalist theory of nature grounded in an understanding of the interdependence of processes and objects in the environment," and his books were the first "fully developed, fully argued and fully evidenced critiques of the European impact on tropical nature"; see Grove, *Green Imperialism: Colonial Expansion, Tropical Island Edens and the Origins of Environmentalism, 1600–1860* (Cambridge: Cambridge University Press, 1995), 247–53.

7. Kant, "To Eternal Peace," in *The Philosophy of Kant*, ed. Carl J. Friedrich (New York: Modern Library, 1949), 447, 442. See Ottmar Ette: "At the age of twenty, Humboldt conceived a scientific program focusing universal history and global connections guided by a 'cosmopolitan intention' in the sense of Immanuel Kant. This cosmopolitan dimension in his thought and investigation will be—even in a sense that differed from the Kantian conceptions that had an early impact on his own conceptions—a *leitmotiv* in the writings of this *citoyen de l'univers*, right up to his last best-seller, *Cosmos*" ("Scientist as Weltbürger," *Northeastern Naturalist* 169).

8. Ibid., 448; *PN* 1:195. Scruton points out that Kant's essay became the basis for the idea of a "league" or "federation of nations," and in the words of one of his commentators, "Kant's theory has been embodied in twentieth-century structures of trans-national legislation." See Scruton, *Kant*, 115. For a closer reading of Kantian cosmopolitanism in the geographic context of Captain Cook and Herder, see Brian W. Richardson's very useful book, *Longitude and Empire: How Captain Cook's Voyages Changed the World* (Vancouver: University of British Columbia Press, 2005), esp. 124–30.

9. *PN* 1:136, 137–47.

10. *PN* 1:180–83.

11. *PN* 1:201, 233.

12. *PN* 1:262, 277, 292–93.

13. *PN* 2:42–47.

14. *PN* 2:176; 3:36; Bruhns 1:265. In a letter to his brother Wilhelm, Humboldt would remark that the Caribs, who impressed him as the smartest and toughest of all the tribes, "are a race of the largest and strongest people I have ever seen. This tribe alone disproves Raynal's and Pauw's wild theories of weakness and degeneration in the peoples of the New World. A fully grown Carib is like a cast-iron Hercules." See Hanno Beck, "Alexander von Humboldt: Letters from His Travels," in *Alexander von Humboldt: 1769/1859* (Bonn: Internationes, 1969), 138.

15. *PN* 2:22.

16. *PN* 2:239–40; 3:177–79.

17. *PN* 2:192; 3:5; 3:230–40; Steven W. Hackel, *Children of Coyote, Missionaries of Saint Fran-cis: Indian-Spanish Relations in Colonial California, 1769–1850* (Chapel Hill: University of North Carolina Press, 2005), 3.

18. *PN* 5:578–79; 5:611.

19. *PN* 3:207, 230, 158.

20. *PN* 3:248; 5:430–32. See also 6:362. Compare the language of the Peace Commission Re-port of 1868, which set U.S. government policy for the next century: eliminate Indian languages, blot out tribal distinctions, "fuse them into one homogenous mass. Uniformity of language will do this—nothing else will." Quoted in Lydia Maria Child, *Hobomok and Other Writings on Indians*, ed. Carolyn L. Karcher (New Brunswick: Rutgers University Press, 1986), 214.

21. *PN* 3:211–12; 5:273.

22. *PN* 4:511–12.

23. *PN* 3:16; 3:213–15. Humboldt is well aware that this number for the population of Ameri-can Indians represents a huge decline from pre-Columbian levels. His point is that the "van-ishing Indian" is a myth that wishes away peoples who are in fact very much alive and need to be included in policy decisions. Bartolomé de las Casas, the conquistador who emigrated to Hispaniola in 1502 and became a Dominican friar (and one of Humboldt's sources), estimated the total pre-Columbian population of the Americas at forty million, a number that is on the low side of current estimates. This question remains extremely controversial. See James Axtell, *Beyond 1492: Encounters in Colonial North America* (Oxford: Oxford University Press, 1992), 203–4; Charles C. Mann, *1491: New Revelations of the Americas before Columbus* (New York: Vintage, 2005, 2006), 147.

24. *PN* 3:219. The concept of "The Casiquiare Exchange" was first proposed by Lee Ster-renberg in a paper delivered at the conference of the Society for Literature and Science, Atlanta, GA, 5–8 October 2000; I remain deeply indebted to Lee's vision for this concept, which has stayed with me ever since. For an extended study and documentation of indigenous earthworks in the Amazon, past and present, see Hugh Raffles, *In Amazonia: A Natural History* (Prince-ton: Princeton University Press, 2002); for the Casiquiare's possible origin in Arawak labor see Raffles 219n69.

25. *PN* 3:321, 355.

26. *PN* 3:197, 414–15. España was executed on 8 May 1799. On the day that Venezuelan in-dependence was declared, 5 July 1811, two of his children unfolded their country's new flag on the spot where he died.

27. *PN* 4:1–2.

28. *PN* 4:143. See Grove, *Green Imperialism*, esp. 365–79.

29. *PN* 4:69–70, 317–18. Humboldt speculates that if horses had been native to South Amer-ica, it would have developed a quite different government and history.

30. *PN* 4:414–17.

31. *PN* 5:406–7.

32. *PN* 5:592; 4:421–23. Humboldt remarks on the frequency of houses in the Andes in his journal: see Benjamin Villegas, *The Route of Humboldt: Venezuela and Colombia*, 2 vols. (Bo-gotá: Villegas Editores, 1994), 2:96.

33. *PN* 3:512; 5:290–91.

34. *PN* 2:287–94. For nations as "imagined communities," see Anderson, *Imagined Com-munities*. Tupac Amaru, frequently mentioned in Humboldt's writings, was also featured in an

epic poem, *The Inca*, by Helen Maria Williams, and a play, *The Incas*, by the Jacobin revolutionary John Thelwall; in 1811 Tupac Amaru's name was taken up by Uruguayan revolutionaries (Anderson 154). What was Humboldt's role, direct and indirect, in helping postcolonial Creole Americans identify themselves as a nation by claiming indigenous figures of heroic resistance (from King Philip on), languages, legends, names, and archaeology as symbolic centers of a distinctive "American" identity, even as they were destroying the populations of living Indians? To what extent did the rise of New World nationalism require American nationals to suppress Humboldt's cosmopolitanism, his respect for and interest in living Indian peoples, in order to more easily appropriate them symbolically? Students interested in this question should begin, but not end, with Pratt's *Imperial Eyes*. One understudied aspect of this question is the degree to which Humboldt may have provided tools for indigenous resistance, too, from his day to our own—one possible reason he is to this day celebrated as a folk hero in Cuba, Mexico, and South America.

35. Benigno Trigo, "Walking Backward to the Future: Time, Travel, and Race," in *Subjects of Crisis: Race and Gender as Disease in Latin America* (Hanover: Wesleyan University Press, 2000), 21.

36. *PN* 3:46–51; 4:216–17; 2:198; 3:120–28; 4:475–93.

37. *PN* 5:81–84, 387–93, 818.

38. *PN* 4:531–34.

39. *PN* 4:542; 5:215, 233–38.

40. Villegas, *Route of Humboldt*, 2:158; *PN* 3:173, 241. It is possible that the *alcalde*'s answer reflected not his desire to please, but an alternative, native worldview that favors cyclical patterns over Western linearity; see Donald L. Fixico, *The American Indian Mind in a Linear World: American Indian Studies and Traditional Knowledge* (New York: Routledge, 2003).

41. *PN* 5:132–33.

42. As Hugh Raffles observes, "Explorer-scientists were vulnerable and dependent, a resource as well as a burden. The lack of direct coercive sanctions available to the naturalists, their acute physical vulnerability on sparsely inhabited, poorly mapped, and unpredictable rivers, and the generalized labor shortage with which foreign travelers were confronted, all gave local workers unusual relative strength," though "positive support was as frequent as obstruction" (139–40). It would be interesting to examine why so many local workers did voluntarily work for, and with, Humboldt and the explorers who followed him. Jorge Cañizares-Esguerra remarks that the Indians "portrayed by La Condamine as submissive and stupid" actively subverted his expedition, destroying or moving his navigational markers, refusing to work for the French, or agreeing to work, then stranding them in the wilderness. See "Spanish America: From Baroque to Modern Colonial Science," *The Cambridge History of Science*, vol. 4, *Eighteenth-Century Science*, ed. Roy Porter (Cambridge: Cambridge University Press, 2003), 718–38, 735. James C. Scott analyzes what he calls "everyday forms of peasant resistance. . . . the ordinary weapons of relatively powerless groups: foot-dragging, dissimulation, desertion, false compliance, pilfering, feigned ignorance, slander, arson, sabotage, and so on," by which the lowest classes have historically defended their interests "against both conservative and progressive orders." These are precisely the forms of resistance that Humboldt encountered, and sought, not always successfully, to understand. As Scott observes, and Humboldt recognized, "Formal, organized political activity, even if clandestine and revolutionary, is typically the preserve of the middle class and the intelligentsia"; see his *Weapons of the Weak: Everyday Forms of Peasant Resistance* (New Haven, CT: Yale University Press, 1985), xv–xvi.

43. *PN* 4:549; 5:256.

44. Harry Liebersohn, *The Travelers' World: Europe to the Pacific* (Cambridge: Harvard University Press, 2006), 302–4.

45. *PN* 2:278; 5:614–15.

46. *PN* 4:496–97.

47. Bruhns 1:321–22; *PN* 3:271–72; *Researches* 1:280.

48. *PN* 6:347–63. For confirmation of the Indian languages' lack of barbarism he cites an 1823 essay by Wilhelm von Humboldt.

49. *PN* 5:620; *Views* 189–90.

50. *PN* 5:427; 4:512–21. Humboldt was a great admirer of the tall, handsome, intelligent and once-nomadic Carib Indians, who originated the custom of painting and had a tremendous influence on the interior tribes, and so, he thought, particularly deserved the attention of the historian. The "profound obscurity" of the history of the vast country to the east of the Andes means that "whatever in this country relates to the preponderance of one nation over others, to distant migrations, to the physiognomical features which denote a foreign race, excite in us a lively interest"; see *PN* 6:9–15.

51. *PN* 5:156, 254–55, 230–31, 267–68, 244–45, 275–76, 528.

52. *PN* 5:533–34, 363, 640–61.

53. *PN* 5:533, 422, 428–29.

54. *PN* 5:285–86; 3:74.

55. *PN* 5:86–116, 127–28, 164–65.

56. *PN* 5:199, 309–10, 270–71.

57. *PN* 5:420–21, 439–44. Depopulation was not confined to the rivers. Back on the Venezuelan shore, Humboldt exclaimed, "What a desert coast! Not one light announced a fisherman's hut. . . . Yet, in the time of Columbus, this territory was inhabited, even along the shore." He noted that wells and erosion still exposed stone hatchets and copper utensils from pre-Columbian times (*PN* 7:323–24). One hundred eighty miles is approximate; the Casiquiare is 328 kilometers long.

58. *PN* 5:446–48; *Views* 195.

59. *PN* 5:695, 693.

60. *PN* 5:864; 7:248; 1:l–li. For the word "commerce" as Humboldt uses it, see Muthu 253.

61. Crosby 137, 144; Kellner 50; Botting 132; Douglas Day, "Humboldt and the Casiquiare: Modes of Travel Writing," *Review: Latin American Literature and Arts* 47 (Fall 1993), 7–8. A controversial movement in Venezuela to identify and protect isolated tribes from all contact has recently (in 2006) met approval by the UN and is spreading to other South American countries, though prospects are dubious. See Claudio Angelo, "Prime Directive for the Last Americans," *Scientific American*, May 2007, 40–41; for its connections to Humboldt's legacy, see Geóg. Diógenes Edgildo Palau, "Los Pueblos Indigenas del Amazonas Venezolano desde Humboldt a Nuestros Dias: Preservacion del Ambiente y Manejo Sostenible," *Northeastern Naturalist* 8.1 (2001): 135–56. By contrast, Brazil is moving toward a policy of controlled and sustainable development in the Amazon, in hopes that better management will protect the forest. Both are Humboldtian solutions, but the point may be moot: two hundred years later, pressures that Humboldt could not imagine are encroaching on his wilderness, for deforestation and global warming is drying the rainforest, which is starting to burn, a process that will result in open savannah. The effect this will have on the native peoples is suggested by recent work by anthropologists, ethnographers, and archaeologists who are today documenting the complex

adaptations native peoples have made to their homelands, and the degree to which the Amazon and Orinoco basins were terraformed and maintained by an intensive custodial regimen of gardening and wild cultivation, a dynamic and long-established historical ecology disrupted by European invasion: see for example Raffles; William Balée, ed., *Advances in Historical Ecology* (New York: Columbia University Press, 1998).

62. *Views* 420; letter from Humboldt to Delambre, posted from Lima, 25 November 1802, in Bruhns 1:317; *PN* 7:468. Measuring the transit of Mercury allowed the determination of the longitude of Lima and therefore of the entire southwest of South America (*Views* 420). For Humboldt's interests in the Andes, see Karl S. Zimmerer, "Humboldt's Nodes and Modes of Interdisciplinary Environmental Science in the Andean World," *Geographical Review* 96.3 (July 2006): 335–60. For Humboldt and Creole and indigenous science, see Jorge Cañizares-Esguerra, "How Derivative Was Humboldt?," in *Nature, Empire, and Nation: Explorations of the History of Science in the Iberian World* (Stanford: Stanford University Press, 2006), 112–28; Pratt, *Imperial Eyes.*

63. *PN* 5:593–602. Humboldt is haunted by these mysterious figures, for as he reflects, "The more a country is destitute of remembrances of generations that are extinct, the more important it becomes to follow the least traces of what appears to be monumental" (*PN* 5:601). He returns to these rock carvings in *Views* (147–51, 164–65), where he expresses relief that Indians refused to help a later archaeologist, Robert Schomburgk, remove a sample of the carvings; the would-be despoiler had to content himself with drawings. Humboldt is pleased to note that Schomburgk's editor appended a request "that no traveller belonging to a civilized nation will in future attempt the destruction of these monuments of the unprotected Indians" (*Views* 149): evidently Humboldt's effort to urge protection and understanding of Indian monuments was having an effect.

64. *Researches* 1:63–67; Villegas 2:134. In his journal Humboldt notes that while the chair-bearer strides "extremely straight and erect," the passenger rides "aslant forward, mak[ing] a helpless, wretched figure" (Villegas 2:134). For a very different approach to *Passage of Quindiu* and various other Humboldt engravings, see Trigo, esp. 22–29.

65. *Researches* 1:87, 97.

66. Ibid. 1:120–21, 117, 122–23.

67. Ibid. 1:79, 237–38; Bruhns 1:311–15, 2:10. Edward Whymper, the great pioneer of mountaineering, attempted to trace Humboldt's route up Chimborazo in 1879, ascending Cotopaxi and several other Andean volcanoes for good measure. Whymper's stated goals, like Humboldt's, were scientific (he was particularly interested in the effect of high altitude on human physiology), but he also exemplified the growing fascination with mountaineering as a heroic exploit with no scientific purpose. See *Travels amongst the Great Andes of the Equator* (1892), ed. F. S. Smythe (London: John Lehmann, 1949), 71–72 and passim. Descriptions of climate zones banding mountainsides became ubiquitous in nineteenth-century literature: for a nice example see Mark Twain, who in *Roughing It* imagined standing atop Hawaii's Mauna Loa volcano and nibbling a snowball while looking down on "sections devoted to productions that thrive in the temperate zone alone," and beyond to palms and other tropical vegetation: "He could see all the climes of the world at a single glance of the eye, and that glance would only pass over a distance of four or five miles as the bird flies!" (New York: Penguin, 1981), 498.

68. *Researches* 1:77, 80. Fittingly, this falls was the first major landscape of South America painted by Frederic Church; see John R. Howat, *Frederick Church* (New Haven: Yale University Press, 2005), 50–51.

69. *Researches* 1:124.

70. Ibid. 1:251–53, 2:73–75, 1:144.

71. Ibid. 1:143.

72. Ibid. 1:247–49.

73. Pérez-Mejía 45; *PE* 4:282. See for example Bruhns 1:322: "Before setting out [from Rio-bamba, Peru] I visited the extensive sulphur mines at Tiscan. The rebel Indians conceived the idea of setting fire to these sulphur works, after the earthquake of 1797; certainly the most horrible plan ever devised even by a people driven to despair. They hoped by this means to produce an eruption by which the whole province of Allausi should be destroyed." See also *PN* 7:281–84; 5:683–84. For a more detailed discussion of Humboldt and rebellion, see chapter 4. For a more subtle analysis of the politics of peasant rebellion, see Scott.

74. *Views* 411–15. Humboldt thinks the reddish stains on the execution rock are hornblende and peroxide, and points out that Atahualpa was strangled, not beheaded, leaving no blood (*Views* 410–11). In his journal Humboldt noted the contempt of the Honda Indians for the Europeans who "mine" Andean tombs and burials for gold ornaments: "To this very day the Indians consider this search for *guacas* [Indian tombs] an *impium opus* [desecration]. They mock the Europeans' pursuit of gold, which upsets the tranquility of the dead. Living in a piece of heaven in which every agricultural task is blessed with countless years of rich harvests, they prefer to turn the earth over and make off with a bone or an earthen pot or poisoned arrows. But the pursuit of gold is a European affliction verging on delirium" (Villegas 2:61).

Chapter Three

1. Ingo Schwarz, "Alexander von Humboldt's Visit to Washington and Philadelphia," *Alexander von Humboldt's Natural History Legacy and Its Relevance for Today*, special issue, *Northeastern Naturalist* 8.1 (2001): 43.

2. Herman R. Friis, "Alexander von Humboldts Besuch in den Vereinigten Staaten von Amerika vom 20. Mai bis zum 30. Juni 1804," in *Alexander von Humboldt: Studien zu seiner universalen Geisteshaltung*, ed. Joachim H. Schultze (Berlin: Verlag Walter de Gruyter, 1959), 148–50, 147; Schwarz, *Alexander von Humboldt*, 87–88, 484; Friis, "Humboldt's Besuch," 150. The number on the customs form is difficult to read; this is Schwarz's interpretation. Madison's passport, written six weeks later, says forty boxes (Schwarz, *Alexander von Humboldt*, 496).

3. Schwarz, "Humboldt's Visit," 44–45; de Terra, "Humboldt's Correspondence," 787; Thomas Jefferson, *Notes on the State of Virginia*, in *Writings* (New York: Library of America, 1984), 167. Humboldt had studied in Hamburg with Christoph Daniel Ebeling, Germany's leading expert on North America. Jefferson's *Notes* was bound with "Draught of a Fundamental Constitution for the Commonwealth of Virginia," "Notes on the Establishment of a Money Unit, and of a Coinage for the United States," and "An Act for establishing Religious Freedom, passed in the assembly of Virginia in the beginning of the year 1786" (Schwarz, *Alexander von Humboldt*, 20). Humboldt was voted an "extraordinary" member of the Academy in 1800, and in 1805 he was voted an "ordinary" member (Schwarz, "Humboldt's Visit," 46).

4. Schwarz, *Alexander von Humboldt*, 92, 513–14; John Bachman, "Von Humboldt" (1869), in *Tributes and Memories* (Boston: Sanctuary Publishing, 1914), 82, 85. Humboldt also took the precaution of writing a shorter and more formal letter to James Madison in which he emphasized his desire to see not just the spectacle of the Andes and the grandeur of nature, but also "the spectacle of a free people worthy of a great destiny" (de Terra, "Correspondence," 796). Philadelphia's leading intellectuals showered Jefferson with further letters vouching for

Humboldt: "If you can spare time, you will be much pleased with the information he has to communicate," wrote Caspar Wistar, an APS vice president; the APS's other vice president, Dr. Benjamin Smith Barton, added that Humboldt was "one of the most intelligent and active philosophers of our times." John Vaughan, APS treasurer, also wrote Jefferson. See Friis, "Humboldts Besuch," 157–59; and his "Baron Alexander von Humboldt's Visit to Washington, D.C., June 1 through June 13, 1804," *Records of the Columbia Historical Society of Washington, 1960–62*, 13.

5. Charles Willson Peale, *The Selected Papers of Charles Willson Peale*, vol. 2, pt. 2 (New Haven: Yale University Press, 1988), 681n1 (the records do not mention whether de la Cruz accompanied them, though one assumes that he did); Friis, "Humboldts Besuch," 168–69; and "Baron von Humboldt," 11. Late one night after dining with the Madisons, Humboldt came into Peale's room to tell him that he had, indeed, extracted a pledge from Jefferson to nationalize Peale's Museum (Peale 694). While Jefferson did deposit Lewis and Clark's collections with Peale, he was unable to keep his promise. Peale's collection was sold to P. T. Barnum in 1850, and Barnum's New York museum was destroyed by fire in 1865.

6. Schwarz, *Alexander von Humboldt*, 104; Peale 691–97. It was probably Rembrandt Peale who, to advertise the polygraph, had a poem printed in the *Federal Gazette*: "Pois'd by the spiral chord above, / The obedient pens in concert move. / Triumph of art! Amaz'd I view, / A transcript fair of all I drew" (Peale 710–11).

7. Mathewson, "Humboldt's Image," 419; Peale 693; Henry Adams, *The Life of Albert Gallatin* (Philadelphia: Lippincott, 1879), 323–34. Adams judged that Gallatin was "one of the best talkers in America, and perhaps the best-informed man in the country" (*History of the United States of America during the Administrations of Thomas Jefferson* [New York: Library of America, 1986], 130). For Humboldt's statistical tables on New Spain, see Schwarz, *Alexander von Humboldt*, 484–95. In his letter to Jefferson, Wilkinson regrets that illness prevented him from attending the dinner, for he had wished to ask Humboldt certain questions in order to determine the accuracy of his answers (Friis, "Baron von Humboldt," 179–80). Evidently Wilkinson was resistant to Humboldt's charms, though given his plans, one can imagine how pleased he was to score a secret copy of Humboldt's map. According to Adams, by 1804 Wilkinson had been secretly in the employ of Carlos IV for roughly eighteen years, at an annual salary of two thousand dollars (839).

8. Henry Adams emphasized that without Toussaint Louverture's resistance, Napoleon would never have been willing to let Louisiana go, a fact that Americans have forgotten: "The prejudice of race alone blinded the American people to the debt they owed to the desperate courage of five hundred thousand Haytian negroes who would not be enslaved" (Adams, *History*, 316).

9. Schwarz, *Alexander von Humboldt*, 92–93, 130; for Humboldt's evaluation of Texas, see the Humboldt Digital Library Project, "Alexander von Humboldt in Washington (1804): Humboldt on the 'Texas' Territories," http://www2.ku.edu/~maxkade/humboldt/contents.htm (accessed 11 January 2009). Jefferson also worried how Lewis and Clark would draw accurate maps without a chronometer, and he asked Humboldt whether one could derive longitude from lunar observations. The answer was yes, such a method had been used before, but it depended on skill and luck with the weather. In actuality, Lewis and Clark's measurements were off, and here as elsewhere the scientific results of their expedition were scanty.

10. Friis, "Baron von Humboldt," 20–23, 30, 34–35. The Jefferson anecdote was originally narrated by Margaret Bayard Smith. Dolley Madison has little to say about Bonpland and

Montúfar, noting only that Humboldt had with him "a train of philosophers, who, though clever and entertaining, did not compare with the Baron" (Friis, "Baron von Humboldt," 23).

11. For more on Ellicott see Susan Faye Cannon, *Science in Culture: The Early Victorian Period* (New York: Dawson and Science History Publications, 1978), 99–102. Cannon argues that by the time of Humboldt's visit, Ellicott, who was working on the problem of measuring longitude using lunar observations, was fully as accurate and sophisticated in his surveying methods as anyone in Europe. By 1806 Ellicott concluded that "the major source of inaccuracies, for the longitude at any rate, was *in astronomical theory*, not in his instruments or techniques" (99). This connected Ellicott directly with work in astronomical theory, particularly the theory of error, being pursued in Paris in the years after Humboldt's return to Europe.

12. The Rush/Humboldt influence went both ways: soon after Humboldt's visit, American physicians, largely trained in Philadelphia, started taking Humboldt's climate theories to California, which by 1850 became the mecca of healthful climate, attracting to its golden shores a steady stream of doctors and patients. See J. B. deC. M. Saunders, "The Influence of Alexander von Humboldt on the Medicine of Western America," *Proceedings of the XXIII Congress of the History of Medicine* (London, 2–9 September 1972): 523–28; and his *Humboldtian Physicians in California* (Davis: University of California, 1971); see also Linda Nash, *Inescapable Ecologies: A History of Environment, Disease, and Knowledge* (Berkeley: University of California Press, 2006), 29–32; and Nicolaas Rupke, "Humboldtian Medicine," *Medical History* 40 (1996): 293–310. For Peale's portrait, as well as complete documentation of Humboldt iconography, see Halina Nelkin, *Alexander von Humboldt: His Portraits and Their Artists: A Documentary Iconography* (Berlin: Dietrich Reimer Verlag, 1980), 61.

13. Schwarz, *Alexander von Humboldt*, 496; de Terra, "Correspondence," 802; Schwarz, *Alexander von Humboldt*, 497–513. Humboldt's abstract was published in *Literary Magazine and American Register for 1804*, Philadelphia, vol. 2 (1804): 321–27. Humboldt was right to be worried about the passport: the British were contemptuous of the protection Madison claimed under the American flag, and searches and seizures of American ships and property, including impressment of American sailors, was so rampant that war with England seemed likely. See Adams, *History*, 571–72, 589, 660.

14. Friis, "Humboldts Besuch," 193; de Terra, "Correspondence," 796–97; Helmut de Terra, "Studies of the Documentation of Alexander von Humboldt: The Philadelphia Abstract of Humboldt's American Travels; Humboldt Portraits and Sculpture in the United States," *Proceedings of the American Philosophical Society* 102.6 (December 1958), 562; de Terra, "Correspondence," 789; Schwarz, *Alexander von Humboldt*, 96–97 (translation by Aaron Sachs, in *Humboldt Current*, 6). For the desire of prominent Germans, in Goethe's words, "nach Amerika fliehen wollte," see Hinrich C. Seeba, "Cultural History: An American Refuge for a German Idea," in *German Culture in Nineteenth-Century America: Reception, Adaptation, Transformation*, ed. Lynne Tatlock and Matt Erlin (Rochester, NY: Camden House, 2005), 8–9.

15. De Terra, "Correspondence," 798.

16. Bruhns 2:5, 1:354–55. A student of the American botanist John Torrey ran into Bonpland in South America and took a picture of him, which Torrey thoughtfully forwarded to Humboldt. See Schwarz, *Alexander von Humboldt*, 321, 342–43.

17. Bruhns 1:347. Francis Lieber recalled an anecdote that circulated widely about the Humboldt-Arago friendship: one day the king, Humboldt, and Niebuhr were talking politics, and Niebuhr started abusing Arago, "who, it is well known, was a very advanced Republican of the Gallican School, an uncompromising French democrat." The king abominated

republicanism, "yet when Neibuhr had finished, Humboldt said, with a sweetness which I vividly remember: 'Still this monster is the dearest friend I have in France'" (Hanno Beck, ed., *Gespräche Alexander von Humboldts* [Berlin: Akademie-Verlag, 1959], 200).

18. Agassiz, *Address*, 21. Agassiz's cost estimate was made in 1869. If he is to be believed, and one uses the standard rough-and-ready multiplier of twenty, the sum would today total a staggering $5 million, with sets retailing for $40,000. The Prussian government subvented publication of four complete sets, distributed to the university libraries of Berlin, Breslau, Halle, and Bonn.

19. De Terra, *Humboldt*, 323, 330; Nicolaas Rupke, "Alexander von Humboldt and Revolution: A Geography of Reception of the Varnhagen von Ense Correspondence," in *Geography and Revolution*, ed. David N. Livingstone and Charles W. J. Withers (Chicago: University of Chicago Press, 2005), 336–50; Botting 245; Bruhns 1:373–88, 2:209–11. In *Views*, Humboldt remarked, "Proximity to princes is apt to rob the most intellectual of their spirit and freedom" (382). One presumes he spoke from experience.

20. According to Susan Faye Cannon the link from Germany to Britain was provided by Charles Babbage; see Cannon 93. Australia soon joined in: see R. W. Home, "Humboldtian Science Revisited: An Australian Case Study," *History of Science* 33 (March 1995): 1–22. Humboldt's collaborative venture was the direct ancestor of the International Geophysical Year of 1957–58; see Botting 254. For Humboldt and geophysics, see also W. Schröder and K.-H. Weiderkehr, "Geomagnetic Research in the 19th century," *Acta Geodaetica et Geophysica Humbarica* 37.4 (2002): 445–66. Humboldt had lectured on global climate change in 1827 (Bruhns 2:104); see also *Views* 362 and *PN* 2:82.

21. Bruhns 2:155–56; Botting 256; Humboldt, *Letters to Varnhagen*, 41–42; Botting 268; de Terra, *Humboldt*, 345–46; Bruhns 2:161–62, 340–48. A charming portrait of Caroline von Humboldt (a force in her own right), and of the Humboldt family circle, is given by Mary Janes Ingham in "A Half-hour with the Humboldts," *Ladies Repository: A Monthly Periodical Devoted to Literature, Arts, and Religion* 26.5 (May 1866): 292–95.

22. Botting 273; Bruhns 2:348.

23. Ingo Schwarz, *From Alexander von Humboldt's Correspondence with Thomas Jefferson and Albert Gallatin* (Berlin: Alexander von Humboldt Research Center, 1991), 7; Jorge Cañizares-Esguerra, "Spanish America," 737. All of Humboldt's known correspondence with citizens of the United States has been painstakingly gathered, edited, and published by Ingo Schwarz in his *Alexander von Humboldt und die Vereinigten Staaten*. Additional letters continue to surface. Helmut de Terra has translated and published Humboldt's correspondence with Jefferson, Madison, and Gallatin ("Correspondence"). Most of Humboldt's correspondence is in French or German; Humboldt read English fluently but was uncomfortable writing it.

24. Schwarz, *Alexander von Humboldt*, 109, 100–11.

25. Ibid., 19–22, 120, 130–32. Jefferson's statement, written in 1813, stands at the head of an emergent policy claiming that the entire Western hemisphere lay within the sphere of influence of the United States. Hence "America" refers both to the United States and, via this prolepsis, this nation's presumed legal and logical extension to the entire hemisphere. Humboldt echoed Jefferson's sentiments when he praised the United States' generous aid to Venezuela (five ships laden with flour, "to be distributed among the poorest inhabitants") sent after the Caracas earthquake of 1812: "a valuable pledge of the mutual benevolence, that ought for ever to unite the nations of both Americas" (*PN* 4:55).

26. Ibid., 122, 131–32, 112, 113–14. On 2 March 1807, Congress passed a bill making the importation of slaves illegal. This did not, as Humboldt hoped, end slavery in the United States: slaves

continued to be smuggled in until 1860, and they were openly bought and sold on the domestic market. Humboldt's passage reads: "The number of African slaves in the United States amounts to more than a million, and constitute a sixth part of the whole population. The southern states, whose influence is increased since the acquisition of Louisiana, very inconsiderately increase the annual importation of these negroes. It is not yet in the power of Congress, nor the chief of the confederation (a magistrate [Jefferson] whose name is dear to the true friends of humanity), to oppose this augmentation, and to spare by that means much distress to the generations to come" (*PE* 1:15).

27. Ibid., 118, 126–27, 159, 186.

28. Ibid., 133; Adams, *Gallatin*, 565. The one exception to the several-year silence was an 1811 letter of introduction from Humboldt to Gallatin for the excellent Correa de Serra, together with a request for tobacco seeds (124). In this same period Humboldt became friends with John Quincy Adams, head of the peace delegation and the American minister to England, a friendship that would soon bear fruit when, as president, Adams supported the ambitious plan of another Humboldt acolyte, J. N. Reynolds, to mount the first major United States Exploring Expedition.

29. Schwarz, *Alexander von Humboldt*, 516–17; Ingo Schwarz, "Transatlantic Communication in the 19th Century: Aspects of the Correspondence between Alexander von Humboldt and George Ticknor," *Asclepio: Revista de Historia de la Medicina y de la Ciencia* 55.2 (2004): 29. As Schwarz details, Ticknor had planned the excursion for years, preparing by translating Goethe's *Sorrows of Young Werther* and rallying his friends for financial support. Robert D. Richardson, Jr., says Ticknor had gone to Germany after reading de Staël's *Germany*, which had "a huge effect on American learning": *Emerson: The Mind on Fire* (Berkeley: University of California Press, 1995), 53. Ticknor met the dying de Staël, who roused herself to tell him that "you are the vanguard of the human race. You are the world's future"; quoted in Maria Fairweather, *Madame de Staël* (London: Robinson, 2006), 462.

30. Robert D. Richardson, Jr., *Henry David Thoreau: A Life of the Mind* (Berkeley: University of California Press, 1986), 13; Schwarz, *Alexander von Humboldt*, 522, 232–33, 238; Schwarz, "Ticknor," 32; Stephens, "An Hour with Humboldt," 152. Schwarz notes that although the liberal Unitarian Ticknor's efforts to reform Harvard according to the German university model established at Berlin by Wilhelm von Humboldt did not succeed until after the Civil War, he not only paved the way but inspired fellow reformers such as James Marsh at Vermont and Francis Wayland at Brown.

31. Schwarz, *Alexander von Humboldt*, 405, 565–66, 443–46, 454–55. Ticknor had Humboldt's letter printed in the *Boston Courier* for 9 June 1858, from which it was widely reprinted.

32. Quoted in Schwarz, "Ticknor," 26. See my *Emerson's Life in Science: The Culture of Truth* (Ithaca: Cornell University Press, 2003). The one thing that divided them was slavery, for Ticknor favored compromise to preserve the Union, where Humboldt was ardent that the United States must abolish slavery at any cost.

33. Philip F. Gura, *American Transcendentalism: A History* (New York: Farrar, Straus and Giroux, 2007), 27; Edward Everett, "Humboldt's Works," *North American Review* 16 (January 1823): 2, 10, 25. Gura states that Everett met Wilhelm von Humboldt in Paris, which is incorrect; he met Alexander von Humboldt in Paris, who gave him a letter of introduction to Wilhelm, then in London. For Everett's Goethe review see *North American Review* 4 (January 1817): 217–62. In 1859, while declining an invitation to speak at a memorial in honor of Humboldt, Everett proudly reminded everyone that it was he who, all those years before in the *North American Review* for 1823, had originally testified to Humboldt's "transcendent merits as a philosophical

traveler and student of nature, both in detail and as one vast system." See Schwarz, *Alexander von Humboldt,* 577.

34. Irving, *Letters* (Boston: Twayne, 1979), 2:211, 219; Ingo Schwarz, "The Second Discoverer of the New World and the First American Literary Ambassador to the Old World: Alexander von Humboldt and Washington Irving," *Acta Historica Leopoldina* 27 (1997): 91. The Humboldt quotation is from *Examen Critique,* quoted in Schwarz, "Irving," 93 (translation by Mary Beth Stein). The original invitation was actually to translate a trove of documents; see John Harmon McElroy's introduction to Washington Irving, *The Life and Voyages of Christopher Columbus* (Boston: Twayne, 1981), xviii–xx. McElroy states that Irving's book became the standard biography of Columbus for the rest of the century, and one of the century's best known biographies of any figure (xxii, xxv).

35. Beck, *Gespräche,* 69 ("saloons" in original text). Pickering thanked Wilhelm for his "powerful stimulus" to studies of American Indian languages, which, he says, have excited little interest in the United States: "As Mr. Du Ponceau [who would soon become president of the APS] justly observes . . . 'We have to go to the German universities to become acquainted with our own country.'" The entire correspondence, which starts in 1821, continues to Wilhelm's death, and runs to roughly 150 manuscript pages, is preserved in the Rare Book Room of the Boston Public Library. For Pickering's understudied role in the development of American linguistics and his influence on Wilhelm von Humboldt and vice versa, see Julie Tetel Andresen, *Linguistics in America, 1769–1924: A Critical History* (New York: Routledge, 1990), 105–10. Wilhelm von Humboldt's side of the correspondence has been published in Germany, and some of Pickering's letters to Wilhelm von Humboldt appear in his 1887 biography, but surprisingly, the full correspondence has apparently never been published; it offers a trove of material on the history of U.S. American linguistics and early studies of American Indian languages.

36. Schwarz, *Alexander von Humboldt,* 164–67. Morse's report had been commissioned by the federal government—which, as Howe remarks, did not follow his advice: see Daniel Walker Howe, *What Hath God Wrought: The Transformation of America, 1815–1848* (Oxford: Oxford University Press, 2007), 342. In a letter to Pickering (which was reprinted in American newspapers), Wilhelm von Humboldt praised Jedidiah Morse's careful details, and more, his "exacting and just" information regarding the high degree of civilization reached by the Five Tribes. Morse's attention to geography helped suggest, he thought, what the Indians owed "to nature, and what, consequently, they possessed, perhaps, too, in a superior degree, before they knew the Europeans, from that they have since acquired. I have been surprised to see, by their speeches, their replies, and even slight sketches of poetry, with what talents nature herself has endowed them" ("Extract of a Letter from Baron Wm. Humboldt, to the Hon. Mr. Pickering of Salem," *Christian Spectator,* 1 September 1823, 495–96). As for Morse's son Samuel F. B. Morse, Kent Mathewson points out that the extent to which his "interactions with Humboldt influenced his global vision of telegraphy's potential is worth examining" ("Humboldt's Image," 418).

37. Everett, "Humboldt's Works," 20. Humboldt outlined his views in the *Political Essay* as well: "The predilections manifested by certain tribes for the cultivation of certain plants, indicates most frequently either an identity of race, or ancient communications between men who live under different climates. In this view the vegetables, like the languages and physiognomy of nations, may become historical monuments" (*PE* 3:437).

38. See Bruhns 2:391. The network of reciprocal circulation evidenced here, which features what Leon Jackson would call embedded authorial economies of exchange, would respond particularly well to the approach he advocates in his new book, *The Business of Letters: Authorial Economies in Antebellum America* (Stanford: Stanford University Press, 2008). Humboldt's in-

terest in the flow of precious metals resulted in his book *The Fluctuations of Gold*, trans. William Maude (New York: Burt Franklin, 1971), and dozens of the letters in Schwarz's collection turn on Humboldt's quest for information on the distribution and mining of gold and platinum in the United States. Humboldt was so identified with the proposed canal across the Isthmus of Panama that at least one American suggested it be erected as "a great international monument to his great name" and called "simply: HUMBOLDT CANAL." ("A Letter from Baron Humboldt," *Columbia Banner* [Columbia, SC], 1853, APS film 570.6–11. The writer, identified only as F.L., was likely Francis Lieber). For Humboldt on the Panama Canal, see *Views* 433.

39. Nicolaas Rupke, "A Geography of Enlightenment: The Critical Reception of Alexander von Humboldt's Mexico Work," in *Geography and Enlightenment*, ed. David N. Livingstone and Charles W. J. Withers (Chicago: University of Chicago Press, 1999), 335. As Rupke rather slyly adds, "These bibliometric facts would appear to narrow the basis on which Pratt puts forward her thesis that during the early part of the nineteenth century, Spanish America was reinvented from Humboldt's writings in terms of the Romantic images of luxuriant tropical forests, snow-capped volcanoes, and vast interior plains." Edward Everett noted in 1823 that "Mexico is certainly our most important frontier neighbor, and it may be very essential to us, for aught we know, to be acquainted with the state of his imperial majesty of Anahuac's dominions. It is not so far by land from New Orleans to the city of Mexico, as from New Orleans to Eastport. . . . Had M. de Humboldt written nothing but this, his name would have stood among the first philosophers of the day" ("Humboldt's Works," 13).

40. *PE* 1:52–53, 46.

41. *PE* 1:xi, cvii. Michael Dettelbach states that Humboldt was not particularly unusual or innovative in his mapping methods, but he was unusual in self-consciously writing about them (personal communication). Since Humboldt was not a professional mapmaker, he may have been more aware of dilemmas the professionals took for granted.

42. *PE* 1:11–12; 3:454 (see also 3:46–48, 92–95). European entrepreneurs marketed extracts from Humboldt's analyses of mines as investment guides. When the resulting boom went bust, Humboldt was blamed. He defended himself by insisting he never offered, nor intended to offer, investment advice, nor did he ever attempt to profit in any way from his knowledge. Humboldt's description of cloth factories is quite detailed: "Free men, Indians, and people of colour, are confounded with the criminals distributed by justice among the manufactories, in order to be compelled to work. All appear half naked, covered with rags, meagre, and deformed. Every workshop resembles a dark prison. The doors, which are double, remain constantly shut, and the workmen are not permitted to quit the house. Those who are married, are only allowed to see their families on Sundays. All are unmercifully flogged, if they commit the smallest tres-pass on the order established in the manufactory." How, he wonders, can the Indians submit to such conditions? They are forced by the stratagem of advancing potential workers a small sum, getting them indebted, then shutting them up under the pretense that they are paying off the debt. "Let us hope that a government friendly to the people, will turn their attention to a species of oppression so contrary to humanity, the laws of the country, and the progress of Mexican in-dustry" (3:463–65). Conditions in many of today's *maquiladoras* would make Humboldt doubt his faith in progress.

43. *PE* 4:282, 270. The Mexican Congress named Humboldt and Bonpland honorary citi-zens on 29 September 1827; see Schwarz, *Alexander von Humboldt*, 248n7. In a later volume of *Personal Narrative*, Humboldt expresses more concern: as the United States advances toward Mexico, it will there find "a European people of another race, other manners, and a different worship." Would the relatively "feeble" population resist its powerful northern neighbor, "or

will it be enveloped by the torrent of the east, and transformed into an Anglo-American state, like the inhabitants of Lower-Louisiana? The future will soon solve this problem" (*PN* 6:307).

44. William H. Goetzmann, *Exploration and Empire: The Explorer and the Scientist in the Winning of the American West* (1966; New York: Norton, 1978), xiii; Beck, *Gespräche*, 235. Humboldt may have been splitting the difference: as Daniel Walker Howe's recent history of this period details, the Mexican War had rapidly become unpopular, especially among Whigs; even some Democrats who supported it hesitated to annex all of Mexico, which would have meant incorporating mixed-race Mexicans into the American population. President Polk's negotiator, Nicholas Trist, refused Polk's demand to require still more territory of Mexico, who eventually accepted the original terms; whether Bancroft refers to Polk's original demands or to his new, more ambitious demands is unclear. Humboldt's assumption that America would bring liberal institutions to the annexed territory was not fulfilled; Mexicans were denied U.S. citizenship until 1912, and its tribal Indians were exterminated. See Howe 731–811. Howe also makes the point that Zachary Taylor's military achievements were brilliant but underappreciated (752); interestingly, Richard Henry Stoddard and John Lloyd Stephens both visited Humboldt about this time, and reported that he was amusing the king by replaying with him Zachary Taylor's battle tactics (Stoddard, *The Life, Travels, and Books of Alexander von Humboldt* [New York: Rudd and Carleton, 1859], 443; Stephens, "An Hour with Humboldt," 152). For an illuminating comparison with Canada, which according to Suzanne Zeller also used Humboldt's science to imperial purpose, see her article "Humboldt and the Habitability of Canada's Great Northwest," *Geographical Review* 96.3 (July 2006): 382–98.

45. Schwarz, *Alexander von Humboldt*, 113; and de Terra, "Correspondence," 790; *PE* 2:230; Goetzmann, *Exploration*, 78–79. As C. Gregory Crampton shows, the authority of Humboldt was so great that features he hypothesized were taken as gospel, leading many an explorer astray ("Humboldt's Utah, 1811," *Utah Historical Quarterly* 26:3 [1958], 268–81).

46. Schwarz, *Alexander von Humboldt*, 132; Jefferson, "Instructions to Meriwether Lewis," *The Annals of America* (Chicago: Britannica, 1968), 4:160–64; Goetzmann, *Exploration*, 6–7. Meriwether Lewis's mysterious death in 1809 was not the only reason for the expedition's lackluster success as a scientific enterprise. As William Stanton observes, an expedition's use to science depends as much on libraries, museums, and collections as on "keelboats and trinkets for the Indians," and while Congress was happy to fund keelboats and trinkets, they had no interest in libraries and museums. Some of the collections found their way to Peale's Museum (to be destroyed in a fire), but the botanical collection went to a German living in England. The journals were not published in their entirety until the twentieth century "and not then at public expense"; *The Great United States Exploring Expedition* (Berkeley: University of California Press, 1975), 5. It took decades of hard lobbying and spectacular results to solidify government support for exploration science.

47. William H. Goetzmann, *New Lands, New Men: America and the Second Great Age of Discovery* (New York: Viking, 1986), 53–54; Susan Faye Cannon, "Humboldtian Science," in *Science in Culture*, 105, 77. For the influence of Creole natural scientists on Humboldt, see Cañizares-Esguerra, "How Derivative Was Humboldt?," 112–28.

48. I first proposed this mnemonic in *Seeing New Worlds*, 98–102.

49. Andreas Daum makes the important point that the discussion of "Humboldtian science" has been carried out almost exclusively in English-speaking countries, and urgently needs to be connected with German scholarship; see his review of Horst Feidler and Ulrike Leitner, eds., *Alexander von Humboldt Schriftden: Bibliographaie der selbstandig erschienenen Werke, Beitrage*

zur Alexander-von-Humboldt-Forshung, no. 20 (Berlin: Akademie Verlag, 2000), in *Journal of the History of Biology* 33.3 (2000): 592. Major articles on Humboldtian science include Home; Goetzmann; Cannon; Buttimer; Dettelbach, "Humboldtian Science"; and Malcolm Nicolson, "Alexander von Humboldt, Humboldtian Science and the Origins of the Study of Vegetation," *History of Science* 25 (1987): 167–94.

50. On Humboldt's innovative forms of graphic representation, see Anne Godlewska, "From Enlightenment Vision to Modern Science? Humboldt's Visual Thinking," in *Geography and Enlightenment*, ed. David N. Livingstone and Charles W. J. Withers (Chicago: University of Chicago Press, 1999), 236–75; quotations are from 241, 253. For the isothermal zodiacal belt, see Goetzmann, *New Lands*, 57–59. The person most directly responsible for turning Humboldt's climate theories to imperial use was William Gilpin, with the help of Senator Thomas Hart Benton: as Saunders says, Gilpin "seized upon the ideas of harmony of nature and the isothermal zodiac to give a new dimension of quasi-scientific rationalisation to the policy of Western expansionism" (*Humboldtian Physicians*, 8). The standard treatment is Henry Nash Smith, *Virgin Land: The American West as Symbol and Myth* (New York: Vintage Books, 1950), 42–46.

51. Goetzmann, *New Lands*, 178.

52. Ibid., 117–19; Goetzmann, *Exploration*, 50–51; see also Arlen J. Large, "The Humboldt Connection," *We Proceeded On* 16.4 (November 1990): 4–12.

53. Quoted in Goetzmann, *Exploration*, 58; Goetzmann, *New Lands*, 120–26; *Exploration*, 58–62.

54. Duke of Württemberg (Paul Wilhelm), *Travels in North America, 1822–1824* (Norman: University of Oklahoma Press, 1973), 36–37, xxv, 193–94; Sonja Schierle, "Introduction: Travels in the Interior of North America," in *The American Indian: Karl Bodmer, Maximilian Prinz zu Wied* (Cologne, Germany: Taschen, 2005), 14.

55. Ben W. Huseman, *Wild River, Timeless Canyons: Balduin Möllhausen's Watercolors of the Colorado* (Tucson: University of Arizona Press, 1995), 20. Some have suggested that Caroline was actually Humboldt's illegitimate daughter; in any case, the marriage made Möllhausen into Humboldt's virtual, if not actual, son-in-law.

56. Alexander von Humboldt, preface to *Diary of a Journey from the Mississippi to the Coasts of the Pacific with a United States Government Expedition*, by Baldwin Möllhausen, 2 vols. (London: Longman, Brown, Green, Longmans, and Roberts, 1858), 1:xvii; quoted in Huseman 65; Goetzmann, *New Lands*, 158. O. Harris King has recently examined Möllhausen's fifth novel, *Die Mandanen-Waise* (1865), an autobiographical work set half in Germany and half on the American prairie in the winter of 1852. It presents a pastiche of liberal German themes: the young hero is moved by the revolutionary energies of young Germany to rebel against the oppressive government. He is captured, sentenced to life imprisonment, and escapes to America in search of a new identity, where he is lost on the prairie and rescued from death by the Oto Indians. In an abandoned Indian village he finds a young girl, lone survivor of the Mandan tribe destroyed by smallpox; the two fall in love and marry, and the novel ends happily. As King observes, this novel does important cultural work in mapping out "two of the main issues affecting German society at the time: the increasing emigration of many Germans to the United States and the lack of freedoms at home, two aspects of German life that were intimately related" (unpublished manuscript of 7 May 2008).

57. For Karl May, see Jace Weaver, *Other Words: American Indian Literature, Law, and Culture* (Norman: University of Oklahoma Press, 2001), 69–90. Matti Bunzl and H. Glenn Penny note that "the German interest in non-Europeans—in their cultures, their religions, their

physiognomy, their physiology, and their history" —was bound up in traditions much richer than "a simple colonialist drive," an insight that might provide a starting point for further discussion ("Introduction," 9).

58. Goetzmann, *New Lands*, 162; Ron Tyler, "Karl Bodmer and the American West," in *Karl Bodmer's North American Prints* (Lincoln: University of Nebraska Press, 2004), 1–45; William J. Orr, "Karl Bodmer: The Artist's Life," in *Karl Bodmer's America*, by Karl Bodmer (Lincoln: University of Nebraska Press, 1984), 349–76; William H. Goetzmann, "Introduction: The Man Who Stopped to Paint America," in *Karl Bodmer's America*, 1–23.

59. Goetzmann, *New Lands*, 164–68; Goetzmann, *Exploration*, 184–91; Schwarz, *Alexander von Humboldt*, 377; George Catlin, *Letters and Notes on the Manners, Customs, and Conditions of North American Indians*, 2 vols. (1844; repr. New York: Dover, 1973), 1:vii–xiv. For the Catlin quotation on Humboldt, and his frontispiece illustration, see his *Notes of Eight Years Travels and Residence in Europe*, 2 vols. (New York: Burgess, Stringer, 1848), vol. 1: title page, 246.

60. Martha Coleman Bray, *Joseph Nicollet and His Map*, 2nd ed. (Philadelphia: American Philosophical Society, 1994), 35, 43–44. For Engelmann, see Andreas Daum, "Die Natur als 'Kosmos,'" 254–55.

61. Goetzmann, "Introduction," 7.

62. Bray 46, 9–10; Goetzmann, *Exploration*, 231; William H. Goetzmann, *Army Exploration in the American West, 1803–1863* (New Haven: Yale University Press, 1959), 12–14; Edmund Bray and Martha Colemen Bray, eds., *Joseph N. Nicollet on the Plains and Prairies: The Expeditions of 1838–39 with Journals, Letters, and Notes on the Dakota Indians* (St. Paul: Minnesota Historical Society Press, 1993), 78.

63. Bray and Bray 231, 15; Goetzmann, *Exploration*, 242. According to Sally Denton, Nicollet added the accent to Frémont's name to instill in his young assistant a reverence for his French heritage: *Passion and Principle: John and Jessie Frémont, the Couple Whose Love Shaped Nineteenth-Century America* (New York: Bloomsbury, 2007), 18.

64. Quoted in Bray and Bray 17.

65. Ibid., 76, 81, 40. The opinion of Bray and Bray is corroborated by an experienced ethnographic fieldworker, who judges that Nicollet's description of how to listen to tribal elders sounds like "an ethnographer of the post-Boasian era" or a twentieth-century folklorist documenting oral traditions in their natural context (Robert E. Walls, personal communication).

66. Quoted in ibid., 36; ibid., 199–201, Nicollet's emphasis.

67. Goetzmann, *Exploration*, 256. Humboldt was greatly impressed by "the admirable labours of the talented French astronomer Nicollet, unhappily lost to science by a premature death," and he devotes half a page to explaining the importance of Nicollet's map of the upper Mississippi (*Views* 39).

68. Goetzmann, *Army Exploration*, 70.

69. My account is heavily indebted to Stanton; see Sachs, *Humboldt Current*, 20–22, 115–76; Goetzmann, *New Lands*, 266. See also Nathaniel Philbrick's *Sea of Glory* (New York: Viking, 2003), which explores why, especially after the California gold strike of 1848, the Wilkes expedition sank from memory.

70. J. N. Reynolds, "An Address on the Subject of a Surveying and Exploring Expedition to the Pacific Ocean and the South Seas." *Southern Literary Messenger*, January 1837.

71. Ralph Waldo Emerson, *The Letters of Ralph Waldo Emerson*, ed. Ralph L. Rusk and Eleanor Tilton, 10 vols. (New York: Columbia University Press, 1939–95), 3:123. For the role of the Smithsonian Institution in creating and maintaining the Humboldt network, see Andreas

Daum, " 'The Next Great Task of Civilization': International Exchange in Popular Science, the German-American Case, 1850–1900," in *The Mechanics of Internationalism: Culture, Society, and Politics from the 1840s to the First World War*, ed. Martin H. Geyer and Johannes Paulmann (Oxford: Oxford University Press, 2001), 285–319.

72. For Jessie Frémont, see Sandra Harbert Petrulionis, "Jessie Ann Benton Frémont," *Legacy* 18.2 (2001): 232–38; and Denton. For American women in natural science during this period see Tina Gianquitto, *"Good Observers of Nature": American Women and the Scientific Study of the Natural World, 1820–1885* (Athens: University of Georgia Press, 2007).

73. Goetzmann, *New Lands*, 299. In his 1869 address honoring Humboldt's centennial, Bache declared that it was Humboldt who saved the Coast Survey when funding for it was about to be cut; see de Terra, "Documentation," 140. Maury's plea to Humboldt reads: "Pray will you not lend me the powerful aid which a word from you would have in favour of a Main General Meteorological Conference of one that should take Cognicion of the land as well as the sea—and aim at the establishment of a universal system of observations" (Schwarz, *Alexander von Humboldt*, 326). Maury is a fascinating figure; for more on his relationship with Humboldt, see Kortum and Schwarz 157–85.

74. William Lewis Herndon and Lardner Gibbon, *Exploration of the Valley of the Amazon Made under Direction of the Navy Department*, 2 vols. (Washington: Robert Armstrong, 1854), 1:159, 20–21, 58, 189.

75. Ibid. 1:224, 277, 337; *New York Daily Times*, 17 February 1854, p. 8. My thanks to Brad Dean for passing this article along to me. For some reason, Brazil never issued the invitation, but according to Goetzmann "a sizeable contingent of unreconstructed Confederates" resettled some three hundred miles up the Amazon, in Santarém, Brazil (*New Lands*, 340). Herndon's attitude toward Indians was not shared by his traveling companion, midshipman Lardner Gibbon, who accompanied Herndon to South America and traveled on a parallel course from Bolivia. Gibbon enjoyed meeting Indians, and he made an observation similar to Humboldt's: "With political affairs the Indian has little or nothing to do. When the Creoles side off on the level plains of Bolivia and fight the battles of their country, the Indians seat themselves on the brows of the hills around, and quietly witness changes or continuance of administration. They seem to be the philosophers of the country, and to take the world very easy. After the struggle is over, they come down and pursue their daily occupations under the new constitution, laws, and powers that be" (2:115).

76. Henry David Thoreau, "Life without Principle," in *Reform Papers*, ed. Wendell Glick (Princeton: Princeton University Press, 1973), 176–77; Schwarz, *Alexander von Humboldt*, 338; quoted in Cheryl Walker, *Indian Nation: Native American Literature and Nineteenth-Century Nationalisms* (Durham: Duke University Press, 1997), 144, from Gae Whitney Canfield, *Sarah Winnemucca of the Northern Paintes* (Norman: University of Oklahoma Press, 1983), 49.

77. Bray and Bray 66–67; David M. Emmons, "Theories of Increased Rainfall and the Timber Culture Act of 1873," *Forest History* 15 (1971): 6–14.

78. Henry Adams quoted in Jonathan Arac, "Narrative Forms," in *The Cambridge History of American Literature*, vol. 2, *Prose Writing, 1820–1865*, ed. Sacvan Bercovitch and Cyrus R. K. Patell (Cambridge: Cambridge University Press, 1995), 776. For European liberalism see Colin Heywood, "Society," in *The Nineteenth Century*, ed. T. C. W. Blanning (Oxford: Oxford University Press, 2000), 51–61; for Manifest Destiny as a manifestation of anxiety rather than confidence, see Thomas R. Hietala's *Manifest Design: American Exceptionalism and Empire*, rev. ed. (Ithaca: Cornell University Press, 2003).

79. David S. Reynolds, *John Brown, Abolitionist* (New York: Knopf, 2005), 143; Oppitz; de Terra, *Humboldt*, 355; and Schwarz, *Alexander von Humboldt*, 457–58; Louis Agassiz, *Address*, 90–91. The other names proposed for Nevada were "Esmeralda" and "Washoe."

80. Humboldt, *Letters to Varnhagen*, 305. He also was blunt with Benjamin Silliman during a visit in 1851; in a letter to Silliman he wrote, "I have moral reasons to fear the immeasurable aggrandizement of your confederacy—the temptations to the abuse of power, dangerous to the Union . . . [and to the] distinct individual character of the other populations of America." Yet he could still acknowledge "the great advantages" that "peaceful conquests" by knowledge would bring, even as it "superimposes, not without violence, new classes of population upon the indigenous races which are in a course of rapid extinction." Somewhere out beyond these conflicts, he descried an epoch "when a high degree of civilization, and institutions free, firm and peaceful" would make Mexico City, Bogotá, and Quito resemble "in their institutions" New York, Boston, and Philadelphia (Silliman, "Baron von Humboldt," 320–21).

Interchapter

1. Ralph Waldo Emerson, "The American Scholar," in *Collected Works of Ralph Waldo Emerson*, ed. Alfred R. Ferguson et al., 6 vols. to date (Cambridge: Harvard University Press, 1971–), 1:52; hereafter cited as *CW*. For the "planetary" Emerson, begin with Wai Chee Dimock, "Deep Time: American Literature and World History," *American Literary History* 13.4 (2001): 755–75. As Paul Giles recently observed, "American literature should be seen as no longer bound to the inner workings of any particular country or imagined organic community but instead as interwoven systematically with traversals between national territory and intercontinental space" ("Transnationalism and Classic American Literature," *PMLA* 118.1 [January 2003]: 63). See also Larzer Ziff, who has recently observed, "The conscious struggle of American writers to create a national literature demanded of them that they confront a world beyond the national boundaries" (*Return Passages: Great American Travel Writing, 1780–1910* [New Haven: Yale University Press, 2000], 14.)

2. Henry David Thoreau, *Walden*, ed. J. Lyndon Shanley (Princeton: Princeton University Press, 1971), 298.

3. Eric Foner, *Tom Paine and Revolutionary America* (New York: Oxford University Press, 2005), 81, 78; Adams, *History*, 101; Thoreau, *Writings: Journal* 4:421. Of Gallatin, Adams notes that "he and his friends hoped to make [the revolution] the most radical that had occurred since the downfall of the Roman empire" (*History*, 111).

4. Irving, *Columbus*, 10, 569; Walt Whitman, "Prayer of Columbus," *Poetry and Prose*, 540–42.

5. J. N. Reynolds, "Mocha Dick, or The White Whale of the Pacific: A Leaf from a Manuscript Journal," *Arthur Gordon Pym, Benito Cereno, and Related Writings*, ed. John Seelye (Philadelphia: J. B. Lippincott, 1967), 288–89; Whitman, *Poetry and Prose*, 532; Emerson, *Collected Works*, 3:6, 18, 5; Whitman, *Poetry and Prose*, 534–35.

6. Nathaniel Hawthorne, *Twice-told Tales* (Ohio: Ohio State University Press, 1974), 62, 54; Ziff 16; 58–117.

7. Parkman quoted in Howe 782.

8. Wilhelm von Humboldt, "On the Historian's Task" (1821), in *The Theory and Practice of History Leopold von Ranke*, ed. Georg G. Iggers and Konrad von Moltke (Indianapolis: Bobbs-Merrill, 1973), 5–8. This move toward historicism calls for a much broader exploration. A good

starting point would be Andreas Daum, "*Wissenschaft* and Knowledge," in *Germany 1800–1870*, ed. Jonathan Sperber (Oxford: Oxford University Press, 2004), 137–61.

9. Francis Parkman, *France and England in North America*, 2 vols. (New York: Library of America, 1982), 1:16; Wilhelm von Humboldt, "Task," 10–11; Parkman, *France and England* 1:13–14.

10. William Hickling Prescott, *History of the Conquest of Mexico* and *History of the Conquest of Peru* (New York: Modern Library, 1936), 275–78.

11. Readers interested in exploring the fascinating question of the racial dimorphism represented by most of these dyads, and their overwhelmingly masculine and homoerotic gendering, will find a large literature on the subject; they should begin with the classic and widely reprinted essay by Leslie Fiedler, "Come Back to the Raft Ag'in, Huck Honey!" *Partisan Review* 25 (1948): 664–71. See also Benedict Anderson's comment that such dyads may rather represent "an eroticized nationalism" (202–3). Sometimes, as with Thoreau and Muir, the "band" of witnesses is compressed to a single person (even though Thoreau, for one, was often, like Humboldt, joined by companions whose presence is suppressed); however, in earlier treatments the adventurer or explorer is seldom without a visible support structure of companions, guides, and assistants.

12. Alexander von Humboldt, *Personal Narrative of Travels to the Equinoctial Regions of America, During the Years 1799–1804*, trans. and ed. by Thomasina Ross (London: Henry G. Bohn, 1852), 3:271.

13. Washington Irving, *Astoria, or Anecdotes of an Enterprise Beyond the Rocky Mountains*, ed. Edgeley W. Todd (Norman: University of Oklahoma, 1964), 113; Emerson, *CW* 6:1–2, 12.

14. Edgar Allan Poe, *The Narrative of Arthur Gordon Pym of Nantucket* (1838), ed. Richard Kopley (New York: Penguin, 1999), 194–45, 217. For Poe and Reynolds, see Aubrey Starke, "Poe's Friend Reynolds," *American Literature* 11.2 (May 1939), 152–59. In his maiden voyage, Pym and his friend Augustus fall unconscious, to be run over, rescued, and revived by the whaling ship *Penguin*, the same name as Reynolds's own ship.

15. This reading is complicated by Tommo's violent reaction to the prospect of being tattooed, or marked as nonwhite, and his growing conviction that the Typee, like Humboldt's gentle cannibal assistant, do in fact eat their enemies. Again, a full reading of this novel would take us beyond the scope of the current project.

16. Irving, *Columbus*, 529–42, 535, 540.

17. Washington Irving, *The Sketch Book* (1819), in *History, Tales, and Sketches*, ed. James W. Tuttleton (New York: Library of America, 1983), 1028.

18. Margaret Fuller, *The Essential Margaret Fuller*, ed. Jeffrey Steele (New Brunswick: Rutgers University Press, 1992), 89–90, 100, 175, 181. For Fuller on the Indian, see Charles Capper, *Margaret Fuller: An American Romantic Life; The Public Years* (Oxford: Oxford University Press, 2007), 148–53.

19. Capper 435.

20. Henry David Thoreau, *Cape Cod*, ed. Joseph J. Moldenhauer (Princeton: Princeton University Press, 1988), 98; Emerson, *CW* 6:17.

21. Reynolds, "Mocha Dick," 284.

22. John Charles Frémont, *Report of the Exploring Expedition to the Rocky Mountains in the year 1842, and to Oregon and North California in the years 1843–44* (Washington: Gales and Seaton, 1845), 62; James Fenimore Cooper, *The Prairie* (Albany: State University of New York Press, 1985), 250. In Michael A. Bryson's telling of this incident, Frémont's barometer becomes a broken sword/phallus: see *Visions of the Land: Science, Literature, and the American Environment*

from the Era of Exploration to the Age of Ecology (Charlottesville: University Press of Virginia, 2002), 30–31.

23. Hermann Melville, *Moby-Dick, or The Whale* (1851; New York: Penguin, 1992), 26; James V. Werner, "'Ground-Moles' and Cosmic Flaneurs: Poe, Humboldt, and Nineteenth-Century Science," *Edgar Allan Poe Review* 3.1 (Spring 2002), 45–64; Henry David Thoreau, *The Journal of Henry David Thoreau* (1906),14 vols., ed. Bradford Torrey and Francis Allen (New York: Dover, 1962), 5:45 [23 March 1853], 8:314 [28 April 1856].

24. Emerson, *CW* 1:60, 7.

25. Emerson, *CW* 1:54.

26. Cooper, *Prairie*, 70–72. Cooper's Dr. Bat bears a certain resemblance to the botanist Thomas Nuttall, as described in Irving's *Astoria*: "He was a zealous botantist [*sic*], and all his enthusiasm was awakened at beholding a new world, as it were, opening upon him in the boundless prairies, clad in the vernal and variegated robe of unknown flowers. . . . Every plant or flower of a rare or unknown species was eagerly seized as a prize." The Canadian voyageurs "were extremely puzzled by this passion for collecting what they considered mere useless weeds," making merry with the botanist treasuring his specimens "as some whimsical kind of madman" (170–71).

27. Thoreau, *Journal*, 9:157–58 [12 April 1856]; Henry David Thoreau, "Natural History of Massachusetts," in *Wild Apples and Other Natural History Essays*, ed. William Rossi (Athens: University of Georgia Press, 2002): 23.

28. Thoreau, *Journal*, 11:153 [8 September 1858]; Wilhelm von Humboldt, "Task," 8, emphasis added; Thoreau, *Writings: Journal*, 4:421 [2 April 1852].

29. Thoreau, *Writings: Journal*, 3:44 [5 January 1850]; Thoreau, "Natural History of Massachusetts," 23.

30. Stanton, *Exploring Expedition*, 308. For Hawthorne, see James R. Mellow, *Nathaniel Hawthorne in His Times* (Boston: Houghton Mifflin, 1980), 84–86; Arlin Turner, *Nathaniel Hawthorne: A Biography* (New York: Oxford University Press, 1980), 91–92.

31. Charles Wilkes, *Narrative of the United States Exploring Expedition*, 5 vols. (Philadelphia: Lea and Blanchard, 1845), 1:xxxiv; Jay Leyda, *The Melville Log: A Documentary Life of Herman Melville, 1819–1891* (New York: Harcourt, Brace, 1951), 241; Samuel Clemens, *Mark Twain's Autobiography* (New York: Harper and Brothers, 1924), 2:120–21. For Mark Twain as a travel writer, see Ziff 170–221.

32. Henry David Thoreau, "Intelligence," *Dial* 3.1 (July 1842): 132–33; Emerson, *Letters*, 3:123; Thoreau, *Walden*, 321–22.

33. Thomas De Quincey, "The Poetry of Pope," *Collected Writings of Thomas De Quincey* (London: A. C. Black, 1897), 11:54–56.

34. Lance Newman, *Our Common Dwelling: Henry Thoreau, Transcendentalism, and the Class Politics of Nature* (New York: Palgrave Macmillan, 2005); Jeffrey Myers, *Converging Stories: Race, Ecology, and Environmental Justice in American Literature* (Athens: University of Georgia Press, 2005). For the depopulation of the national parks, see Mark David Spence, *Dispossessing the Wilderness: Indian Removal and the Making of the National Parks* (New York: Oxford University Press, 1999).

Chapter Four

1. For Jefferson's original draft of the Declaration of Independence and his stated reasons for altering it, see Thomas Jefferson, *Writings*, 18–24.

2. Reginald Horsman, *Race and Manifest Destiny: The Origins of American Racial Anglo-Saxonism* (Cambridge: Harvard University Press, 1981), 101.

3. C. Loring Brace, *"Race" Is a Four-Letter Word: The Genesis of the Concept* (New York: Oxford, 2005), 46; Alexander von Humboldt, *Cosmos: Sketch of a Physical Description of the Universe*, trans. Elizabeth Sabine, 2 vols., 5th ed. (London: Longman, Brown, Green, and Longmans, 1849), 1:355; Stephen Jay Gould, *The Mismeasure of Man* (New York: Norton, 1981), 38.

4. Johann Friedrich Blumenbach, *The Anthropological Treatises of Johann Friedrich Blumenbach* (1865), trans. Thomas Bendyshe (Boston: Longwood Press, 1978), 264, 305, 312.

5. Ibid., 301; *PN* 7:463–64, 260.

6. *Cosmos* 1:358, 358n; *PN* 7:282–83.

7. *PN* 6:124; 3:438–41; see *PE* 1:200–203 for more detail.

8. *PN* 3:430–32.

9. *PE* 1:184, 188, 199–203.

10. *PE* 1:235–42.

11. *PE* 1:243–46. An extraordinary series of casta paintings is reproduced in M. Concepción García Sáiz, "Miguel Cabrera," *The Arts in Latin America, 1492–1820*, ed. Joseph N. Newland et al. (Philadelphia: Philadelphia Museum of Art, 2006), 402–9. See also Ilona Katzew, "Casta Painting: Identity and Social Stratification in Colonial Mexico," in *New World Orders: Casta Painting and Colonial Latin America*, ed. Ilona Katzew (New York: Americas Society Art Gallery, 1996), 8–29.

12. Anderson 150–53; *PE* 1:245–46; *Researches* 1:66; *PE* 1:245–47. For more on these "certificates of legal 'whiteness'" see Katzew 12.

13. David Brion Davis, *Inhuman Bondage: The Rise and Fall of Slavery in the New World* (Oxford: Oxford University Press, 2006), 177, 192.

14. *PN* 4:432–33; *PE* 1:257–60.

15. *PE* 1:262; *PN* 7:267.

16. *PN* 7:269; Davis 159, 270–71; *PN* 7:334–35.

17. Erwin H. Ackerknecht, "George Forster, Alexander von Humboldt, and Ethnology," *Isis* 46 (1955): 83–95.

18. Matti Bunzl, "Franz Boas and the Humboldtian Tradition," in *Volksgeist as Method and Ethic: Essays on Boasian Ethnography and the German Anthropological Tradition*, ed. George W. Stocking, Jr., vol. 8 of *History of Anthropology* (Madison: University of Wisconsin Press, 1996), 24, 29, 31. It was de Staël who originated the comment about Wilhelm von Humboldt's intellectual preeminence; Mary Janes Ingham notes that his wife Caroline von Humboldt was one of de Staël's closest friends (294).

19. *Cosmos* 1:49; Bunzl, "Franz Boas," 39. To assert, like Humboldt, that Indians were historical peoples had subversive implications in the United States, where American government's abrogation of Indian treaty rights was based on the projection of Indians out of history altogether to a static, timeless, primitive past. This point is well made by Maureen Konkle in *Writing Indian Nations: Native Intellectuals and the Politics of Historiography, 1827–1863* (Chapel Hill: University of North Carolina Press, 2004): when nineteenth-century native intellectuals like William Apess argued that tribes had both complex political institutions and a civil history, they were insisting that Indians had political and historical autonomy, which meant the right to negotiate treaties—and the right to demand that those treaties be respected.

20. Wilhelm von Humboldt, "Task," 23; Bunzl, "Franz Boas," 31; Wilhelm von Humboldt, *Essays on Language*, ed. T. Harden and D. Farrelly (Frankfurt am Main: Peter Lang, 1997), 26–27.

21. See Bunzl, "Franz Boas," 33, 34; John E. Joseph, "A Matter of *Consequenz*: Humboldt, Race and the Genius of the Chinese Language," *Historiographia Linguistica* 26.1/2 (1999): 119, 143.

22. Quoted in Joseph 136; Wilhelm von Humboldt, *Language*, 7–8, 68, 85.

23. *Cosmos* 1:358–59; Wilhelm von Humboldt, *Language*, 138–39. Both Joseph and Bunzl seek to rescue Wilhelm von Humboldt from imputations of racism, derived from decontextualized misreadings of his often difficult prose, and by repetition of misunderstandings perpetuated by his ideological opponents.

24. Bunzl, "Franz Boas," 31 and passim.

25. Eloise Quiñones Keber, "Humboldt and Aztec Art," *Colonial Latin American Review* 5.2 (1996): 277–97; Halina Nelkin, *Humboldtiana at Harvard* (Cambridge: Widener Library, Harvard University, 1976), 27. For Humboldt as the founder of New World landscape archaeology, see Kent Mathewson, "Alexander von Humboldt and the Origins of Landscape Archaeology," *Journal of Geography* 85.2 (March–April 1986): 50–56.

26. Adams, *Gallatin*, 672–77; Raymond Walters, Jr., *Albert Gallatin: Jeffersonian Financier and Diplomat* (New York: Macmillan, 1957), 354.

27. Humboldt, *Essay on the Geography of Plants*, 54.

28. H. F. Augstein, *James Cowles Prichard's Anthropology: Remaking the Science of Man in Early Nineteenth Century Britain* (Amsterdam-Atlanta, GA: Rodopi, 1999), 79; James Cowles Prichard, *The Natural History of Man*, 2 vols. (London: H. Baillière, 1855), 2:714; Augstein, *Prichard's Anthropology*, 145–47.

29. Quoted in Stanton, *Exploring Expedition*, 340.

30. Ibid., 340, 346–47. Charles Pickering seems to have been taking after his uncle John Pickering. According to Julie Tetel Andresen, when Lewis Cass—about to become President Jackson's secretary of war, in which role he would direct the annihilation of Black Hawk and the Seminoles—asserted in 1826 that Indians were barely capable of thinking, John Pickering shot back a reply under the pen name Kass-ti-ga-tor-skee, or "the Feathered Arrow" (127).

31. Robert Knox, *The Races of Men: A Fragment* (Philadelphia: Lea and Blanchard, 1850), 13, 7, 13.

32. Robert E. Bieder, *Science Encounters the Indian, 1820–1880: The Early Years of American Ethnology* (Norman: University of Oklahoma Press, 1986), 91, 70; Gould, *Mismeasure*, 50–69. Brace defends Morton against Gould's accusations in *"Race,"* 88–90.

33. Lester D. Stephens, *Science, Race, and Religion in the American South: John Bachman and the Charleston Circle of Naturalists, 1815–1895* (Chapel Hill: University of North Carolina Press, 2000), 171; Josiah Nott, *American Journal of Science* [*Silliman's Journal*] 6 (1843): 252–53.

34. Elizabeth Cary Agassiz, *Louis Agassiz: His Life and Correspondence*, 2 vols. (Boston: Houghton Mifflin, 1887), 2:417; Agassiz's letter translated and quoted by Gould, *Mismeasure*, 45. As Gould points out, this passage was expurgated from Elizabeth Cary Agassiz's standard biography of her husband. Gould believed he was the first to publish it; a much briefer and less explicit excerpt was printed in Edward Lurie, *Louis Agassiz: A Life in Science* (Chicago: University of Chicago Press, 1960), 257.

35. Agassiz, "The Diversity of Origin of the Human Races," *Christian Examiner* 49 (July 1850): 110–12, 118, 135–42. Morton agreed: after reading *Cosmos* he wrote to Gliddon that while Humboldt's word *désolante* (translated as "depressing") "is true in sentiment and in morals . . . it is wholly inapplicable to physical reality." That reality was that nonwhite races were inferior, a fact science must acknowledge: "Let us search out the truth, and reconcile it afterwards." Quoted in Richard H. Popkin, "The Philosophical Bases of Modern Racism," in *Philosophy and*

the Civilizing Arts, ed. Craig Walton and John P. Anton (Athens: Ohio University Press, 1974), 151–52.

36. Popkin 133; William Stanton, *The Leopard's Spots: Scientific Attitudes toward Race in America, 1815–1859* (Chicago: University of Chicago Press, 1960), 162–63; Agassiz, "Sketch of the Natural Provinces of the Animal World and Their Relation to the Different Types of Man," in *Types of Mankind*, by Josiah C. Nott and George R. Gliddon, 8th ed. (Philadelphia: Lippincott, 1857), lxxiv.

37. Arnold Guyot, *The Earth and Man: Lectures on Comparative Physical Geography, in Its Relation to the History of Mankind*, trans. Cornelius Felton (Boston: Gould, Kendall, and Lincoln, 1849), 307, 296–97, emphasis added. In 1858 James Henry Hammond, the governor of South Carolina, paraphrased Guyot's thesis in a speech to the U.S. Senate: "In all social systems there must be a class to do the menial duties, to perform the drudgery of life. . . . Its requisites are vigor, docility, fidelity. Such a class you must have, or you would not have that other class which leads progress, civilization, and refinement. It constitutes the very mud-sill of society and political government, and you might as well attempt to build a house in the air, as to build either one or the other, except on this mud-sill" (quoted in Davis 189). As Davis comments, "Southern leaders considered themselves exceptionally fortunate in having found a race that had been created to be a mud-sill."

38. Stephens, *Science*, 173; for reproductions of the photographs, see Brian Wallis, "Black Bodies, White Science: Louis Agassiz's Slave Daguerreotypes," *American Art* 9 (Summer 1995): 38–61.

39. Stanton, *Leopard's Spots*, 75. Stanton notes that Asa Gray supported Bachman's position but thought public engagement futile. Gallatin had of course died in 1849.

40. John Bachman, *The Doctrine of the Unity of the Human Race Examined on the Principles of Science* (Charleston, SC: C. Canning, 1850), 209, 211–12.

41. Peter McCandless, "The Political Evolution of John Bachman: From New York Yankee to South Carolina Secessionist," *The South Carolina Historical Magazine* 108.1 (January 2007), 13; Simms quoted in Eric Sundquist, "Literature of Expansion and Race," *Cambridge History of American Literature*, vol. 2, *Prose Writing, 1820–1865* (Cambridge: Cambridge University Press, 1995), 262; John Winthrop, "A Model of Christian Charity," in *The Norton Anthology of American Literature*, vol. A, *Beginnings to 1820*, ed. Wayne Franklin, Philip F. Gura, and Arnold Krupat (New York: Norton, 2007), 148; Bachman, *Unity*, 8.

42. Bachman, *Unity*, 167–69, 306–7. At the end of his tribute to Humboldt, Parker asked his audience to remember "that Mr. Humboldt is one of the men of science whom American churchlings have sought to brand with the name of 'Atheist.' . . . Certainly, this great philosopher is not one of those who are continually crying 'Lord, Lord!' But when did he fail to do the duty of a man?" (*Liberator*, 9 July 1858).

43. Peter W. Becker, "Lieber's Place in History," in *Francis Lieber and the Culture of the Mind*, ed. Charles R. Mack and Henry H. Lesesne (Columbia: University of South Carolina Press, 2005), 5–6; Schwarz, *Alexander von Humboldt*, 424–25. Lieber's letter to Humboldt reads: "You may have heard about my resignation in South Carolina. Why I resigned would certainly lead too far and would be barely interesting for you. But I could not resist telling you about my election to this honorable position. It is the urge of a son of Germany to notify his Humboldt. Shall I add that my 'Civil Liberty and Self-Government,' as well as my 'Political Ethics' have several times been cited as authorities in difficult constitutional cases by the judges at the Supreme Court?" (trans. by Catherina Wuetig).

44. Becker, "Lieber's Place," 6.

45. Franz Boas, *The Mind of Primitive Man*, rev. ed. (New York: Macmillan, 1938), v.

46. Schwarz, *Alexander von Humboldt*, 97.

47. *Liberator*, 20 November 1863, 187; Schwarz, *Alexander von Humboldt*, 461–62. The Rare Book Room of the Boston Public Library possesses one of these cards; Humboldt's original letter is photographically reproduced on one side in microscopic text, with the translation provided on the verso. Matthews invented, manufactured, and distributed the modern soda fountain.

48. *PN* 7:107.

49. *PN* 7:223.

50. *PN* 7:273, 156–57, 152–54; quoted in Sundquist 744; see Harriet Beecher Stowe, *Uncle Tom's Cabin*, in *Three Novels* (New York: Library of America, 1982), 395.

51. *PN* 7:208–9, 103–4 (emphasis in original); 5:683–84; 7:236–37.

52. *PN* 7:260.

53. Vera Kutzinsky has explored the many distortions in the several English translations of Humboldt's essay on Cuba, including Thrasher's, in some detail; she and Ottmar Ette are leading a research team that is undertaking new translations of Humboldt's American works, forthcoming from the University of Chicago Press. Humboldt's *Political Essay on the Island of Cuba* will be the first volume in this series. See "Translating Humboldt's *On the Island of Cuba*," paper delivered at "Alexander von Humboldt and the Hemisphere," Vanderbilt University, January 15–17, 2009.

54. Alexander von Humboldt, *The Island of Cuba: A Political Essay*, trans. J. S. Thrasher (1856; Princeton: Markus Wiener Publishers, 2001), 21.

55. Schoenwalt, "Alexander von Humboldt und die Vereinigten Staaten von Amerika," 458; Schwarz, *Alexander von Humboldt*, 452; Stoddard 461; Humboldt, *Letters to Varnhagen*, letter 208, p. 373. For Humboldt on slavery, see Philip S. Foner, "Alexander von Humboldt on Slavery in America," *Science and Society* 47.3 (Fall 1983): 330–42.

56. Humboldt, *Island of Cuba*, 43–44.

57. Ibid., 44–65.

58. *PN* 7:6, Humboldt, *Island of Cuba*, 77, 187, 251 n23.

59. *PN* 7:260–64.

60. *PN* 7:269–70, 281–84.

61. Schwarz, *Alexander von Humboldt*, 365–67 (including a photograph of Thrasher's letter), 561–62; Humboldt, "Letter from Humboldt," in *Letters on American Slavery from Victor Hugo, de Tocqueville, Emile de Gerardin, Carnot, Passy, Mazzini, Humboldt, O. Lafayette—&c.* (Boston: American Anti-Slavery Society, 1860): 17–19. The letter was widely reprinted in U.S. newspapers, and appears, together with remarks on Webster quoted below (see n. 65), in *Letters on American Slavery* (Boston: American Anti-Slavery Society, 1860), 17–20. The anonymous editor comments, "These were the opinions of Baron Humboldt, a Christian philosopher of world-wide reknown, whose views of men and of nations went further to establish their character, than any man now living. As Humboldt thought, the Christian world would think."

62. Schwarz, *Alexander von Humboldt*, 563; Humboldt, *Letters to Varnhagen*, 325–26; quoted in Foner, "Humboldt on Slavery," 339. To a friend Humboldt wrote that this comment "had caused a great outburst of anger among the defenders of slavery in the United States" (338).

63. Schwarz, *Alexander von Humboldt*, 434–35, 276–77, 387; Humboldt, *Letters to Varnhagen*, 326; Foner, "Humboldt on Slavery," 340. See John Bigelow, *Memoir of the Life and Public Services of John Charles Fremont* (New York: Derby and Jackson, 1856): "To Alexander von Hum-

boldt, this memoir of one whose genius he was among the first to discover and acknowledge, is respectfully inscribed by the Author." The thirty-two-page campaign flyer, *Life of Col. Fremont* (New York: 1856), ends with testimonials from supporters in Charleston, South Carolina, and from Humboldt, including a description of the impressive medal: "Of fine gold, massive, more than double the size of the American double eagle, and of exquisite workmanship" (30–31). Horace Greeley printed and distributed 150,000 copies of the campaign booklet.

64. Bayard Taylor, "Alexander von Humboldt," in *At Home and Abroad* (New York: Putnam 1860), 357; Humboldt, *Letters to Varnhagen*, 339 (letter 187); Schwarz, *Alexander von Humboldt*, 572; Bruhns 2:254.

65. Cornelius Conway Felton, *Familiar Letters from Europe* (Boston: Ticknor and Fields, 1865), 63–64; also in Schwarz, *Alexander von Humboldt*, 551–52 (emphasis in original).

66. *Liberator*, 11 May 1855; *New York Evening Post*, 10 May 1855.

67. Manuscript, Boston Public Library Ms. A.9.2 vol.1, no.11 (my profound thanks to Sandy Petrulionis for passing this previously unknown letter along to me, and for pointing out the two previously cited Humboldt letters reprinted in *Liberator*); Foner, "Humboldt on Slavery," 334, 331.

68. *Liberator*, 9 July 1858. On 17 June 1858 Garrison wrote to Parker, "I am delighted to know that you are willing to say something in merited praise of that wonderful old man, Humboldt, at our gathering at Framingham. . . . We will have your tribute carefully reported, printed, and transmitted to that scientific prodigy and veteran philanthropist. Perhaps you will draw up a resolution, to be adopted by acclamation on the occasion" (William Lloyd Garrison to Theodore Parker, Boston Public Library Rare Book Room Ms. A.1.1 vol. 5, p. 63).

Chapter Five

Epigraph: In the original Spanish, Carrera Andrade's poem reads: "Comprende y venera al objeto. / Penetra en ese orbe secreto / y sea la flor tu amuleto." Translated by Charles Bergman; my thanks to Charles for acquainting me with Carrera Andrade's work.

1. Richard V. Francaviglia, *Mapping and the Imagination in the Great Basin: A Cartographic History* (Reno: University of Nevada Press, 2005), 3, 9.

2. Peter Kropotkin, "Modern Science and Anarchism," in *Kropotkin's Revolutionary Pamphlets: A Collection of Writings by Peter Kropotkin*, ed. Roger N. Baldwin (New York: Dover, 1970), 147, 152; Bowen 262–63; Marvin Harris, *The Rise of Anthropological Theory* (New York: Harper and Row, 1968), 250.

3. Franz Boas, "The Background of My Early Thinking," in *The Shaping of American Anthropology, 1883–1911: A Franz Boas Reader*, ed. George W. Stocking, Jr. (New York: Basic Books, 1974), 41–42; Douglas Cole, *Franz Boas: The Early Years, 1858–1906* (Seattle: University of Washington Press, 1999), 79–81. For Boas' Jewish family background, see Cole 13–14.

4. Cole 99–133, Harris, *Rise*, 254; Boas on Humboldt quoted in Cole 26; Franz Boas, "The Study of Geography," in *Race, Language and Culture* (New York: Macmillan, 1966), 639–42.

5. Boas, "Geography," 644–45.

6. Ibid., 644, 646–47; Thoreau, *Journal* 10:165; Carrera Andrade. Jorge Carrera Andrade (1903–78) was a diplomat and humanitarian as well as a poet; his work is virtually unknown in the United States, although in Latin America he is acclaimed as one of their greatest poets. His poem "Humboldt" ends: "The hoofbeat of your mule / awakened nations / you forged a new world / and raised the standard of light / in the night of the centuries" (Carrera Andrade, *Selected Poems* 230–33).

7. See Harris, *Rise*, 255; Cole 128; Bunzl, "Franz Boas," 19; Bunzl and Penney, "Introduction," 22; Boas, *Primitive Man*, 272. For more on Boas see Bunzl, "Franz Boas," particularly 18–19, 64–73; Harris, *Rise*, 254–92. Recently Bunzl has joined several other anthropologists to argue at length for a "neo-Boasian" anthropology for the twenty-first century; see *American Anthropologist* 106.4 (September 2004): 433–94.

8. However, a biography had appeared: William McGillivray's *The Travels and Researches of Alexander von Humboldt* (1832; New York: Harper, 1833).

9. Stoddard 432–33.

10. Humboldt, *Essay on the Geography of Plants*, 56–57.

11. Bruhns 2:112, 308; 1:87–88, 114; 1:173–75; quoted in Buttimer 114. For Humboldt's *Cosmos* and popular science, see E. R. Brann, *Alexander von Humboldt: Patron of Science* (Madison, WI: E. R. Brann, 1954), 9–10; Bruhns 2:123–24; and Andreas Daum, "Science, Politics, and Religion: Humboldtian Thinking and the Transformations of Civil Society in Germany, 1830–1870," in *Science and Civil Society*, ed. Lynn K. Nyhart and Thomas H. Broman, *Osiris* 17 (2002): 107–40.

12. Humboldt, *Letters to Varnhagen*, 35–39.

13. Bruhns 2:366; Rupke, Review of Petra Werner, *Himmel und Erde. Alexander von Humboldt und sein Kosmos* (Berlin: Akademie Verlag, 2004), in *Annals of Science* 62.4 (October 2005): 554; Stoddard 437, 475. Despite Stoddard's claim, *Kosmos* clearly was never finished. The fifth and final volume, which is mostly index, has never been translated into English.

14. Quoted in de Terra, *Humboldt*, 359; Cedric Hentschel, "Alexander von Humboldt's Synthesis of Literature and Science," in *Alexander von Humboldt, 1769/1969* (Bonn: InterNationes, 1969), 124; Humboldt, *Letters to Varnhagen*, 197.

15. *Cosmos* 1:50, 54–55, 23. For a useful analysis of *Kosmos*, see Gisela Brude-Firnau, "Alexander von Humboldt's Sociopolitical Intentions: Science and Poetics," in *Traditions of Experiment from the Enlightenment to the Present* (Ann Arbor: University of Michigan Press, 1991), 45–61.

16. *Cosmos* 1:47, 53 (emphasis added), 77, 51, 53.

17. *Cosmos* 1:24, 68, 79–80, 69–71. Not everyone was convinced by Humboldt's neologism: a London reviewer grumbled, "We venture to think his introduction of the word *Cosmos* into our vocabulary, and the word itself, after all, indefinite." Review of *Cosmos*, vol. 1, *Quarterly Review* 77 (December 1845): 164.

18. *Cosmos* 1:79; Bowen 257.

19. *Cosmos* 1:24, 3:9, 5.

20. *Cosmos* 1:55; "Alexander von Humboldt's Cosmos," *Broadway Journal*, 12 July 1845; *Cosmos* 1:49; 3:10–11; Thoreau, *Walden*, 320.

21. *Cosmos* 1:49.

22. John Burroughs, "Science and Literature," in *Indoor Studies*, vol. 8 of *The Writings of John Burroughs* (Boston: Houghton Mifflin, 1904), 49–74, 64–65; Kropotkin, *Memoirs*, 211.

23. *Cosmos* 1:76, 162–63. Humboldt is referring to his associates, some of them his friends, in the German school of *Naturphilosophie*, with which he is sometimes erroneously categorized.

24. *PN* 3:90–91; *Cosmos* 1:37–38. Humboldt did worry about offending Schelling and Hegel, as his letters to Varnhagen made clear; he intended his aspersions for their followers, the sort who would practice "Chemistry, without so much as wetting one's fingers," and write such delicious nonsense as, "The diamond is a pebble arrived at consciousness" (101–7).

25. *Cosmos* 1:40–41; Thoreau, *Journal*, 4:416–21 (2 April 1852).

26. *Cosmos* 1:25, 26 (the expression in quotations is taken from Bernardin de Saint Pierre's *Paul and Virginia*); David Kenosian, "Speaking of Nature," in *Alexander von Humboldt: From the Americas to the Cosmos*, ed. Raymond Erickson et al. (New York: Bildner Center, City Uni-

versity of New York online publication), 506, http://web.gc.cuny.edu/bildnercenter/publications/humboldt.pdf (accessed 11 January 2009).

27. *Views* 217–18; Humboldt, *Essay on the Geography of Plants* 55; *PN* 4:133–34. In *Views* Humboldt notes an interesting example of bioregional imprinting on personal character: on his first sight of a pine forest, Montúfar, who had been born in Quito and had therefore never seen needle-leafed trees, felt that the trees were leafless, "and because we were journeying towards the cold north, he thought he recognised already, in the extreme contraction of the organs, the impoverishing influence of the pole" (328–29).

28. Steigerwald, "Figuring Nature," 69–70; *Cosmos* 1:34; *PN* 4:419; *Views* 192; *Cosmos* 2:81. Many commentators presume a connection between Humboldt and the similar precepts of the British art critic John Ruskin. However, when asked if he had drawn on Humboldt's *Cosmos*, Ruskin huffed defensively that he had glanced at it and tossed it aside: "Certainly I owe it absolutely nothing." (*Modern Painters*, 5 vols. [London: George Allen, 1897], 3:361). Ruskin in general found American landscapes too historically shallow to be interesting. Tropical landscapes struck him as grotesque, incapable of developing the mind or heart: "It would be difficult to conceive of groves less fit for academic purposes than those mentioned by Humboldt, into which no one can enter except under a stout wooden shield, to avoid the chance of being killed by the fall of a nut" (*Modern Painters*, 5:151–52). Bernard Smith makes a case for the influence of Humboldt on Ruskin in *European Vision and the South Pacific* (New Haven: Yale University Press, 1985), 205–6.

29. *Cosmos* 1:72.

30. *Cosmos* 1:72.

31. Michel-Rolph Trouillot, *Silencing the Past: Power and the Production of History* (Boston: Beacon Press, 1995), 2.

32. Trouillot 16.

33. There is a scientific field called "historical ecology," which according to one of its founders, William Balée, "seeks a synthetic understanding of human/environmental interactions within specific societal, biological, and regional contexts. In other words, the focus of historical ecology is a relationship, not an organism, species, society—not a 'thing.'" As the name of the field suggests, its practitioners follow changing ecological relationships, putting humans as key ecological actors at the center of explanations of ecological change across historical time. Its pioneers have done much of their empirical work in Humboldt's old haunts, the Amazon and Orinoco river basins, and they trace the roots of their field to Boas and, at least implicitly, to Humboldt. For an overview, start with William Balée, ed., *Advances in Historical Ecology*; the quotation from Balée appears as an epigraph on p. 213.

34. *Cosmos* 1:83; *PN* 1:195; 4:505–6; 3:105–6.

35. *Cosmos* 1:25; *Views* 153–54, 219; *PN* 1:185; 5:139 (emphasis added). Humboldt's line of metaphors bears comparison with the work of George Lakoff and Mark Johnson, starting with their classic work *Metaphors We Live By* (Chicago: University of Chicago Press, 1980).

36. *Cosmos* 1:26; Coleridge, "Dejection: An Ode," *Poetical Works*, ed. Ernest Hartley Coleridge (Oxford: Oxford University Press, 1967, 1973), 365; Bruhns 1:88; *Views* 173.

37. Bowen 255–57. I would argue that in this respect Humboldt's philosophy anticipates that of the French sociologist of science Bruno Latour. A Latourian analysis of Humboldtian science would be a very productive project.

38. *Cosmos* 1:76; first quotation as translated by Bowen 257 (nowhere is the need for a modern translation of *Cosmos* more evident than in this passage). I have treated the problem of Humboldt and scientific objectivity in "The Birth of the Two Cultures," 247–58.

39. *Cosmos* 1:56; Thoreau, *Walden*, 175; *Cosmos* 1:76. The connections between Humboldt and Charles Sanders Peirce remain to be explored; for a promising start, see Bradley Ray King, "The Entanglement of the Rainbow: Henry David Thoreau, Charles Sanders Peirce, and the Meanings of Nature in Nineteenth-Century America" (MA thesis, University of South Carolina, 2008).

40. *Cosmos* 3:9 (as translated by Bowen 257); *Cosmos* 3:7; Hacking quoted in Tristram R. Kidder, "The Rat That Ate Louisiana: Aspects of Historical Ecology in the Mississippi River Delta," in Balée, *Advances*, 144; Bowen 272; *Cosmos* 1:311; Bowen 274.

41. Villegas 2:138.

42. Kant, "Idea for a Universal History," in *The Philosophy of Kant*, ed. Carl J. Friedrich (New York: Modern Library, 1949), 116; *Cosmos* 1:38.

43. Andrew Wilton, "The Sublime in the Old World and the New," in *American Sublime: Landscape Painting in the United States, 1820–1880*, by Andrew Wilton and Tim Barringer (Princeton: Princeton University Press, 2002), 15. For Humboldt and the dating of rocks, see *A Geognostical Essay on the Superposition of Rocks, in Both Hemispheres* (London: Longman, Hurst, Rees, Orme, Brown, and Green, 1823).

44. Gould, "Church, Humboldt, and Darwin," 97, 104–5; *Cosmos* 1:24, 8; Charles Darwin, *On the Origin of Species*, facsimile of 1st ed. (1859; Cambridge: Harvard University Press, 1964), 489–90.

45. *PN* 4:421–22; Richards, *Romantic Conception*, 551.

46. Peter Hanns Reill, "The Legacy of the 'Scientific Revolution': Science and the Enlightenment," *The Cambridge History of Science*, vol. 4, *Eighteenth-Century Science*, ed. Roy Porter (Cambridge: Cambridge University Press, 2003), 41; *PN* 4:403–4; 3:512; *Views* 21; *Cosmos* 1:42; quoted in Bowen 254.

47. *Cosmos* 1:84, 154. How Humboldt would have enjoyed the Hubble telescope images!

48. *Cosmos* 1:88–89, 136–37, 149–50; emphasis in original.

49. *Cosmos* 1:160, 155; 156; 256; Thoreau, *Walden*, 306, 309.

50. *Cosmos* 1:270–72, 339–40, 355; Humboldt is quoting his brother Wilhelm.

51. *Cosmos* 1:342, *Views* 210; *Cosmos* 1:341–42, 297–98, *Essay on the Geography of Plants* 52; *Cosmos* 1:299–301, 55 (quoting Schelling).

52. *Cosmos* 1:359.

53. The allusion is to Wallace Stevens's poem "The Death of a Soldier."

54. *Cosmos* 1:40, 2:20.

55. *Cosmos* 2:20, 62; Cheryll Glotfelty, "Introduction: Literary Studies in an Age of Environmental Crisis," in *The Ecocriticism Reader*, ed. Cheryll Glotfelty and Harold Fromm (Athens: University of Georgia Press, 1996), xix; Clarence Glacken, *Traces on the Rhodian Shore* (Berkeley: University of California Press, 1967), 12. As this book was in press, Alice Jenkins published an article that supports my argument for giving *Cosmos* a place in the history of ecocriticism, though she identifies Humboldt's belief in the urgency of exploiting natural resources as "a challenge to ecocritical theory," and Humboldt's legacy as, hence, "ambivalent." Given that Humboldt tried to imagine a balanced and equitable distribution of natural resources, it might be more fair to his thinking, if equally anachronistic, to suggest that he anticipated and sought what we would now call "sustainable" resource use. See Alice Jenkins, "Alexander von Humboldt's *Kosmos* and the Beginnings of Ecocriticism," *Interdisciplinary Studies in Literature and the Environment* 14.2 (Summer 2007), 91, 97–98.

56. *Cosmos* 2:32, 38, 50–57.

57. *Cosmos* 2:66–68, 75–77, 79, 81.

58. *Cosmos* 2:93–95, 98.

59. *Cosmos* 2:100–101, 103.

60. *Cosmos* 2:108–11.

61. *Cosmos* 2:112–16. Interestingly, on page 112 Otté mistranslates "several" as "one," attributing to Humboldt belief in "one common point of radiation"; however, the larger context makes Humboldt's actual meaning clear. Michael Dettelbach discusses the mistranslation at length, speculating that it was a "'correction'" of Humboldt intended to soften his imputation that Adam and Eve were purely mythical and to avoid association with polygenesis (*Cosmos* 2: xxxix).

62. *Cosmos* 2:114, 187, 116.

63. *Cosmos* 2:141, 187, 201, 209–21. My point about his staging the Mediterranean as an origin, but not *the* origin, is similar to Ottmar Ette's contrast of "the *European* with the *Eurocentric* perspective." Ette argues (referring to the beginning of Humboldt's *Personal Narrative*) that "it is not a preset objectivity but rather the awareness and conscious incorporation of one's own origins that make an adequate perception of otherness possible at all. Humboldt, it seems to me, was giving a clear signal by beginning his account of his travels with a description of his 'transformation' into a European" ("Transatlantic Perceptions," 165–97).

64. *Cosmos* 2:271, 297.

65. *Cosmos* 2:301–2, 356.

66. Bowen 256–59.

67. *Cosmos* 3:11, 24–25.

Chapter Six

1. Ibid.

2. Richardson, *Thoreau*, 13; Ralph Waldo Emerson, *Journals and Miscellaneous Notebooks*, ed. William Gilman et al., 16 vols. (Cambridge: Harvard University Press, 1960–1982), 1:57 (hereafter cited as *JMN*); Emerson, *Letters*, 7:120, 133.

3. *Letters* 3:77; Ralph Waldo Emerson, *The Complete Works of Ralph Waldo Emerson*, ed. Edward Waldo Emerson, 12 vols. (Boston: Houghton Mifflin, 1903–4), 7:323, 11:391; *JMN* 11:157; *CW* 4:58.

4. Ralph Waldo Emerson, *The Early Lectures of Ralph Waldo Emerson, 1833–42*, ed. Stephen E. Whicher, Robert E. Spiller, and Wallace E. Williams, 3 vols. (Cambridge: Harvard University Press, 1959–1972), 1:39, 3:107, 1:18, 1:78, 2:172 (hereafter cited as *EL*); Emerson, "American Scholar," *CW* 1:69; *JMN* 5:238, 10:98.

5. *JMN* 8:401; *EL* 2:68, 200–201; *EL* 3:213. Compare Henry T. Tuckerman, who makes a similar argument in "The Naturalist. Humboldt," in *Characteristics of Literature Illustrated by the Genius of Distinguished Writers*, 2nd ser. (Philadelphia: Lindsay and Blakiston, 1851), 56–77. Where Humboldt in *Cosmos* would call for a "natural history of the mind," Emerson would spend his last year working on "the natural history of the intellect."

6. *EL* 1:26; Emily Dickinson, *The Poems of Emily Dickinson: Reading Edition*, ed. R. W. Franklin (Cambridge: Harvard University Press, 1999), poem 598, p. 269; Ralph Waldo Emerson, *The Later Lectures of Ralph Waldo Emerson, 1843–1871*, ed. Ronald A. Bosco and Joel Myerson, 2 vols. (Athens: University of Georgia Press, 2001), 1:93, hereafter cited as *LL*; *CW* 3:105. I have explored Emerson's science, which I argue was central to his thinking from the 1820s on, in *Emerson's Life in Science*.

7. *JMN* 4:331, 5:246, 258; *CW* 1:20; *EL* 1:100; *LL* 2:343; *CW* 1:23. After reading Humboldt, Thoreau in "Walking" commented similarly: "We have to be told that the Greeks called the world κόσμος, Beauty, or Order, but we do not see clearly why they did so, and we esteem it at best only a curious philological fact" (*Wild Apples*, 88.)

8. *CW* 3:6; *JMN* 4:329.

9. Emerson, *Letters* 3:77; *JMN* 8:193, 7:227, 10:38, 11:285; Helen R. Deese, "Alcott's Conversations on the Transcendentalists: The Record of Caroline Dall," *American Literature* 60.1 (March 1988): 24. Emerson was grateful to Humboldt for defending Emerson's brother-in-law Charles T. Jackson's controversial claim to have been the first to discover anaesthesia; see *Letters* 8:606.

10. *Letters* 8:52. It is unclear whether Emerson was reading and circulating the original German text or the pirated and inferior translation by Prichard that was soon superceded by Sabine's and Otté's. As his next letter makes clear, he was reading Humboldt in preparation for delivering the series of lectures that was eventually published as *Representative Men* (*Letters* 8:53).

11. *JMN* 10:270; *CW* 6:50–51; *LL* 1:120. In this passage Emerson is comparing *Cosmos* with the writings of Mary Somerville, a frequent pairing in contemporary reviews. Somerville admired Humboldt and modeled her book *Physical Geography* (1848) on his work; for his part, Humboldt admired Somerville as well, assisted her work, and likely found one source of inspiration for *Cosmos* in her book *On the Connexion of the Physical Sciences* (1834). See Kathryn A. Neeley, *Mary Somerville: Science, Illumination, and the Female Mind* (Cambridge: Cambridge University Press, 2001).

12. Emerson, *Letters* 6:86, 9:354; *JMN* 16:160–61; *CW* 3:12.

13. Quoted in Roland W. Nelson, "Apparatus for a Definitive Edition of Poe's *Eureka*," in *Studies in the American Renaissance, 1978*, ed. Joel Myerson (Boston: Twayne, 1978): 186; Harold Beaver in Edgar Allan Poe, *The Science Fiction of Edgar Allan Poe*, ed. Harold Beaver (New York: Penguin, 1976), 397–98; Edgar Allan Poe, *Poetry and Tales* (New York: Library of America, 1984), 1258.

14. Quoted by Harold Beaver 396, 395; Nelson 161–62, 180.

15. Poe, *Poetry and Tales*, 1257–59; Glen A. Omans, "'Intellect, Taste, and the Moral Sense': Poe's Debt to Immanuel Kant," in *Studies in the American Renaissance, 1980*, ed. Joel Myerson (Boston: Twayne, 1980), 142. Henry A. Pochman offers Poe's translation of Humboldt, which he compares to the two others then available in print, as proof of Poe's fluency in German: it is "a faithful rendition of the substance and form of the original, such as would be made by one thoroughly conversant with the original German, and hence feeling himself free to make changes not essential to the sense for the sake of a good literary translation" (392).

16. Poe, *Poetry and Tales*, 1261–62.

17. Ibid., 1272–75 (see also 1305, 1327–28); 1353–54. Curiously, Poe's theory here repeats with some precision the theory of life Humboldt offered in his allegory "Vital Force, or The Rhodian Genius," published in Schiller's *Horen* in 1795 and reprinted in all three editions of *Ansichten* even though Humboldt repudiated it soon after. It is quite possible Poe read this piece, but he does not refer to it.

18. *Cosmos* 3:14–15.

19. Poe, *Poetry and Tales*, 1282, 1306, 1334–35, 1346–47; see *Cosmos* 1:146.

20. Poe, *Poetry and Tales*, 1312, 1341–42. This sentiment is not confined to poets: compare James Watson's comment, when he first imagined the molecular structure of DNA, that it was too pretty not to be true (*The Double Helix* [New York: Norton, 1980], 120). The metaphor of

organic wholeness, in which constituent elements are reciprocally part and whole, goes back through Kant to Plato. See my *Emerson's Life in Science*, 110–26. In the terms discussed there, Poe can be seen as "gnomic," very similar to Emerson, and equally indebted to Coleridge and Kant.

21. Poe, *Poetry and Tales*, 1352, 1355, 1356.

22. Ibid., 1356, 1358–59.

23. Dometa Wiegand, "Alexander von Humboldt and Samuel Taylor Coleridge: The Intersection of Science and Poetry," *Coleridge Bulletin: The Journal of the Friends of Coleridge* 20 (Winter 2002), 111, 107, 113. That Coleridge joins us on Humboldt's bridge is something of a surprise, for it was he who originally erected and popularized, in English, the distinction between "objective" and "subjective" that made such a bridge necessary.

24. Henry David Thoreau, *A Week on the Concord and Merrimack Rivers*, ed. Karl F. Hovde et al. (Princeton: Princeton University Press, 1983), 382; Thoreau, *Natural History Essays*, ed. Robert Sattelmeyer (Salt Lake City: Peregrine Smith, 1980), 31.

25. Henry David Thoreau, *The Maine Woods*, ed. Joseph J. Moldenhauer (Princeton: Princeton University Press, 1972), 70, 64, 71. I have explored Thoreau's turn to Humboldt at length in *Seeing New Worlds*. On Thoreau's reading in the literature of travel and exploration, see John Aldrich Christie, *Thoreau as World Traveler* (New York: Columbia University Press and American Geographical Society, 1965). Thoreau wasn't entirely won over by all the reports of the exploring expeditions. In his journal, he grumbled, "A good book is not made in the cheap and offhand manner of many of our scientific Reports, ushered in by the message of the President communicating it to Congress . . . the bulk of the book being a journal of a picnic or sporting expedition by a brevet Lieutenant-Colonel . . . followed by an appendix on the palaeontology of the route by a distinguished savant who was not there, the last illustrated by very finely executed engravings of some old broken shells picked up on the road" (*Journal* 11:456).

26. Thoreau, *Journal*, 12:77–78; Bradley P. Dean, "Natural History, Romanticism, and Thoreau," in *American Wilderness: A New History*, ed. Michael Lewis (Oxford: Oxford University Press, 2007), 82.

27. Thoreau, *Writings: Journal*, 3:331, 5:233, 6:269; *Cosmos* 2:68; Thoreau, *Natural History Essays*, 28; Thoreau, *Week*, 325; *Writings: Journal* 4:356–57.

28. *Walden* 130, 297–98, 20–21, 323.

29. *Walden* 4, 321, 225; *Writings: Journal* 4:467–68.

30. *Walden* 286–91, 138, 97; *Writings: Journal* 5:378; *Walden* 320, 290.

31. *Walden* 290–91, *Journal* 9:44–45.

32. *Writings: Journal* 8:98; *Journal* 10:164–65.

33. *Writings: Journal* 5:469–70; Henry David Thoreau, *The Correspondence of Henry David Thoreau*, ed. Walter Harding and Carl Bode (New York: New York University Press, 1958), 309–10.

34. *Journal* 10:467–68; 14:146–47. For Thoreau and ecology, see Frank Egerton and Laura Dassow Walls, "Rethinking Thoreau and the History of American Ecology," *Concord Saunterer*, n.s., 5 (Fall 1997): 4–20; Michael Benjamin Berger, *Thoreau's Late Career and "The Dispersion of Seeds"* (Rochester, NY: Camden House, 2000).

35. Dean, "Natural History," 86; *Journal* 14:146–47. Both manuscripts have been edited and published by Bradley P. Dean: see Henry David Thoreau, *Faith in a Seed* (Covelo, CA: Island Press, 1993), and *Wild Fruits* (New York: Norton, 2000).

36. Walls, *Seeing New Worlds*, 4. To summarize the comparison I devised two contrasting lists of characteristic features which are, I think, still useful (see 60–61, 85–86).

37. Barbara Novak, *Nature and Culture: American Landscape and Painting, 1825–1875* (New York: Oxford University Press, 1980), 160, 9, 66–67; Wilton and Barringer 56, 220, 29; Novak 71. The quoted passage originally appeared in *Art-Journal*, December 1865, 265. For more on Humboldt and Church, see Howat 44–47 and passim.

38. Thoreau, *Natural History Essays*, 171; *Journal* 14:3.

39. Edward S. Casey, *Representing Place: Landscape Painting and Maps* (Minneapolis: University of Minnesota Press, 2002), 274.

40. Casey 216–24.

41. The work involved in creating hachure lines was immense, and the dying Nicollet, who could no longer do the work himself, agonized over this feature of his own great map, which he thought was fatally compromised; his techniques were passed along most successfully not, in fact, to Frémont but to William H. Emory. The story is told in Bray, *Joseph Nicollet*, 267–71. For Thoreau's original maps of Walden Pond, see Robert F. Stowell, *A Thoreau Gazetteer* (Princeton: Princeton University Press, 1970), 5–9; as noted there, a friend of Emerson's told him he thought Thoreau's "survey and map of the pond were not real, but a caricature of the Coast Surveys." All of Thoreau's maps and surveys are available online; see the website for the Concord Free Public Library.

42. Humboldt, *Researches*, 1:238–39; Casey 170; Humboldt, *Researches*, 1:238–39. E. L. Youmans observed that Humboldt's illustrations of scenery were "the first examples of landscapes adhering strictly to the truth of natural history" ("Alexander von Humboldt," *Appleton's Journal* 2.26 [1869]: 182–83).

43. Pérez-Mejía, *Geography*, 42; Whitman, *Poetry and Prose*, 987; Casey 249. For Humboldt's influence on pictorial images of the Americas, see Renate Löschner, "The Influence of Alexander von Humboldt on Illustrations of America," *Alexander von Humboldt: Life and Work*, ed. Wolfgang-Hagen Hein (Ingelheim am Rheim: C. H. Boehringer Sohn, 1987): 283–300. See also Rigby, *Topographies*, 76–78, for more on the way Humboldtian aesthetics disclose or "give" themselves "to a perceiving subject, who is, in turn, subtly altered in the encounter."

44. W. J. T. Mitchell, "Imperial Landscape," in *Landscape and Power*, ed. W. J. T. Mitchell (Chicago: University of Chicago, 1994), 1, 10; Wilton and Barringer 59, 56.

45. Foner, *Tom Paine*, 111; Mark Stoll, "Religion 'Irradiates' the Wilderness," in *American Wilderness: A New History*, ed. Michael Lewis (Oxford: Oxford University Press, 2007), 37.

46. Novak 20, 23–24; Pratt, *Imperial*, 153. Novak notes that panoramas were disposable art; wound and unwound on cylinders, they simply wore out, and few survive.

47. Alexis de Tocqueville, *Democracy in America*, trans. Harvey C. Mansfield and Delba Winthrop (Chicago: University of Chicago Press, 2000), 460–62. Humboldt and Tocqueville seem to have had a rather chilly relationship.

48. Dickinson, poem 557, p. 252.

49. "L. of G.'s Purport," in Whitman, *Poetry and Prose*, 652–53, 50, 210; David S. Reynolds, *Walt Whitman's America: A Cultural Biography* (New York: Knopf, 1995), 244; Charles Eliot Norton, "Whitman's Leaves of Grass," *Putnam's Monthly* 6 (September 1855): 321–23, in *Walt Whitman: The Contemporary Reviews*, ed. Kenneth M. Price (Cambridge: Cambridge University Press, 1996), 18.

50. Reynolds, *Walt Whitman's*, 244–45; Pochman, *German Culture in America*, 467; Whitman, *Poetry and Prose*, 516–17; Reynolds, *Walt Whitman's*, 245–46.

51. *CW* 3:22; Whitman, *Poetry and Prose*, 5, 15, 18, 24–25 (note that "the United States" was a plural noun until after the Civil War). For a good discussion of Whitman and science, including particularly this passage, see Robert J. Scholnick, " 'The Password Primeval': Whitman's Use of

Science in 'Song of Myself,'" in *Studies in the American Renaissance 1986* (Charlottesville: University Press of Virginia, 1986), 385–425.

52. Whitman, *Poetry and Prose*, 926, 984–85, 987–88; Thoreau, *Journal*, 9:44–45.

53. Horsman 303; David S. Reynolds, "Politics and Poetry: Leaves of Grass and the Social Crisis of the 1850s," in *The Cambridge Companion to Walt Whitman*, ed. Ezra Greenspan (Cambridge: Cambridge University Press, 1995), 88–89.

54. Whitman, *Poetry and Prose*, 990, 993.

55. Tocqueville 464; Whitman, *Poetry and Prose*, 42–43. One suspects Tocqueville might have applied his next observation to Poe: "We have seen, moreover, that in democratic peoples the sources of poetry are beautiful but not abundant. In the end one soon exhausts them. Not finding more material for the ideal in the real and true, poets leave them entirely and create monsters."

56. Thoreau, *Journal*, 9:45; *Walden* 100, 97–98; Whitman, *Poetry and Prose*, 539; *CW* 4:91.

57. Greg Garrard, *Ecocriticism* (New York: Routledge, 2005), 108; Susan Fenimore Cooper, *Essays on Nature and Landscape*, ed. Rochelle Johnson and Daniel Patterson (Athens: University of Georgia Press, 2002), 12; Michael G. Ziser, "*Walden* and the Georgic Mode," in *More Day to Dawn: Thoreau's Walden for the Twenty-First Century*, ed. Sandra Harbert Petrulionis and Laura Dassow Walls (Amherst: University of Massachusetts Press, 2006), 175, 185.

58. Susan Fenimore Cooper, *Rural Hours*, ed. Rochelle Johnson and Daniel Patterson (Athens: University of Georgia Press, 1998), 142; Thoreau, *Natural History Essays*, 259; George Perkins Marsh, *Man and Nature, or, Physical Geography as Modified by Human Action* (1864), ed. David Lowenthal (Cambridge: Harvard University Press, 1965), 42–43; Cooper, *Rural Hours*, 56.

59. Lawrence Buell, *Writing for an Endangered World* (Cambridge: Harvard University Press, 2001), 170–71; Lawrence Buell, *The Environmental Imagination: Thoreau, Nature Writing, and the Formation of American Culture* (Cambridge: Harvard University Press, 1995), 405–8, 266; Cooper, *Rural Hours*, 56, xix.

60. *Rural Hours* 128, 149, 190, 311, 314; *Essays* 83–86.

61. *Rural Hours* 119, 139, 133, 180.

62. *Rural Hours* xviii, 208.

63. *Rural Hours* 152; Tina Gianquitto, "The Noble Designs of Nature: God, Science, and the Picturesque in Susan Fenimore Cooper's *Rural Hours*," in *Susan Fenimore Cooper: New Essays on Rural Hours and Other Works*, ed. Rochelle Johnson and Daniel Patterson (Athens: University of Georgia Press, 2001), 186, 184, 172 (quoting Novak 6).

64. *Rural Hours* 53, 75; *Essays* 6; *Rural Hours* 81.

65. Buell, *Environmental Imagination*, 407; Cooper, *Essays*, 37; *Cosmos* 2:38–39; Cooper, *Essays*, 38, 32, 39.

66. *Essays* 40, 42–43; *Rural Hours* 57, 302–3.

67. Rochelle Johnson, *Passions for Nature: Nineteenth-Century America's Aesthetics of Alienation* (Athens: Georgia University Press, forthcoming), [MSS pp. 23, 300–301].

68. *Rural Hours* 117, 153; *Essays* 114.

69. Quoted in Steven J. Holmes, *The Young John Muir: An Environmental Biography* (Madison: University of Wisconsin Press, 1999), 6.

70. John Muir, *Nature Writings* (New York: Library of America, 1997), 18, 83, 101, 129; Holmes 56, 121, 128. Muir's description of the ritual tradition of slaughtering songbirds is quite shocking today: "They divided into two squads, and, choosing leaders, scattered through the woods in different directions, and the party that killed the greatest number enjoyed a supper at the expense of the other. The whole neighborhood seemed to enjoy the shameful sport especially the

farmers afraid of their crops. . . . All the blessed company of mere songbirds, warblers, robins, thrushes, orioles, with nuthatches, chickadees, blue jays, woodpeckers, etc., counted only one head each. The heads of the birds were hastily wrung off and thrust into the game-bags to be counted. . . . The blood-stained bags of the best slayers were soon bulging full. Then at a given hour all had to stop and repair to the town, empty their dripping sacks, count the heads, and go rejoicing to their dinner." Muir says he refused to participate in "these abominable head-hunts" (Muir 83).

71. Holmes 130–33, 142.

72. Ibid., 161, 179. Though published too late to be included in the present volume, Paul Outka's provocative argument that the American sublime emerged as a reaction to the emotional trauma of slavery—an argument that, as he shows, pertains with particular force to John Muir—deserves to be considered here; see his *Race and Nature from Transcendentalism to the Harlem Renaissance* (New York: Palgrave Macmillan, 2008), 155–70.

73. Holmes 181; Muir 826.

74. Saunders, *Humboldtian Physicians*; Holmes 198, 236–37; *Cosmos* 3:7; Holmes 224; Muir 304. Muir did, near the end of his life, find his way at long last to South America: from 1911 to 1912 he steamed a thousand miles up the Amazon to the Rio Negro, dropped down the coast to Argentina, then traveled overland to the Chilean Andes, finishing up with a tour around Africa via the Canary Islands. His travel journal has been published as *John Muir's Last Journey: South to the Amazon and East to Africa*, ed. Michael P. Branch (Washington, DC: Island Press, 2001).

75. Sachs, *Humboldt Current*, 311, 313, 317, 330–31; Catlin 1:261–62. For instances of Muir's racism against Indians, see Muir 281–82, 285, 729. On Indian removal from the national parks, see Spence; Robert H. Keller and Michael F. Turek, *American Indians and National Parks* (Tucson: University of Arizona Press, 1998), especially 232–40.

76. Sachs, *Humboldt Current*, 334; Wilcomb E. Washburn, *The Cosmos Club of Washington: A Centennial History, 1878–1978* (Washington, DC: Cosmos Club, 1978), xiii. It seems extremely unlikely that the founders did not have Humboldt's *Cosmos* in mind, but I been unable to confirm this. Its historians state, "In their infinite wisdom and foresight the founders chose the name *Cosmos*, which uniquely signifies the nature of the Club and its breadth and diversity" (Washburn xii). Donald Worster, Powell's biographer, notes that "rational order and harmony, the universal principles, were written into its very name" (*A River Running West: The Life of John Wesley Powell* [Oxford: Oxford University Press, 2001]). For the report of the Harriman expedition, see John Burroughs, John Muir et al., *Alaska: The Harriman Expedition, 1899* (New York: Dover, 1986).

77. Muir 161, Holmes 228, 230; emphasis in original.

78. David Lowenthal, *George Perkins Marsh: Prophet of Conservation* (Seattle: University of Washington Press, 2000), xv; Marsh, *Man and Nature*, xxi–xxii. For the introduction of camels see *PN* 4:184–85; and George Perkins Marsh, *So Great a Vision: The Conservation Writings of George Perkins Marsh*, ed. Stephen C. Trombulak (Hanover, NH: University Press of New England, 2001), 24–33. Trombulak notes that the thirty-four camels brought from Egypt for the purpose, and sold or released during the Civil War, did not establish a viable population, though feral camels were occasionally sighted in the American desert through the early 1900s (25).

79. Marsh, *Man and Nature*, 43; Marsh, *Vision*, 17; Marsh, *Man and Nature*, 42–43.

80. Marsh, *Man and Nature*, 79, 80, 82, 91.

81. Marsh, *Vision*, 82–83, 84–86.

82. Ibid. 74–75, 96.

83. Marsh, *Man and Nature*, 52, 9–10, 12–13.

84. Ibid., 53; *CW* 1:8; 36.

85. Ibid., 38–39.

86. Ibid., 91–92.

87. Ibid., 203–4, 280.

88. Prescott 275–78; Lowenthal, *Marsh*, 416–19; David Lowenthal, *The Past Is a Foreign Country* (Cambridge: Cambridge University Press, 1985), xvi (quoting L. P. Hartley).

Epilogue

1. Bruhns 2:353; Botting 277; Kellner 226–27, 219; Botting 278; Bruhns 2:395; John Lloyd Stephens 151; Taylor 357–58; Francis Lieber, "Alexander von Humboldt: Address before the American Geographical Society, 1859," in *Miscellaneous Writings*, vol. 1, *Reminiscences, Addresses, and Essays* (Philadelphia: Lippincott, 1881), 390. It was widely known that Humboldt was deeply in debt to his "servant," Siefert, an imperious soul who managed the household, and his master, with an iron hand. Humboldt willed virtually all his worldly goods to Siefert, and extracted a pledge from the king to pay his debts upon his death (Bruhns 2:402–9).

2. Humboldt, *Letters to Varnhagen*, 358, 402; Schwartz 574, de Terra, *Humboldt*, 368.

3. Botting 283–84, Bruhns 2:411–12; Lieber, "Alexander von Humboldt," 402–3.

4. Asa Gray, "Biographical Sketches: Brown and Humboldt," (1859), in *Scientific Papers of Asa Gray*, vol. 2 (Boston: Houghton, Mifflin, 1889), 288; Everett, "Alexander von Humboldt," 173; William Rounseville Alger, *Lessons for Mankind, from the Life and Death of Humboldt* (Cambridge, MA: Welch, Bigelow, 1859), 18; Louis Agassiz, "Eulogy by Professor Agassiz upon Baron von Humboldt," *Littell's Living Age*, 3rd ser., no. 63 (11 June 1859), 646–48; Lieber, "Alexander von Humboldt," 390, 396. During a recess of the American Oriental Society meeting of 18 May 1859, word came by telegraph of Humboldt's death, which Emerson announced as soon as the meeting resumed. Cornelius Felton proposed, and the society unanimously passed, a resolution to pay tribute to their honorary member (*Proceedings at the Annual Meeting of the American Oriental Society*, 18 May 1859). Bachman would say, years later, that wherever science contemplated the history of the earth, "wherever the tides of ocean and of air, the rush of mighty rivers and the stillness of unbounded plains proclaim the laws which made this globe a habitable world; wherever forests wave, decked with exuberant foliage, laden with many-hued and fragrant flowers, and fruits of luscious taste, and teeming with throngs of beasts, birds and insects . . . the name of Humboldt stands confessed" ("Von Humboldt," 82).

5. Sandra Nichols, "Why Was Humboldt Forgotten in the United States?" *Geographical Review* 96.3 (July 2006), 401–4; Andreas Daum, "Celebrating Humanism in St. Louis: The Origins of the Humboldt Statue in Tower Grove Park, 1859–1878," *Gateway Heritage* 15.2 (Fall 1994), 50; Cora Lee Nollendorfs, "Alexander von Humboldt Centennial Celebrations in the United States: Controversies Concerning His Work," *Monatshefte* 80.1 (Spring 1988): 59–66; *New York Times*, 14 and 15 September 1869. For 1869 Humboldt celebrations in Germany, see Denise Phillips, "Building Humboldt's Legacy: The Humboldt Memorials of 1869 in Germany," *Northeastern Naturalist* 8.1 (2001): 21–32. According to her, in these commemorations "Humboldt functioned as a synecdoche for natural science as a whole," with speakers using his image to argue for the importance of science (31).

6. The event is documented in detail in Anon., *Celebration of the Humboldt Centennial and Opening of the Iowa Institute of Science and Arts, at Dubuque, Iowa, September 14th, 1869* (Dubuque: Daily Times, 1869).

7. Bachman, "Von Humboldt," 85.

8. John Bachman, "The Humboldt Festival," *Charleston [S.C.] Daily Courier*, 15 September 1869, 4; Anon. 15. Bachman lived for another five years. McCrady's attempt to revive science in the South failed, and in 1873 Agassiz hired him to work at his Museum of Comparative Zoology. Soon after he arrived, Agassiz died, and after several difficult years in Boston making himself unpopular with Northern students, McCrady moved to the University of the South in Sewanee, Tennessee, where he spent the rest of his short life. See Stephens, *Science, Race, and Religion*, 146–64, 224, 227–59.

9. Anon. 17, 27, 32, 16.

10. Harriet Martineau, "Alexander von Humboldt: Died May 6th, 1859," in *Biographical Sketches, 1852–1875*, 4th ed. (London: MacMillan, 1876), 282–83; Everett, "Alexander von Humboldt," 176; Francis Ellingwood Abbot, *An Oration Delivered at the Centennial Celebration of the Birth of Alexander von Humboldt, September 14, 1869* (Toledo, OH: Weekly Express, 1869), 13; Alger, *Lessons* 23.

11. Humboldt, *Letters to Varnhagen*, 122–23, 112–13, 128–29, 339, 181–82, 340–41. The minister General Leopold von Gerlach, leader of the reactionary party, was a favorite target for Humboldt. Varnhagen noted in his diary that "when the King was with Humboldt [who was recovering from his stroke], Schoenlein said to the latter, that he would not be able for some time to stand firmly on his left side, to which Humboldt rejoined: 'For all that, it will not be necessary for me to sit on the right with Gerlach' " (Ibid., 358).

12. Letter in "Humboldt Centenary 1869," scrapbook in the Boston Public Library Rare Book Collection; Agassiz, *Address*, 108, 46. The members of the organizing committee were Rev. R. C. Waterston, Jeffries Wyman, N. B. Shurtleff, Thomas Wentworth Higginson, Samuel Kneeland, and Samuel H. Scudder. Details of the event are drawn from the scrapbook and from the proceedings that were printed with Agassiz's address. Those familiar with Mozart's opera will appreciate the significance of their selecting *The Magic Flute*'s "Chorus of Priests." The two Humboldt portraits are reproduced in Nelkin, *Alexander von Humboldt*, 54, 143.

13. Agassiz, *Address*, 87, 88, 100. Jackson did not mention that Humboldt had defended him when Morton tried to claim credit for Jackson's great discovery, the use of ether as an anaesthesia, although Emerson, his brother-in-law, remembered this in a letter to Agassiz in which he remarked that Humboldt had arranged for the Prussian Eagle to be awarded to Jackson (*Letters* 8:606).

14. Agassiz, *Address*, 70–71, 53–56, 92.

15. Max Rodenbeck, "How Terrible Is It?" *New York Review of Books*, 30 November 2006, 34; Karl Heinzen, *The True Character of Humboldt: Oration, Delivered at the German Humboldt Festival, in Boston* (Indianapolis: Association for the Propagation of Radical Principles, 1869), 3–4, 5, 13, emphasis in original.

16. Foner, "Humboldt on Slavery," 334; Heinzen 15; Robert G. Ingersoll, "Humboldt: The Universe is Governed by Law," in *The Gods and Other Lectures* (Peoria, Illinois: C. P. Farrell, 1878), 112, 115, 117.

17. Samuel Osgood, "Humboldt and Young Germany," *Appleton's Journal of Literature, Science, and Art* 2.26 (25 September 1869), 180.

18. Londa Schiebinger, "Agnotology and Exotic Abortifacients," *Proceedings of the American Philosophical Society* 149.3 (September 2005), 320; Bruhns 2:359. I am indebted to a conversation with Bob Habich for the insight into postbellum memorials.

19. Zeller 394; Agassiz, *Address*, 5–6; Kellner 234; Whitman, *Poetry and Prose*, 26.

20. Brading 534; Elizabeth Cary Agassiz 1:227–29; Zeller 394. A formative instance of Agassiz's removing himself from his text in favor of the speech of facts may be found in his *Essay on the*

Classification of Fishes, the work that made his scientific reputation—and that he dedicated to Humboldt: "Such facts [of comparative anatomy and comparative embryology] *loudly proclaim* principles which science has not discussed, but which palaeontological researches *place before the eyes* of the observer with increasing persistency; I mean the relation of the creation to the Creator. . . . More than 1,500 species of fossil fishes with which I have become acquainted *say to me* that the species do not pass gradually from one to the other, but appear and disappear suddenly without direct relations with their predecessors. . . . Are not these facts manifestations of a thought as rich as it is powerful, acts of an intelligence as sublime as provident?" (Quoted in Arnold "Memoir of Louis Agassiz, 1807–1878" [1878], in *Biographical Memoirs* [Washington, DC: National Academy of Sciences, 1886], 58; emphasis added.) I have explored Agassiz, Thoreau, and the formation of scientific objectivity at length in "Textbooks and Texts from the Brooks: Inventing Scientific Authority in America," *American Quarterly* 49.1 (March 1997): 1–25.

21. Edward O. Wilson, *Consilience: The Unity of Knowledge* (Cambridge: Harvard University Press, 1998), 54–55 and passim. William Whewell coined the word "consilience" in 1840. For more on Whewell and Wilson, see my essays, "Consilience Revisted," *ebr: Electronic Book Review* (27 December 1999: http://altx.com/ebr/ebr10/10wal.htm); and "Seeking Common Ground: Integrating the Sciences and the Humanities," in *Coming into Contact: Explorations in Ecocritical Theory and Practice*, ed. Annie Merrill Ingram et al. (Athens: University of Georgia Press, 2007): 199–208. Wilson's stellar work as a naturalist, his ability to bring science to the public through the media, and his drive to save the panoply of nature's biological diversity are clearly in the Humboldt tradition. His autobiography indicates that his turn toward reductionism in the 1960s, at a Harvard dominated by James Watson and the new biochemistry of life, was a salvage project responding to the endangerment of natural history. Lest natural history be overwritten altogether, Wilson turned it into a natural science by founding the reductionist field of sociobiology (see Edward O. Wilson, *Naturalist* [Washington, DC: Shearwater Press, 1994], esp. 225).

22. As Kent Mathewson argues, aspects of Humboldt's vision and approach persisted in American geography well beyond his death and remain powerful today; see "Humboldt's Image and Influence." Sachs, *Humboldt Current*, 340–42, 351; Lewis Mumford, *The Myth of the Machine: The Pentagon of Power* (New York: Harcourt Brace Jovanovich, 1970), 351; de Terra, *Humboldt*, 321–22.

23. Nichols, "Why?" 406–8; Theresa M. Kelley and Paula R. Feldman, "Introduction," *Romantic Women Writers: Voices and Countervoices*, ed. Paula R. Feldman and Theresa M. Kelley (Hanover: University Press of New England, 1995), 3.

24. Martineau 283; Ed Folsom, "Database as Genre: The Epic Transformation of Archives," *PMLA* 122.5 (October 2007), 1573. For digital Humboldts, see the electronic publication *Humboldt im Netz* and the Alexander von Humboldt Digital Library. Currently Rex Clark is leading a project to rediscover Humboldt as a literary writer, and his colleague, Oliver Lubrich, has written persuasive literary analyses of Humboldt's works: see his "Alexander von Humboldt," 360–87; and "In the Realm of Ambivalence: Alexander von Humboldt's Discourse on Cuba," *German Studies Review* 26.1 (February 2003), 63–80. For an earlier treatment, see Robert Van Dusen, *The Literary Ambitions and Achievements of Alexander von Humboldt* (Bern: Herbert Lang, 1971).

25. Nichols 408–11; Jörg Nagler, "From Culture to *Kultur*: Changing American Perceptions of Imperial Germany, 1870–1914," in *Transatlantic Images and Perceptions: Germany and America since 1776*, ed. David E. Barclay and Elisabeth Glaser-Schmidt (Cambridge: Cambridge University Press, 1997), 131–70 passim; Sachs, *Humboldt Current*, 339. By contrast, Kirsten

Belgum suggests that Humboldt was available to U.S. Americans because he could be viewed as cosmopolitan, *not* German; see "Reading Alexander von Humboldt: Cosmopolitan Naturalist with an American Spirit," in *German Culture in Nineteenth-Century America: Reception, Adaptation, Transformation,* ed. Lynn Tatlock and Matt Erlin (Rochester, NY: Camden House, 2005), 107–27. I am mindful that the American capacity for xenophobia has, since Humboldt's death, been demonstrated in various ways against several nations and ethnic groups, including but not limited to Jewish immigrants who were refused entry from the 1880s through the 1940s, Japanese Americans whose property was seized and who were interned in concentration camps during World War II, the French nation when it refused to join the United States in the war on Iraq ("liberty fries" briefly replaced "french fries" on fast-food menus and French's mustard saw sales plummet), Muslims following 9/11, and Latin Americans as immigration across the United States' southern border became increasingly contested.

26. Henry David Thoreau, "Slavery in Massachusetts," in *Reform Papers,* ed. Glick, 108. For *Cotopaxi* as an allegory of the Civil War, see Rebecca Bedell, *The Anatomy of Nature: Geology and American Landscape Painting, 1825–1875* (Princeton: Princeton University Press, 2001), 81–83.

27. The painting itself bears witness to this fragility: though meticulously restored, a careful observer will note a long tear extending from the upper-right corner clear through to the top of the waterfall. For generations this painting was a castoff, gathering dust and contempt until it was rediscovered, repaired, and given pride of place in San Francisco's De Young Museum. See Isabel Breskin, "A Vision of Unity in Disunited Times," in *Masterworks of American Painting at the De Young,* ed. Timothy Anglin Burgard (San Francisco: Fine Arts Museums of San Francisco, 2005), 110. Breskin suggests that "*Rainy Season in the Tropics* can be read as Church's paean to Humboldt's ideas" (112).

28. Wai Chee Dimock, "The Planetary Dead: Margaret Fuller, Ancient Egypt, Italian Revolution," *ESQ* 50.1–3 (2004): 35, 43, 46; Wai Chee Dimock, "A Theory of Resonance," *PMLA* 112.5 (October 1997): 1061.

Bibliography

Alexander and Wilhelm von Humboldt, Primary and Secondary

Abbot, Francis Ellingwood. *An Oration Delivered at the Centennial Celebration of the Birth of Alexander von Humboldt, September 14, 1869.* Toledo, OH: Weekly Express, 1869.

Ackerknecht, Erwin H. "George Forster, Alexander von Humboldt, and Ethnology." *Isis* 46 (1955): 83–95.

Agassiz, Louis. *Address Delivered on the Centennial Anniversary of the Birth of Alexander von Humboldt under the Auspices of the Boston Society of Natural History, with an Account of the Evening Reception.* Boston: Boston Society of Natural History, 1869.

———. "Eulogy by Professor Agassiz upon Baron von Humboldt, delivered before the American Academy of Arts and Science, at their annual meeting, on Tuesday, 24 May, 1859." *Littel's Living Age,* 3rd ser., no. 63 (11 June 1859): 642–49.

Aldrich, Robert. "Humboldt and His Friends." In *Colonialism and Homosexuality.* New York: Routledge, 2003. 24–29.

Alger, William Rounseville. *Lessons for Mankind, from the Life and Death of Humboldt.* Cambridge, MA: Welch, Bigelow, 1859.

Anon. *Celebration of the Humboldt Centennial and Opening of the Iowa Institute of Science and Arts, at Dubuque, Iowa, September 14th, 1869.* Dubuque: Daily Times, 1869.

Bachman, John. "The Humboldt Festival." *Charleston [S.C.] Daily Courier,* 15 September 1869, 4–5.

———. "Von Humboldt." 1869. *Tributes and Memories.* Boston: Sanctuary Publishing, 1914. 75–99.

Baron, Frank, and Detlev Doherr. "Exploring the Americas in a Humboldt Digital Library." *Geographical Review* 96.3 (July 2006): 439–51.

Barrett, Paul H., and Alain F. Corcos. "A Letter from Alexander Humboldt to Charles Darwin." *Journal of the History of Medicine and Allied Science* 27.2 (April 1972): 159–72.

Baumgartner, Hans. "Alexander von Humboldt: Remarks on the Meaning of Hypothesis in His Geological Researches." In *Toward a History of Geology,* edited by Cecil J. Schneer. Cambridge: MIT Press, 1967. 19–35.

Beck, Hanno. *Alexander von Humboldt.* 2 vols. Wiesbaden: Steiner, 1959–61.

———. "Alexander von Humboldt: Letters from His Travels." In *Alexander von Humboldt: 1769/1859.* Bonn: Internationes, 1969.

———, ed. *Gespräche Alexander von Humboldts*. Berlin: Akademie-Verlag, 1959.

Belgum, Kirsten. "Reading Alexander von Humboldt: Cosmopolitan Naturalist with an American Spirit." In *German Culture in Nineteenth-Century America: Reception, Adaptation, Transformation*, edited by Lynn Tatlock and Matt Erlin. Rochester, NY: Camden House, 2005. 107–27.

Botting, Douglas. *Humboldt and the Cosmos*. New York: Harper and Row, 1973.

Bowen, Margarita. *Empiricism and Geographical Thought from Francis Bacon to Alexander von Humboldt*. Cambridge: Cambridge University Press, 1981.

Brading, D. A. "Scientific Traveller." In *The First America: The Spanish Monarchy, Creole Patriots, and the Liberal State, 1492–1867*. Cambridge: Cambridge University Press, 1991. 515–34.

Brann, E. R. *Alexander von Humboldt: Patron of Science*. Madison, WI: E. R. Brann, 1954.

———. *The Political Ideas of Alexander von Humboldt: A Brief Preliminary Study*. Madison, WI: E. R. Brann, 1954.

Breskin, Isabel. "A Vision of Unity in Disunited Times." In *Masterworks of American Painting at the De Young*, edited by Timothy Anglin Burgard. San Francisco: Fine Arts Museums of San Francisco, 2005. 110–12.

Browne, C. A. "Alexander von Humboldt as Historian of Science in Latin America." *Isis* 35, pt. 2 (Spring 1944): 134–39.

Brude-Firnau, Gisela. "Alexander von Humboldt's Sociopolitical Intentions: Science and Poetics." In *Traditions of Experiment from the Enlightenment to the Present*, edited by Nancy Kaiser and David E. Wellbery. Ann Arbor: University of Michigan Press, 1991. 45–61.

Bruhns, Karl, ed. *Life of Alexander von Humboldt Compiled in Commemoration of the Centenary of His Birth by J. Löwenberg, Robert Ave-Lallemant, and Alfred Dove*. Translated by Jane and Caroline Lassell. 2 vols. London: Longmans, Green, 1873.

Bunkše, Edmunds V. "Humboldt and an Aesthetic Tradition in Geography." *Geographical Review* 71.2 (April 1981): 127–46.

Bunzl, Matti. "Franz Boas and the Humboldtian Tradition." In *Volksgeist as Method and Ethic: Essays on Boasian Ethnography and the German Anthropological Tradition*, edited by George W. Stocking, Jr. Vol. 8 of *History of Anthropology*. Madison: University of Wisconsin Press, 1996. 17–78.

Buttimer, Anne. "Beyond Humboldtian Science and Goethe's Way of Science: Challenges of Alexander von Humboldt's Geography." *Erdkune* 55.2 (2001): 105–20.

Cañizares-Esguerra, Jorge. "How Derivative Was Humboldt?" In *Nature, Empire, and Nation: Explorations of the History of Science in the Iberian World*. Stanford: Stanford University Press, 2006. 112–28.

Cannon, Susan Faye. "Humboldtian Science." In *Science in Culture: The Early Victorian Period*. New York: Dawson and Science History Publications, 1978. 73–110.

Catlin, George. *Notes of Eight Years Travels and Residence in Europe*. 2 vols. New York: Burgess, Stringer, 1848.

Chambers, David Wade. "Centre Looks at Periphery: Alexander von Humboldt's Account of Mexican Science and Technology." *Journal of Iberian and Latin-American Studies* 2.1 (July 1996): 94–113.

Clark, Rex. "If Humboldt Had a Laptop: Moving Knowledge Networks from Print to Digital Media." *HiN: Alexander von Humboldt im Netz, International Review for Humboldtian Studies* 2.3 (2002): n.p. http://www.uni-potsdam.de/u/romanistik/humboldt/hin/hin3/lindquist.htm.

Crampton, C. Gregory. "Humboldt's Utah, 1811." *Utah Historical Quarterly* 26.3 (1958): 268–81.

Daum, Andreas. "Alexander von Humboldt, die Natur als 'Kosmos' und die Suche nach Einheit." *Berichte zur Wissenschaftsgeschichte* 23 (2000): 243–68.

———. "Celebrating Humanism in St. Louis: The Origins of the Humboldt Statue in Tower Grove Park, 1859–1878." *Gateway Heritage* 15.2 (Fall 1994): 48–58.

———. " 'The Next Great Task of Civilization': International Exchange in Popular Science, the German-American Case, 1850–1900." In *The Mechanics of Internationalism: Culture, Society, and Politics from the 1840s to the First World War*, edited by Martin H. Geyer and Johannes Paulmann. Oxford: Oxford University Press, 2001. 285–319.

———. Review of *Alexander von Humboldt Schriftden: Bibliographaie der selbstandig erschienenen Werke. Beitrage zur Alexander-von-Humboldt-Forshung, no. 20*, ed. Horst Feidler and Ulrike Leitner (Berlin: Akademie Verlag, 2000). *Journal of the History of Biology* 33.3 (2000): 591–92.

———. "Science, Politics, and Religion: Humboldtian Thinking and the Transformations of Civil Society in Germany, 1830–1870." In *Science and Civil Society*, edited by Lynn K. Nyhart and Thomas H. Broman. *Osiris* 17 (2002): 107–40.

———. "*Wissenschaft* and Knowledge." In *Germany 1800–1870*, edited by Jonathan Sperber. Oxford: Oxford University Press, 2004. 137–61.

Day, Douglas. "Humboldt and the Casiquiare: Modes of Travel Writing." *Review: Latin American Literature and Arts* 47 (Fall 1993): 4–8.

De Terra, Helmut. "Alexander von Humboldt's Correspondence with Jefferson, Madison, and Gallatin." *Proceedings of the American Philosophical Society* 103.6 (December 1959): 783–806.

———. *Humboldt: The Life and Times of Alexander von Humboldt, 1769–1859, Explorer, Naturalist, and Humanist*. New York: Knopf, 1955.

———. "Motives and Consequences of Alexander von Humboldt's Visit to the United States (1804)." *Proceedings of the American Philosophical Society* 104.3 (June 1960): 314–16.

———. "Studies of the Documentation of Alexander von Humboldt." *Proceedings of the American Philosophical Society* 102.2 (April 1958): 136–41.

———. "Studies of the Documentation of Alexander von Humboldt: The Philadelphia Abstract of Humboldt's American Travels; Humboldt Portraits and Sculpture in the United States." *Proceedings of the American Philosophical Society* 102.6 (December 1958): 560–89.

Dettelbach, Michael. "Alexander von Humboldt between Enlightenment and Romanticism." *Northeastern Naturalist* 8.1 (2001): 9–20.

———. "Global Physics and Aesthetic Empire: Humboldt's Physical Portrait of the Tropics." In *Visions of Empire*, edited by David Philip Miller and Peter Hans Reill. Cambridge: Cambridge University Press, 1996. 258–301.

———. "Humboldtian Science." In *Cultures of Natural History*, edited by N. Jardine, J. A. Secord, and E. C. Spary. Cambridge: Cambridge University Press, 1996. 287–304.

———. Introduction to *Cosmos: A Sketch of a Physical Description of the Universe*, by Alexander von Humboldt. 2 vols. Translated by E. C. Otté. Baltimore: Johns Hopkins University Press, 1997. 2:vii–xlvii.

———. "The Stimulations of Travel." In *Tropical Visions in an Age of Empire*, edited by Felix Driver and Luciana Martius. Chicago: University of Chicago Press, 2005. 43–58.

Edwards, John S. "Humboldt's South America Today." *AvH-Magazin* 73 (1999): 47–52.

Egerton, Frank N. "Humboldt, Darwin, and Population." *Journal of the History of Biology* 3.2 (Fall 1970): 325–60.

Ette, Ottmar. *Literature on the Move*. Translated by Katharina Vester. Amsterdam: Rodopi, 2003.

———. "The Scientist as Weltbürger: Alexander von Humboldt and the Beginning of Cosmopolitics." *Northeastern Naturalist* 8.1 (2001): 157–82.

———. "Transatlantic Perceptions: A Contrastive Reading of the Travels of Alexander von Humboldt and Fray Servando Teresa de Mier." *Dispositio* 17.42–43 (1992): 165–97.

Everett, Edward. "Alexander von Humboldt." 1859. In *Orations and Speeches on Various Occasions*, vol. 4. Boston: Little, Brown, 1879. 170–77.

———. "Humboldt's Works." *North American Review* 16 (January 1823): 1–30.

Felton, Cornelius Conway. *Familiar Letters from Europe*. Boston: Ticknor and Fields, 1865.

Foner, Philip S. "Alexander von Humboldt on Slavery in America." *Science and Society* 47.3 (Fall 1983): 330–42.

Fränzl, Otto. "Alexander von Humboldt's Holistic World View and Modern Inter- and Transdisciplinary Ecological Research." *Northeastern Naturalist* 8.1 (2001): 57–90.

Friis, Herman R. "Alexander von Humboldts Besuch in den Vereinigten Staaten von Amerika vom 20. Mai bis zum 30. Juni 1804." In *Alexander von Humboldt: Studien zu seiner universalen Geisteshaltung*, edited by Joachim H. Schultze. Berlin: Verlag Walter de Gruyter & Co., 1959. 142–95.

———. "Baron Alexander von Humboldt's Visit to Washington, D.C., June 1 through June 13, 1804." In *Records of the Columbia Historical Society of Washington, 1960–1962*, edited by Francis Coleman Rosenberger. Washington, DC: Columbia Historical Society. 1–35.

Gilman, Daniel Coit. "Humboldt, Ritter, and the New Geography." *New Englander* 18.70 (May 1860): 277–306.

Godlewska, Anne Marie Claire. "From Enlightenment Vision to Modern Science? Humboldt's Visual Thinking." In *Geography and Enlightenment*, edited by David N. Livingstone and Charles W. J. Withers. Chicago: University of Chicago Press, 1999. 236–75.

———. *Geography Unbound: French Geographic Science from Cassini to Humboldt*. Chicago: University of Chicago Press, 1999.

Gould, Stephen Jay. "Church, Humboldt, and Darwin: The Tension and Harmony of Art and Science." In *Frederic Edwin Church*, edited by Franklin Kelly et al. Washington, DC: Smithsonian Institution Press, 1989. 94–107.

Gray, Asa. "Biographical Sketches: Brown and Humboldt." 1859. In *Scientific Papers of Asa Gray*, vol. 2. Boston: Houghton, Mifflin, 1889. 283–88.

Hein, Wolfgang-Hagen, ed. *Alexander von Humboldt: Life and Work*. Ingelheim am Rheim: C. H. Boehringer Sohn, 1987.

———. "Humboldt and Goethe." In *Alexander von Humboldt: Life and Work*, edited by Wolfgang-Hagen Hein. Ingelheim am Rheim: C. H. Boehringer Sohn, 1987. 46–55.

Heinzen, Karl. *The True Character of Humboldt: Oration, Delivered at the German Humboldt Festival, in Boston*. Indianapolis: Association for the Propagation of Radical Principles, 1869.

Helferich, Gerard. *Humboldt's Cosmos: Alexander von Humboldt and the Latin American Journey That Changed the Way We See the World*. New York: Penguin, 2004.

Hentschel, Cedric. "Alexander von Humboldt's Synthesis of Literature and Science." In *Alexander von Humboldt, 1769/1969*. Bonn: InterNationes, 1969. 97–132.

Herschell, John F. W. "Humboldt's Kosmos" (*Edinburgh Review*, January 1848). In *Essays from the Edinburgh and Quarterly Reviews*. London: Longman, Brown, Green, Longmans, and Roberts, 1857. 257–364.

Home, Roderick W. "Humboldtian Science Revisited: An Australian Case Study." *History of Science* 33 (1995): 1–22.

Humboldt, Alexander von. *Abridgement of Humboldt's Statistical Essay on New Spain . . . by a Citizen of Maryland.* Baltimore, 1813.

———. *Cosmos: A Sketch of the Physical Description of the Universe.* Translated by Elise C. Otté. 2 vols. New York: Harper and Brothers, 1850, 1858; Baltimore: Johns Hopkins, 1997.

———. *Cosmos: A Sketch of the Physical Description of the Universe.* Vol. 3, translated by Elise C. Otté. New York: Harper and Brothers, 1858.

———. *Cosmos: Sketch of a Physical Description of the Universe.* Translated by Elizabeth Sabine. 2 vols. 5th ed. London: Longman, Brown, Green, and Longmans, 1849.

———. *Essay on the Geography of Plants.* In *Foundations of Biogeography: Classic Papers with Commentaries,* edited by Mark V. Lomolino et al. Chicago: University of Chicago Press, 2004. 49–57.

———. *The Fluctuations of Gold.* Translated by William Maude. New York: Burt Franklin, 1971.

———. *A Geognostical Essay on the Superposition of Rocks, in Both Hemispheres.* London: Longman, Hurst, Rees, Orme, Brown, and Green, 1823.

———. Introduction to *Meteorological Essays: by François Arago.* London: Longman, Brown, Green, and Longmans, 1855.

———. *The Island of Cuba: A Political Essay.* Translated by John S. Thrasher (1856). Princeton: Markus Wiener Publishers, 2001.

———. "Letter from Humboldt." In *Letters on American Slavery from Victor Hugo, de Tocqueville, Emile de Gerardin, Carnot, Passy, Mazzini, Humboldt, O. Lafayette—&c.* Boston: American Anti-Slavery Society, 1860. 17–19.

———. *Letters of Alexander von Humboldt to Varnhagen von Ense from 1827 to 1858.* New York: Rudd and Carleton, 1860.

———. *Personal Narrative of Travels to the Equinoctial Regions of America, During the Years 1799–1804,* trans. and ed. by Thomasina Ross. 3 vols. London: Henry G. Bohn, 1852.

———. *Personal Narrative of Travels to the Equinoctial Regions of the New Continent, During the Years 1799–1804.* 1814, 1819, 1825. Translated by Helen Maria Williams. 7 vols. in 6 books. London: Longman, Hurst, Rees, Orme, and Brown, 1814–29.

———. *Political Essay on the Kingdom of New Spain.* Translated by John Black. 4 vols. London, 1811; facs. ed., New York: AMS Press, 1966.

———. Preface to *Diary of a Journey from the Mississippi to the Coasts of the Pacific with a United States Government Expedition,* by Baldwin Möllhausen. 2 vols. London: Longman, Brown, Green, Longmans, and Roberts, 1858. 1:xi–xxv.

———. *Researches concerning the Institutions and Monuments of the Ancient Inhabitants of America, with Descriptions and Views of Some of the Most Striking Scenes in the Cordilleras!* Translated by Helen Maria Williams. 2 vols. London, 1814. Facs. ed., Amsterdam: Plenum Publishing, 1972.

———. *Selections from the Works of the Baron de Humboldt, relating to the Climate, Inhabitants, Productions, and Mines of Mexico,* by John Taylor. London: Longman, Hurst, Rees, Orme, Brown, and Green, 1824.

———. *Views of Nature: or Contemplations on the Sublime Phenomena of Creation; with Scientific Illustrations.* Translated by E. C. Otté and Henry G. Bohn. London: Henry G. Bohn, 1850.

———. *Vues des Cordillères, et monumens des peuples indigènes de l'Amérique.* Paris, 1810.

Humboldt, Wilhelm von. *Essays on Language*, ed. T. Harden and D. Farrelly. Frankfurt am Main: Peter Lang, 1997.

———. "Extract of a Letter from Baron Wm. Humboldt, to the Hon. Mr. Pickering of Salem." *Christian Spectator*, 1 September 1823, 495–96.

———. "On the Historian's Task." 1821. In *The Theory and Practice of History: Leopold von Ranke*, edited by Georg G. Iggers and Konrad von Moltke. Indianapolis: Bobbs-Merrill, 1973.

Ingersoll, Robert G. "Humboldt: The Universe is Governed by Law." In *The Gods and Other Lectures*. Peoria, IL: C. P. Farrell, 1878. 93–117.

Ingham, Mary Janes. "A Half-hour with the Humboldts." *Ladies Repository: A Monthly Periodical Devoted to Literature, Arts, and Religion* 26.5 (May 1866): 292–95.

Jahn, Ilsa. "The Influence of Alexander von Humboldt on Young Biologists and Biological Thinking during the XIXth Century." In *Actes du XIe Congrès International d'Histoire des Sciences*. Vol. 4. Wroclaw: Maison de'Édition de l'Académie Polonaise des Sciences, 1965.

Jeffries, Stephen. "Alexander von Humboldt and Ferdinand von Mueller's Argument for the Scientific Botanic Garden." *Historical Records of Australian Science* 11.3 (June 1997): 301–10.

Jenkins, Alice. "Alexander von Humboldt's *Kosmos* and the Beginnings of Ecocriticism." *ISLE: Interdisciplinary Studies in Literature and the Environment* 14.2 (Summer 2007): 89–105.

Joseph, John E. "A Matter of *Consequenz*: Humboldt, Race and the Genius of the Chinese Language." *Historiographia Linguistica* 26 1/2 (1999): 89–148.

Keber, Eloise Quiñones. "Humboldt and Aztec Art." *Colonial Latin American Review* 5.2 (1996): 277–97.

Kellner, Lotte. *Alexander von Humboldt*. London: Oxford University Press, 1963.

Kenosian, David. "Speaking of Nature." In *Alexander von Humboldt: From the Americas to the Cosmos*, ed. Raymond Erickson et al. New York: Bildner Center, City University of New York online publication. http://web.gc.cuny.edu/bildnercenter/publications/humboldt .pdf. 501–08.

Klencke, Hermann, "Alexander von Humboldt: A Biographical Monument." In *Lives of the Brothers Humboldt, Alexander and William*. Translated by Juliette Bauer. London: Ingram, Cooke, 1852. 1–245.

Kohlhepp, Gerb. "Scientific Findings of Alexander von Humboldt's Expedition into the Spanish-American Tropics (1799–1804) from a Geographical Point of View." *Anais da Academia Brasileira de Ciências* 77.2 (2005): 325–42.

Kortum, Gerhard. "Humboldt und das Meer: Eine Ozeanographiegeschichtliche Bestandsaufnahme." *Northeastern Naturalist* 8.1 (2001): 91–108.

Kortum, Gerhard, and Ingo Schwarz. "Alexander von Humboldt and Matthew Fontaine Maury—Two Pioneers of Marine Science." In *Historisch-Meereskundliches Jahrbuch (History of Oceanography Yearbook)*, Band 10. Stralsund: Deutsches Meeresmuseum: 2003–4. 157–85.

Kwa, Chunglin. "Alexander von Humboldt's Invention of the Natural Landscape." *European Legacy* 10.2 (2005): 149–62.

Large, Arlen J. "The Humboldt Connection." *We Proceeded On* 16.4 (November 1990): 4–12.

Leask, Nigel. "Alexander von Humboldt and the Romantic Imagination of America: The Impossibility of Personal Narrative." In *Curiosity and the Aesthetics of Travel Writing, 1770–1840*. Oxford: Oxford University Press, 2002. 243–98.

———. "Darwin's 'Second Sun': Alexander von Humboldt and the Genesis of *The Voyage of the Beagle*." In *Literature, Science, Psychoanalysis, 1830–1970*, edited by Helen Small and Trudi Tate. Oxford: Oxford University Press, 2003. 13–36.

———. "Salons, Alps and Cordilleras: Helen Maria Williams, Alexander von Humboldt, and the Discourse of Romantic Travel." In *Women, Writing, and the Public Sphere, 1700–1830*, edited by Elizabeth Eger et al. Cambridge: Cambridge University Press, 2001. 217–35.

Lieber, Francis. "Address When a Bust of Humboldt Was Placed in the Central Park, New York, 1869." In *Miscellaneous Writings*, vol. 1, *Reminiscences, Addresses, and Essays*. Philadelphia: Lippincott, 1881. 405–10.

———. "Alexander von Humboldt: Address before the American Geographical Society, 1859." In *Miscellaneous Writings*, vol. 1, *Reminiscences, Addresses, and Essays*. Philadelphia: Lippincott, 1881. 389–404.

Lindquist, Jason H. "'Under the influence of an exotic nature . . . national remembrances are insensibly effaced': Threats to the European Subject in Humboldt's *Personal Narrative of Travels to the Equinoctial Regions of the New Continent*." *HiN: Alexander von Humboldt im Netz, International Review for Humboldtian Studies* 5.9 (2004): n.p. http://www.uni-potsdam.de/u/romanistik/humboldt/hin/hin9/lindquist.htm.

Löschner, Renate. "The Influence of Alexander von Humboldt on Illustrations of America." In *Alexander von Humboldt: Life and Work*, edited by Wolfgang-Hagen Hein. Ingelheim am Rhein: C. H. Boehringer Sohn, 1987. 283–300.

Lubrich, Oliver. "Alexander von Humboldt: Revolutionizing Travel Literature." *Monatshefte* 96.3 (2004): 360–87.

———. "In the Realm of Ambivalence: Alexander von Humboldt's Discourse on Cuba." *German Studies Review* 26.1 (February 2003): 63–80.

Macpherson, Anne. "The Human Geography of Alexander von Humboldt." PhD diss., University of California, Berkeley, 1971.

———. "Man Must Will the Good and the Great—on Alexander von Humboldt's 'Political Essay on the Island of Cuba.'" *Revista Geográfica* [Rio de Janeiro, Brazil] 69 (December 1968): 157–65.

Martineau, Harriet. "Alexander von Humboldt: Died May 6th, 1859." In *Biographical Sketches, 1852–1875*. 4th ed. London: MacMillan, 1876. 278–89.

Mathewson, Kent. "Alexander von Humboldt and the Origins of Landscape Archaeology." *Journal of Geography* 85.2 (March–April 1986): 50–56.

———. "Alexander von Humboldt's Image and Influence in North American Geography, 1804–2004." *Geographical Review* 96.3 (July 2006): 416–38.

McGillivray, William. *The Travels and Researches of Alexander von Humboldt*. 1832. New York: Harper, 1833.

McIntyre, Loren. "Humboldt's Way: Pioneer of Modern Geography." *National Geographic* 168.3 (September 1985): 318–51.

Meyer-Abich, Adolf. "Alexander von Humboldt." In *Alexander von Humboldt 1769/1969*. Bonn: Inter Nationes, 1969. 5–94.

———. "Alexander von Humboldt as a Biologist." In *Alexander von Humboldt: Werk und Weltgeltung*. Munich: R. Piper, 1969. 179–96.

Moheit, U. "Alexander von Humboldt and Australia." In *Australia: Studies in the History of the Discovery and Exploration*. Frankfurt, 1994. 33–42.

Nelkin, Halina. *Alexander von Humboldt: His Portraits and Their Artists: A Documentary Iconography*. Berlin: Dietrich Reimer Verlag, 1980.

———. *Humboldtiana at Harvard*. Cambridge: Widener Library, Harvard University, 1976.

Nichols, Sandra. "Why Was Humboldt Forgotten in the United States?" *Geographical Review* 96.3 (July 2006): 399–415.

Nicolson, Malcolm. "Alexander von Humboldt and the Geography of Vegetation." In *Romanticism and the Sciences*, edited by Andrew Cunningham and Nicholas Jardine. Cambridge: Cambridge University Press, 1990. 169–85.

———. "Alexander von Humboldt, Humboldtian Science and the Origins of the Study of Vegetation." *History of Science* 25 (1987): 167–94.

———. "Humboldtian Plant Geography after Humboldt: The Link to Ecology." *British Journal of the History of Science* 29 (1996): 289–310.

Nollendorfs, Cora Lee. "Alexander von Humboldt Centennial Celebrations in the United States: Controversies Concerning His Work." *Monatshefte* 80.1 (Spring 1988): 59–66.

Ochoa, John. "Alexander von Humboldt's Work on Mexico, Cultural Allegory, and the Limits of Vision." In *The Uses of Failure in Mexican Literature and Identity*. Austin: University of Texas Press, 2004. 81–109.

Olesko, Kathryn. "Humboldtian Science." In *The Oxford Companion to the History of Modern Science*, edited by J. L. Heilbron. Oxford: Oxford University Press, 2003.

Oppitz, Ulrich-Dieter. "De Name der Brüder Humboldt in aller Welt." In *Alexander von Humboldt, Werk und Weltgeltung*. Munich: R. Piper, 1969. 277–429.

Osgood, Samuel. "Humboldt and Young Germany." *Appleton's Journal of Literature, Science, and Art* 2.26 (25 September 1869): 178–80.

Palau, Geóg. Diógenes Edgildo. "Los Pueblos Indigenas del Amazonas Venezolano desde Humboldt a Nuestros Dias: Preservacion del Ambiente y Manejo Sostenible." *Northeastern Naturalist* 8.1 (2001): 135–56.

Phillips, Denise. "Building Humboldt's Legacy: The Humboldt Memorials of 1869 in Germany." *Northeastern Naturalist* 8.1 (2001): 21–32.

Pratt, Mary Louise. "Humboldt and the Reinvention of America." In *Amerindian Images and the Legacy of Columbus*, edited by René Jara and Nicholas Spadaccini. Minneapolis: University of Minnesota Press, 1992. 584–606.

Pratt, Mary Louise. "Alexander von Humboldt and the Reinvention of America." In *Imperial Eyes: Travel Writing and Transculturation*. London: Routledge, 1992. 111–43.

Rebok, Sandra. "Two Exponents of the Enlightenment: Transatlantic Communication by Thomas Jefferson and Alexander von Humboldt." *Southern Quarterly* 43 (Summer 2006): 126–52.

Rippy, J. Fred, and E. R. Brann. "Alexander von Humboldt and Simón Bolívar." *American Historical Review* 52.4 (July 1947): 697–703.

Robinson, A. H., and Helen M. Wallis. "Humboldt's Map of Isothermal Lines: A Milestone in Thematic Cartography." *Cartographic Journal* 4.1 (June 1967): 119–23.

Romero-González, Gustavo A. "Alexander von Humboldt's Legacy in Venezuela." *Northeastern Naturalist* 8.1 (2001): 33–42.

Rose, M. A. "Alexander von Humboldt and Australian Art and Exploration." In *The German Experience of Australia, 1833–1938*, edited by I. Harmstorf and P. Schwerdtfeger. Adelaide, 1988. 106–19.

Rupke, Nicolaas. "Alexander von Humboldt: German Naturalist and Traveller." In *Reader's Guide to the History of Science*, edited by Arne Hessenbruch. London: Fitzroy Dearborn, 2000. 356–59.

———. *Alexander von Humboldt: A Metabiography*. Frankfurt am Main: Peter Lang, 2005.

———. "Alexander von Humboldt and Revolution: A Geography of Reception of the Varnhagen von Ense Correspondence." In *Geography and Revolution*, ed. David N. Livingstone and Charles W. J. Withers. Chicago: University of Chicago Press, 2005. 336–50.

——. "A Geography of Enlightenment: The Critical Reception of Alexander von Humboldt's Mexico Work." In *Geography and Enlightenment*, edited by David N. Livingstone and Charles W. J. Withers. Chicago: University of Chicago Press, 1999. 319–39.

——. "Humboldtian Medicine." *Medical History* 40 (1996): 293–310.

——. Introduction to *Cosmos: A Sketch of a Physical Description of the Universe*, by Alexander von Humboldt. 2 vols. Translated by E. C. Otté. Baltimore: Johns Hopkins University Press, 1997. 1:vii–xlii.

——. Review of Petra Werner, *Himmel und Erde. Alexander von Humboldt und sein Kosmos*. Berlin: Akademie Verlag, 2004. *Annals of Science* 62.4 (October 2005): 553–54.

Sachs, Aaron. *The Humboldt Current: Nineteenth-Century Exploration and the Roots of American Environmentalism*. New York: Viking Penguin, 2006.

——. "The Ultimate 'Other': Post-Colonialism and Alexander von Humboldt's Ecological Relationship with Nature." *History and Theory*, theme issue 42 (December 2003): 111–35.

Sarton, George. "Aime Bonpland (1773–1858)." *Isis* 34.5 (Summer 1943): 385–99.

Saunders, J. B. deC. M. *Humboldtian Physicians in California*. Davis: University of California, 1971.

——. "The Influence of Alexander von Humboldt on the Medicine of Western America." *Proceedings of the XXIII Congress of the History of Medicine* (2–9 September 1972): 523–28.

Schoenwaldt, Peter. "Alexander von Humboldt and the USA." In *Alexander von Humboldt: Life and Work*, edited by Wolfgang-Hagen Hein. Ingelheim am Rheim: C. H. Boehringer Sohn, 1987. 273–82.

——. "Alexander von Humboldt und die Vereinigten Staaten von Amerika." In *Alexander von Humboldt: Werk und Weltgeltung*. Munich: R. Piper, 1969. 431–82.

Schwarz, Ingo, ed. *Alexander von Humboldt und die Vereinigten Staaten von Amerika: Briefwechsel*. Berlin: Akademie Verlag, 2004.

——. "Alexander von Humboldt's Visit to Washington and Philadelphia, His Friendship with Jefferson, and His Fascination with the United States." *Northeastern Naturalist* 8.1 (2001): 43–56.

——. *From Alexander von Humboldt's Correspondence with Thomas Jefferson and Albert Gallatin*. Berlin: Alexander von Humboldt Research Center, 1991.

——. "The Second Discoverer of the New World and the First American Literary Ambassador to the Old World: Alexander von Humboldt and Washington Irving." *Acta Historica Leopoldina* 27 (1997): 89–97.

——. "Transatlantic Communication in the 19th Century: Aspects of the Correspondence between Alexander von Humboldt and George Ticknor." *Asclepio: Revista de Historia de la Medicina y de la Ciencia* 55.2 (2004): 25–39.

Silliman, Benjamin. "Baron von Humboldt." In *A Visit to Europe in 1851*. 2 vols. New York: A. S. Barnes, 1856. 2:318–22.

Sluyter, Andrew. "Humboldt's Mexican Texts and Landscapes." *Geographical Review* 96.3 (July 2006): 361–81.

Stearn, W. T. "Humboldt's 'Essai sur la Géographie des Plantes.'" In *Humboldt, Bonpland, Kunth and Tropical American Botany*, edited by William T. Stearn. Germany: J. Cramer, 1968. 121–28.

Steigerwald, Joan. "The Cultural Enframing of Nature: Environmental Histories during the Early German Romantic Period." *Environment and History* 6 (2000): 451–96.

——. "Figuring Nature/Figuring the (Fe)Male: The Frontispiece to Humboldt's Ideas Towards a Geography of Plants." In *Figuring It Out: Science, Gender, and Visual Culture*. Hanover, NH: Dartmouth College Press/University Press of New England, 2006. 54–82.

Stephens, John Lloyd. "An Hour with Alexander von Humboldt." *Littell's Living Age* 15 (October–December 1847): 151–53.

Stillé, Alfred. *Humboldt's Life and Character: An Address before the Linnaean Association of Pennsylvania College.* Philadelphia: Linnaean Association, 1859.

Stoddard, Richard Henry. *The Life, Travels, and Books of Alexander von Humboldt.* New York: Rudd and Carleton, 1859.

Tang, Chenzi. *The Geographic Imagination of Modernity: Geography, Literature, and Philosophy in German Romanticism.* Stanford: Stanford University Press, 2008.

Taylor, Bayard. "Alexander von Humboldt." In *At Home and Abroad.* New York: Putnam, 1860. 351–65.

Tresch, John. "¡Viva la República Cosmica! Or, The Children of Humboldt and Coca-Cola." In *Making Things Public: Atmospheres of Democracy,* edited by Bruno Latour and Peter Weibel. Cambridge: MIT Press, 2005. 352–56.

Trigo, Benigno. "Walking Backward to the Future: Time, Travel, and Race." In *Subjects of Crisis: Race and Gender as Disease in Latin America.* Hanover: Wesleyan University Press, 2000. 16–46.

Tuan, Yi-Fu. "Alexander von Humboldt and His Brother: Portrait of an Ideal Geographer in Our Time." Lecture, UCLA Faculty Center, 7 February 1997.

Tuckerman, Henry T. "The Naturalist. Humboldt." In *Characteristics of Literature Illustrated by the Genius of Distinguished Writers.* 2nd ser. Philadelphia: Lindsay and Blakiston, 1851. 56–77. Repr. from *Godey's Lady's Book* 41 (September 1850): 133–38.

Van Dusen, Robert. *The Literary Ambitions and Achievements of Alexander von Humboldt.* Bern: Herbert Lang, 1971.

Villegas, Benjamin. *The Route of Humboldt: Venezuela and Colombia.* 2 vols. Bogotá:Villegas Editores, 1994.

Vogel, Ursula. "Humboldt and the Romantics: Neither *Hausfrau* nor *Citoyenne*; The Idea of 'Self-Reliant Femininity' in German Romanticism." In *Women in Western Political Philosophy: Kant to Nietzsche,* edited by Ellen Kennedy and Susan Mendus. New York: St. Martins, 1987. 106–26.

von Hagen, Victor Wolfgang. *South America Called Them.* New York: Knopf, 1945.

Walls, Laura Dassow. "The Birth of the Two Cultures." In *Alexander von Humboldt: From the Americas to the Cosmos,* ed. Raymond Erickson et al. New York: Bildner Center, City University of New York online publication. http://web.gc.cuny.edu/bildnercenter/publications/humboldt.pdf. 247–58, accessed 11 January 2009.

———. "'Hero of Knowledge, Be Our Tribute Thine': Alexander von Humboldt in Victorian America." *Northeastern Naturalist* 8.1 (2001): 121–34.

———. "'The Napoleon of Science': Alexander von Humboldt in Antebellum America." *Nineteenth-Century Contexts* 14.1 (1990): 71–98. Repr. in *Nineteenth Century Literature Criticism,* vol. 170, edited by Russel Whitaker. Oklahoma City: Thomson Gale, 2006. 145–56.

———. "Rediscovering Humboldt's Environmental Revolution." *Environmental History* 10.4 (October 2005): 758–60.

———. "Seeing New Worlds: Thoreau and Humboldtian Science." In *Thoreau's World and Ours: A Natural Legacy,* edited by Edmund A. Schofield and Robert Baron. Golden, CO: North American Press, 1993. 55–63.

———. *Seeing New Worlds: Henry David Thoreau and Nineteenth-Century Natural Science.* Madison: University of Wisconsin Press, 1995.

Weisert, John J. "A Young American Visits von Humboldt." *American-German Review* 28.3 (1962): 27–28.

Wiegand, Dometa. "Alexander von Humboldt and Samuel Taylor Coleridge: The Intersection of Science and Poetry." *Coleridge Bulletin: The Journal of the Friends of Coleridge* 20 (Winter 2002): 105–13.

Werner, James V. "'Ground-Moles' and Cosmic Flaneurs: Poe, Humboldt, and Nineteenth-Century Science." *Edgar Allan Poe Review* 3.1 (Spring 2002): 45–65.

Whitaker, Arthur P. "Alexander von Humboldt and Spanish America." *Proceedings of the American Philosophical Society* 104.3 (June 1960): 317–22.

Wilhelmy, Herbert. "Alexander von Humboldt in the Light of His American Journey." *Universitas: A German Review of the Arts and Sciences* 12.1 (1969): 37–52.

Württemberg, Duke of (Paul Wilhelm). *Travels in North America, 1822–1824.* Norman: University of Oklahoma Press, 1973.

Youmans, E. L. "Alexander von Humboldt." *Appleton's Journal of Literature, Science, and Art* 2.26 (1869): 182–83.

Young, Edward J. "Remarks at the Centennial Anniversary of the Birth of Alexander von Humboldt, in Horticultural Hall, Boston, September 14, 1869." *Harvard University Alumni Writings, Class of 1848*, vol. 1, no. 8 (n.d.).

Zantop, Susanne. "The German Columbus." In *Colonial Fantasies: Conquest, Family, and Nation in Precolonial Germany, 1770–1870.* Durham: Duke University Press, 1997. 166–72.

Zeller, Suzanne. "Humboldt and the Habitability of Canada's Great Northwest." *Geographical Review* 96.3 (July 2006): 382–98.

Zimmerer, Karl S. "Humboldt and the History of Environmental Thought." *Geographical Review* 96.3 (July 2006): 456–58.

———. "Humboldt's Nodes and Modes of Interdisciplinary Environmental Science in the Andean World." *Geographical Review* 96.3 (July 2006): 335–60.

All Other Sources, Primary and Secondary

Adams, Henry. *History of the United States of America during the Administrations of Thomas Jefferson.* New York: Library of America, 1986.

———. *The Life of Albert Gallatin.* Philadelphia: Lippincott, 1879.

Agassiz, Elizabeth Cary. *Louis Agassiz: His Life and Correspondence.* 2 vols. Boston: Houghton Mifflin, 1887.

Agassiz, Louis. "Contemplations of God in the Kosmos." *Christian Examiner* 50 (January 1851): 1–17.

———. "The Diversity of Origin of the Human Races." *Christian Examiner* 49 (July 1850): 110–45.

———. "Geographical Distribution of Animals." *Christian Examiner* 48 (March 1850): 181–204.

———. "Sketch of the Natural Provinces of the Animal World and Their Relation to the Different Types of Man." In *Types of Mankind*, by Josiah C. Nott and George R. Gliddon, 8th ed. Philadelphia: Lippincott, 1857. lviii–lxxvi.

Agassiz, Louis, and Elizabeth Cary Agassiz. *A Journey in Brazil.* Boston: Fields, Osgood, & Co., 1869.

Anderson, Benedict. *Imagined Communities.* London: Verso, 1991.

Andresen, Julie Tetel. *Linguistics in America, 1769–1924: A Critical History.* New York: Routledge, 1990.

Arac, Jonathan. "Narrative Forms." In *The Cambridge History of American Literature*. Vol. 2, *Prose Writing, 1820–1865*, edited by Sacvan Bercovitch and Cyrus R. K. Patell. Cambridge: Cambridge University Press, 1995. 605–777.

Augstein, H. F. *James Cowles Prichard's Anthropology: Remaking the Science of Man in Early Nineteenth Century Britain.* Amsterdam: Rodopi, 1999.

Bachman, John. *The Doctrine of the Unity of the Human Race Examined on the Principles of Science.* Charleston, SC: C. Canning, 1850.

Balée, William, ed. *Advances in Historical Ecology.* New York: Columbia University Press, 1998.

Bieder, Robert E. *Science Encounters the Indian, 1820–1880: The Early Years of American Ethnology.* Norman: University of Oklahoma Press, 1986.

Blumenbach, Johann Friedrich. *The Anthropological Treatises of Johann Friedrich Blumenbach.* 1865. Translated by Thomas Bendyshe. Boston: Longwood Press, 1978.

Boas, Franz. "The Background of My Early Thinking." In *The Shaping of American Anthropology, 1883–1911: A Franz Boas Reader*, edited by George W. Stocking, Jr. New York: Basic Books, 1974. 41–42.

———. *The Mind of Primitive Man.* Rev. ed. New York: Macmillan, 1938.

———. *Primitive Art.* 1927. New York: Dover, 1955.

———. "The Study of Geography." In *Race, Language and Culture.* New York: Macmillan, 1966. 639–47.

Brace, C. Loring. *"Race" Is a Four-Letter Word: The Genesis of the Concept.* New York: Oxford, 2005.

Bray, Edmund, and Martha Colemen Bray, eds. *Joseph N. Nicollet on the Plains and Prairies: The Expeditions of 1838–39 with Journals, Letters, and Notes on the Dakota Indians.* St. Paul: Minnesota Historical Society Press, 1993.

Bray, Martha Coleman. *Joseph Nicollet and His Map.* 2nd ed. Philadelphia: American Philosophical Society, 1994.

Buell, Lawrence. *The Environmental Imagination: Thoreau, Nature Writing, and the Formation of American Culture.* Cambridge: Harvard University Press, 1995.

———. *Writing for an Endangered World.* Cambridge: Harvard University Press, 2001.

Bunzl, Matti, and H. Glenn Penny. "Introduction: Rethinking German Anthropology, Colonialism, and Race." In *Worldly Provincialism: German Anthropology in the Age of Empire*, edited by H. Glenn Penny and Matti Bunzl. Ann Arbor: University of Michigan Press, 2003.

Burroughs, John, John Muir, et al. *Alaska: The Harriman Expedition, 1899.* New York: Dover, 1986.

———. *Indoor Studies.* Vol. 8 of *The Writings of John Burroughs.* Boston: Houghton Mifflin, 1904.

Cañizares-Esguerra, Jorge. "Spanish America: From Baroque to Modern Colonial Science." In *The Cambridge History of Science.* Vol. 4, *Eighteenth-Century Science*, edited by Roy Porter. Cambridge: Cambridge University Press, 2003. 718–38.

Capper, Charles. *Margaret Fuller: An American Romantic Life; The Public Years.* Oxford: Oxford University Press, 2007.

Carrera Andrade, Jorge. *Selected Poems of Jorge Carrera Andrade*, translated by H. R. Hays. Albany: State University of New York, 1972.

Casey, Edward S. *Representing Place: Landscape Painting and Maps.* Minneapolis: University of Minnesota Press, 2002.

Catlin, George. *Letters and Notes on the Manners, Customs, and Conditions of North American Indians.* 2 vols. London, 1844; repr. New York: Dover, 1973.

Cole, Douglas. *Franz Boas: The Early Years, 1858–1906.* Seattle: University of Washington Press, 1999.

Cooper, James Fenimore. *The Prairie.* Albany: State University of New York Press, 1985.

Cooper, Susan Fenimore. *Essays on Nature and Landscape,* edited by Rochelle Johnson and Daniel Patterson. Athens: University of Georgia Press, 2002.

———. *Rural Hours,* edited by Rochelle Johnson and Daniel Patterson. Athens: University of Georgia Press, 1998.

Crosby, Alfred W. *Ecological Imperialism: The Biological Expansion of Europe, 900–1900.* Cambridge: Cambridge University Press, 1986.

Darwin, Charles. *On the Origin of Species.* 1859. Facs. of 1st ed. Cambridge: Harvard University Press, 1964.

Davis, David Brion. *Inhuman Bondage: The Rise and Fall of Slavery in the New World.* Oxford: Oxford University Press, 2006.

Dean, Bradley P. "Natural History, Romanticism, and Thoreau." In *American Wilderness: A New History,* edited by Michael Lewis. Oxford: Oxford University Press, 2007. 73–89.

Dickinson, Emily. *The Poems of Emily Dickinson: Reading Edition,* edited by R. W. Franklin. Cambridge: Harvard University Press, 1999.

Denton, Sally. *Passion and Principle: John and Jessie Frémont, the Couple Whose Love Shaped Nineteenth-Century America.* New York: Bloomsbury, 2007.

Dimock, Wai Chee. "Deep Time: American Literature and World History." *American Literary History* 13.4 (2001): 755–75.

———. "The Planetary Dead: Margaret Fuller, Ancient Egypt, Italian Revolution." *ESQ* 50.1–3 (2004): 23–57.

———. "A Theory of Resonance." *PMLA* 112.5 (October 1997): 1060–71.

Elliott, J. H. *The Old World and the New, 1492–1650.* Cambridge: Cambridge University Press, 1992.

Emerson, Ralph Waldo. *The Collected Works of Ralph Waldo Emerson,* edited by Alfred R. Ferguson et al. 6 vols. to date. Cambridge: Harvard University Press, 1971–.

———. *The Complete Works of Ralph Waldo Emerson,* edited by Edward Waldo Emerson. 12 vols. Boston: Houghton Mifflin, 1903–4.

———. *The Early Lectures of Ralph Waldo Emerson, 1833–42,* edited by Stephen E. Whicher, Robert E. Spiller, and Wallace E. Williams. 3 vols. Cambridge: Harvard University Press, 1959–1972.

———. *Essays and Lectures,* edited by Joel Porte. New York: Library of America, 1983.

———. *Journals and Miscellaneous Notebooks of Ralph Waldo Emerson,* edited by William Gilman et al. 16 vols. Cambridge: Harvard University Press, 1960–1982.

———. *The Later Lectures of Ralph Waldo Emerson, 1843–1871,* edited by Ronald A. Bosco and Joel Myerson. 2 vols. Athens: University of Georgia Press, 2001.

———. *The Letters of Ralph Waldo Emerson.* 10 vols, edited by Ralph L. Rusk and Eleanor Tilton. New York: Columbia University Press, 1939–95.

Emmons, David M. "Theories of Increased Rainfall and the Timber Culture Act of 1873." *Forest History* 15 (1971): 6–14.

Emory, William H. *Notes of a Military Reconnaissance, from Fort Leavenworth, in Missouri, to San Diego, in California, including Parts of the Arkansas, Del Norte, and Gila Rivers.* Washington: Wendell and Van Benthuysen, 1848.

Evans, Howard Ensign. *The Natural History of the Long Expedition to the Rocky Mountains.* New York: Oxford, 1997.

Fixico, Donald L. *The American Indian Mind in a Linear World: American Indian Studies and Traditional Knowledge.* New York: Routledge, 2003.

Folsom, Ed. "Database as Genre: The Epic Transformation of Archives." *PMLA* 122.5 (October 2007): 1571–1579.

Foner, Eric. *Tom Paine and Revolutionary America.* Updated ed. New York: Oxford University Press, 2005.

Francaviglia, Richard V. *Mapping and the Imagination in the Great Basin: A Cartographic History.* Reno: University of Nevada Press, 2005.

Frémont, John Charles. *Report of the Exploring Expedition to the Rocky Mountains in the year 1842, and to Oregon and North California in the years 1843–44.* Washington: Gales and Seaton, 1845.

Fuller, Margaret. *The Essential Margaret Fuller,* edited by Jeffrey Steele. New Brunswick: Rutgers University Press, 1992.

García Sáiz, M. Concepción. "Miguel Cabrera." In *The Arts in Latin America, 1492–1820,* edited by Joseph N. Newland et al. Philadelphia: Philadelphia Museum of Art, 2006. 402–9.

Garrard, Greg. *Ecocriticism.* New York: Routledge, 2005.

Gianquitto, Tina. "The Noble Designs of Nature: God, Science, and the Picturesque in Susan Fenimore Cooper's *Rural Hours.*" In *Susan Fenimore Cooper: New Essays on Rural Hours and Other Works,* edited by Rochelle Johnson and Daniel Patterson. Athens: University of Georgia Press, 2001. 169–90.

Goetzmann, William H. *Army Exploration in the American West, 1803–1863.* New Haven: Yale University Press, 1959.

———. *Exploration and Empire: The Explorer and the Scientist in the Winning of the American West.* 1966. New York: Norton, 1978.

———. "Introduction: The Man Who Stopped to Paint America." In *Karl Bodmer's America.* Lincoln: Joslyn Art Museum and University of Nebraska Press, 1984. 1–23.

———. *New Lands, New Men: America and the Second Great Age of Discovery.* New York: Viking, 1986.

———. "Paradigm Lost." In *The Sciences in the American Context: New Perspectives,* edited by Nathan Reingold. Washington, DC: Smithsonian, 1979. 21–34.

Grabbe, Hans-Jürgen. "Weary of Germany—Weary of America: Perceptions of the United States in Nineteenth-Century Germany." In *Transatlantic Images and Perceptions: Germany and America since 1776,* edited by David E. Barclay and Elisabeth Glaser-Schmidt. Cambridge: German Historical Society and Cambridge University Press, 1997. 65–86.

Gould, Stephen Jay. *The Mismeasure of Man.* New York: Norton, 1981.

Graustein, Jeannette E. *Thomas Nuttall, Naturalist: Explorations in America, 1808–1841.* Cambridge: Harvard University Press, 1967.

Grove, Richard H. "Environmental History." In *New Perspectives on Historical Writing.* 2nd ed., edited by Peter Burke. University Park: Pennsylvania State University Press, 2001. 261–82.

———. *Green Imperialism: Colonial Expansion, Tropical Island Edens and the Origins of Environmentalism, 1600–1860.* Cambridge: Cambridge University Press, 1995.

Guha, Ramachandra. *Environmentalism: A Global History.* New York: Longman, 2000.

Gura, Philip F. *American Transcendentalism: A History.* New York: Farrar, Straus and Giroux, 2007.

Guyot, Arnold. *The Earth and Man: Lectures on Comparative Physical Geography, in Its Relation to the History of Mankind.* Translated by Cornelius Felton. Boston: Gould, Kendall, and Lincoln, 1849.

Hackel, Steven W. *Children of Coyote, Missionaries of Saint Francis: Indian-Spanish Relations in Colonial California, 1769–1850.* Chapel Hill: University of North Carolina Press, 2005.

Halpin, Marjorie. Introduction to *George Catlin, Letters and Notes on the Manners, Customs, and Conditions of North American Indians.* 2 vols. London, 1844; repr. New York: Dover, 1973. vii–xiv.

Harris, Marvin. *The Rise of Anthropological Theory.* New York: Harper and Row, 1968.

Harvey, David. "Cosmopolitanism and the Banality of Geographical Evils." *Public Culture* 12.2 (2000): 529–64.

Herder, Johann Gottfried. *Outlines of a Philosophy of the History of Man.* Translated by T. Churchill. 4 vols. 1800; repr., New York: Bergman, n.d.

Herndon, William Lewis, and Lardner Gibbon. *Exploration of the Valley of the Amazon Made under Direction of the Navy Department.* 2 vols. Washington: Robert Armstrong, 1854.

Heywood, Colin. "Society." In *The Nineteenth Century*, edited by T. C. W. Blanning. Oxford: Oxford University Press, 2000. 47–77.

Hietala, Thomas R. *Manifest Design: American Exceptionalism and Empire.* Rev. ed. Ithaca: Cornell University Press, 2003.

Holmes, Steven J. *The Young John Muir: An Environmental Biography.* Madison: University of Wisconsin Press, 1999.

Horsman, Reginald. *Race and Manifest Destiny: The Origins of American Racial Anglo-Saxonism.* Cambridge: Harvard University Press, 1981.

Howat, John K. *Frederic Church.* New Haven: Yale University Press, 2005.

Howe, Daniel Walker. *What Hath God Wrought: The Transformation of America, 1815–1848.* Oxford: Oxford University Press, 2007.

Huseman, Ben. W. *Wild River, Timeless Canyons: Balduin Möllhausen's Watercolors of the Colorado.* Tucson: University of Arizona Press, 1995.

Irving, Washington. *Astoria, or Anecdotes of an Enterprise Beyond the Rocky Mountains*, edited by Edgeley W. Todd. Norman: University of Oklahoma, 1964.

———. *The Life and Voyages of Christopher Columbus*, edited by John Harmon McElroy. Boston: Twayne, 1981.

———. *The Sketch Book.* 1819. In *History, Tales, and Sketches*, edited by James W. Tuttleton. New York: Library of America, 1983. 731–1091.

Jefferson, Thomas. *Writings.* New York: Library of America, 1984.

Johnson, Rochelle. *Passions for Nature: Nineteenth-Century America's Aesthetics of Alienation.* Athens: Georgia University Press, forthcoming.

Johnson, Rochelle, and Daniel Patterson, eds. *Susan Fenimore Cooper: New Essays on Rural Hours and Other Works.* Athens: University of Georgia Press, 2001.

Kant, Immanuel. *Metaphysical Foundations of Natural Science.* 1786. Translated by James Ellington. Indianapolis: Bobbs-Merrill, 1970.

———. *The Philosophy of Kant*, edited by Carl J. Friedrich. New York: Modern Library, 1949.

Katzew, Ilona. "Casta Painting: Identity and Social Stratification in Colonial Mexico." In *New World Orders: Casta Painting and Colonial Latin America*, edited by Ilona Katzew. New York: Americas Society Art Gallery, 1996. 8–29.

Keller, Robert H., and Michael F. Turek. *American Indians and National Parks.* Tucson: University of Arizona Press, 1998.

Kelley, Theresa M., and Paula R. Feldman. Introduction to *Romantic Women Writers: Voices and Countervoices*, edited by Paula R. Feldman and Theresa M. Kelley. Hanover: University Press of New England, 1995.

Kennedy, Deborah. *Helen Maria Williams and the Age of Revolution.* Lewisburg: Bucknell University Press, 2002.

Knox, Robert. *The Races of Men: A Fragment.* Philadelphia: Lea and Blanchard, 1850.

Konkle, Maureen. *Writing Indian Nations: Native Intellectuals and the Politics of Historiography, 1827–1863.* Chapel Hill: University of North Carolina Press, 2004.

Kramer, Fritz L. "Eduard Hahn and the End of the 'Three Stages of Man.'" *Geographical Review* 57.1 (January 1967): 73–89.

Kropotkin, Peter. *Memoirs of a Revolutionist.* Montréal: Black Rose Books, 1989.

———. "Modern Science and Anarchism." In *Kropotkin's Revolutionary Pamphlets: A Collection of Writings by Peter Kropotkin,* edited by Roger N. Baldwin. New York: Dover, 1970. 146–94.

Latour, Bruno. *Pandora's Hope: Essays on the Reality of Science Studies.* Cambridge: Harvard University Press, 1999.

Leyda, Jay. *The Melville Log: A Documentary Life of Herman Melville, 1819–1891.* New York: Harcourt, Brace, 1951.

Liebersohn, Harry. *The Travelers' World: Europe to the Pacific.* Cambridge: Harvard University Press, 2006.

Lowenthal, David. *George Perkins Marsh: Prophet of Conservation.* Seattle: University of Washington Press, 2000.

———. *The Past Is a Foreign Country.* Cambridge: Cambridge University Press, 1985.

Lynch, John. *Simón Bolívar: A Life.* New Haven: Yale University Press, 2006.

Mack, Charles R., and Henry H. Lesesne, eds. *Francis Lieber and the Culture of the Mind.* Columbia: University of South Carolina Press, 2005.

Marsh, George Perkins. *Man and Nature, or, Physical Geography as Modified by Human Action.* 1864, edited by David Lowenthal. Cambridge: Harvard University Press, 1965.

———. *So Great a Vision: The Conservation Writings of George Perkins Marsh,* edited by Stephen C. Trombulak. Hanover, NH: University Press of New England, 2001.

McCandless, Peter. "The Political Evolution of John Bachman: From New York Yankee to South Carolina Secessionist." *South Carolina Historical Magazine* 108.1 (January 2007): 6–31.

Mellow, James R. *Nathaniel Hawthorne in His Times.* Boston: Houghton Mifflin, 1980.

Melville, Herman. *Moby-Dick, or The Whale.* 1851. New York: Penguin, 1992.

Mitchell, W. J. T. "Imperial Landscape." In *Landscape and Power,* edited by W. J. T. Mitchell. Chicago: University of Chicago, 1995. 5–34.

Muir, John. *Nature Writings.* New York: Library of America, 1997.

Muthu, Sankar. *Enlightenment against Empire.* Princeton: Princeton University Press, 2003.

Myers, Jeffrey. *Converging Stories: Race, Ecology, and Environmental Justice in American Literature.* Athens: University of Georgia Press, 2005.

Nagler, Jörg. "From Culture to *Kultur.* Changing American Perceptions of Imperial Germany, 1870–1914." In *Transatlantic Images and Perceptions: Germany and America since 1776,* edited by David E. Barclay and Elisabeth Glaser-Schmidt. Cambridge: Cambridge University Press, 1997. 131–70.

Nelson, Roland W. "Apparatus for a Definitive Edition of Poe's *Eureka.*" In *Studies in the American Renaissance, 1978,* edited by Joel Myerson. Boston: Twayne, 1978. 161–205.

Nott, Josiah C., and George R. Gliddon. *Types of Mankind.* 8th ed. Philadelphia: Lippincott, 1857.

Novak, Barbara. *Nature and Culture: American Landscape and Painting, 1825–1875.* New York: Oxford University Press, 1980.

Omans, Glen A. "'Intellect, Taste, and the Moral Sense': Poe's Debt to Immanuel Kant." In *Studies in the American Renaissance, 1980*, edited by Joel Myerson. Boston: Twayne, 1980. 123–68.

Orr, William J. "Karl Bodmer: The Artist's Life." In *Karl Bodmer's America*, by Karl Bodmer. Lincoln: University of Nebraska Press, 1984. 349–76.

Outka, Paul. *Race and Nature from Transcendentalism to the Harlem Renaissance*. New York: Palgrave Macmillan, 2008.

Pagden, Anthony. *European Encounters with the New World: From Renaissance to Romanticism*. New Haven: Yale University Press, 1993.

Parkman, Francis. *France and England in North America*. 2 vols. New York: Library of America, 1982.

Peale, Charles Willson. *The Selected Papers of Charles Willson Peale*. New Haven: Yale University Press, 1988.

Peirce, Charles S. "The Place of Our Age in the History of Civilization." In *Writings of Charles S. Peirce*. 5 vols. to date, edited by Max H. Fisch et al. Bloomington: Indiana University Press, 1982–. 1:101–14.

Pérez-Mejía, Ángela. *A Geography of Hard Times: Narratives about Travel to South America, 1780–1849*. Translated by Dick Cluster. Albany: State University of New York, 2004.

Pochman, Henry A. *German Culture in America, Philosophical and Literary Influences, 1600–1900*. Madison: University of Wisconsin Press, 1957.

Poe, Edgar Allan. *The Narrative of Arthur Gordon Pym of Nantucket*. 1838. Edited by Richard Kopley. New York: Penguin, 1999.

———. *Poetry and Tales*. New York: Library of America, 1984.

———. *The Science Fiction of Edgar Allan Poe*. Edited by Harold Beaver. New York: Penguin, 1976.

Poole, Deborah. *Vision, Race, and Modernity: A Visual Economy of the Andean Image World*. Princeton: Princeton University Press, 1997.

Popkin, Richard H. "The Philosophical Bases of Modern Racism." In *Philosophy and the Civilizing Arts*, edited by Craig Walton and John P. Anton. Athens: Ohio University Press, 1974. 126–65.

Prescott, William Hickling. *History of the Conquest of Mexico and History of the Conquest of Peru*. New York: Modern Library, 1936.

Prichard, James Cowles. *The Natural History of Man*. 2 vols. London: H. Baillière, 1855.

Raffles, Hugh. *In Amazonia: A Natural History*. Princeton: Princeton University Press, 2002.

Reill, Peter Hanns. "The Legacy of the 'Scientific Revolution': Science and the Enlightenment." In *The Cambridge History of Science*. Vol. 4, *Eighteenth-Century Science*, edited by Roy Porter. Cambridge: Cambridge University Press, 2003. 23–43.

Reynolds, David S. *John Brown, Abolitionist*. New York: Knopf, 2005.

———. "Politics and Poetry: Leaves of Grass and the Social Crisis of the 1850s." In *The Cambridge Companion to Walt Whitman*, edited by Ezra Greenspan. Cambridge: Cambridge University Press, 1995. 66–91.

———. *Walt Whitman's America: A Cultural Biography*. New York: Knopf, 1995.

Reynolds, J. N. "An Address on the Subject of a Surveying and Exploring Expedition to the Pacific Ocean and the South Seas." *Southern Literary Messenger*, January 1837.

———. "Mocha Dick, or The White Whale of the Pacific: A Leaf from a Manuscript Journal." In *Arthur Gordon Pym, Benito Cereno, and Related Writings*, edited by John Seelye. Philadelphia: J. B. Lippincott, 1967. 267–89.

Richards, Robert J. *The Romantic Conception of Life: Science and Philosophy in the Age of Goethe.* Chicago: University of Chicago Press, 2002.

Richardson, Brian W. *Longitude and Empire: How Captain Cook's Voyages Changed the World.* Vancouver: University of British Columbia Press, 2005.

Richardson, Robert D., Jr. *Emerson: The Mind on Fire.* Berkeley: University of California Press, 1995.

———. *Henry David Thoreau: A Life of the Mind.* Berkeley: University of California Press, 1986.

Rigby, Kate. *Topographies of the Sacred: The Poetics of Place in European Romanticism.* Charlottesville: University of Virginia Press, 2004.

Said, Edward W. *Culture and Imperialism.* New York: Random House, 1993.

Saint Pierre, Bernardin de. *Paul and Virginia.* Translated by Helen Maria Williams. New York: Appleton, 1851; repr. Doyleston, PA: Wildside Press, n.d.

Schiebinger, Londa. "Agnotology and Exotic Abortifacients: The Cultural Production of Ignorance in the Eighteenth-Century Atlantic World." *Proceedings of the American Philosophical Society* 149.3 (September 2005): 316–43.

Schierle, Sonja. "Introduction: Travels in the Interior of North America: The Fascination and Reality of Native American Cultures." In *The American Indian: Karl Bodmer, Maximilian Prinz zu Wied.* Cologne, Germany: Taschen, 2005. 9–15.

Scholnick, Robert J. "'The Password Primeval': Whitman's Use of Science in 'Song of Myself.'" In *Studies in the American Renaissance, 1986.* Charlottesville: University Press of Virginia, 1986. 385–425.

Scott, James C. *Weapons of the Weak: Everyday Forms of Peasant Resistance.* New Haven, CT: Yale University Press, 1985.

Scruton, Roger. *Kant: A Very Short Introduction.* Oxford: Oxford University Press, 2001.

Seeba, Hinrich C. "Cultural History: An American Refuge for a German Idea." In *German Culture in Nineteenth-Century America: Reception, Adaptation, Transformation,* edited by Lynne Tatlock and Matt Erlin. Rochester, NY: Camden House, 2005. 3–20.

Smith, Bernard. *European Vision and the South Pacific.* New Haven: Yale University Press, 1985.

Spence, Mark David. *Dispossessing the Wilderness: Indian Removal and the Making of the National Parks.* New York: Oxford University Press, 1999.

Stanton, William. *The Great United States Exploring Expedition.* Berkeley: University of California Press, 1975.

———. *The Leopard's Spots: Scientific Attitudes toward Race in America, 1815–1859.* Chicago: University of Chicago Press, 1960.

Stephens, Lester D. *Science, Race, and Religion in the American South: John Bachman and the Charleston Circle of Naturalists, 1815–1895.* Chapel Hill: University of North Carolina Press, 2000.

Stoll, Mark. "Religion 'Irradiates' the Wilderness." In *American Wilderness: A New History,* edited by Michael Lewis. Oxford: Oxford University Press, 2007. 35–53.

Sundquist, Eric. "Literature of Expansion and Race." In *The Cambridge History of American Literature.* Vol. 2, *Prose Writing, 1820–1865.* Cambridge: Cambridge University Press, 1995. 125–328.

Thoreau, Henry David. *Cape Cod,* edited by Joseph J. Moldenhauer. Princeton: Princeton University Press, 1988.

———. *The Journal of Henry David Thoreau.* 1906. 14 vols. Edited by Bradford Torrey and Francis Allen. New York: Dover, 1962.

———. *The Maine Woods*, edited by Joseph J. Moldenhauer. Princeton: Princeton University Press, 1972.

———. *Natural History Essays*, edited by Robert Sattelmeyer. Salt Lake City: Peregrine Smith, 1980.

———. *Reform Papers*, edited by Wendell Glick. Princeton: Princeton University Press, 1973.

———. *The Writings of Henry David Thoreau: Journal.* 7 vols. to date. Princeton: Princeton University Press, 1981–.

———. *Walden*, edited by J. Lyndon Shanley. Princeton: Princeton University Press, 1971.

———. *A Week on the Concord and Merrimack Rivers*, edited by Karl F. Hovde et al. Princeton: Princeton University Press, 1983.

———. *Wild Apples and Other Natural History Essays*, edited by William Rossi. Athens: University of Georgia Press, 2002.

Tocqueville, Alexis de. *Democracy in America*. Translated by Harvey C. Mansfield and Delba Winthrop. Chicago: University of Chicago Press, 2000.

Trouillot, Michel-Rolph. *Silencing the Past: Power and the Production of History*. Boston: Beacon Press, 1995.

Turner, Arlin. *Nathaniel Hawthorne: A Biography*. New York: Oxford University Press, 1980.

Tyler, Ron. "Karl Bodmer and the American West." In *Karl Bodmer's North American Prints*. Lincoln: University of Nebraska Press, 2004. 1–45.

Walls, Laura Dassow. *Emerson's Life in Science: The Culture of Truth*. Ithaca: Cornell University Press, 2003.

———. "Exploring the World: Volney, Goethe, Humboldt." In *Oxford History of Literary Translation in English*, vol. 4, edited by Peter France and Kenneth Haynes. Oxford: Oxford University Press, 2006. 1036–45.

Wallis, Brian. "Black Bodies, White Science: Louis Agassiz's Slave Daguerreotypes." *American Art* 9 (Summer 1995): 38–61.

Walters, Raymond Jr. *Albert Gallatin: Jeffersonian Financier and Diplomat*. New York: Macmillan, 1957.

Washburn, Wilcomb E. *The Cosmos Club of Washington: A Centennial History, 1878–1978*. Washington, DC: Cosmos Club, 1978.

Weaver, Jace. *Other Words: American Indian Literature, Law, and Culture*. Norman: University of Oklahoma Press, 2001.

White, Richard. *"It's Your Misfortune and None of My Own": A New History of the American West*. Norman: University of Oklahoma Press, 1991.

Whitman, Walt. *Leaves of Grass*, edited by Sculley Bradley and Harold W. Blodgett. New York: Norton, 1973.

———. *Poetry and Prose*. New York: Library of America, 1982.

Wilkes, Charles. *Narrative of the United States Exploring Expedition*. 5 vols. Philadelphia: Lea and Blanchard, 1845.

Wilton, Andrew, and Tim Barringer. *American Sublime: Landscape Painting in the United States, 1820–1880*. Princeton: Princeton University Press, 2002.

Worster, Donald. *Nature's Economy: A History of Ecological Ideas*. Cambridge: Cambridge University Press, 1977.

Ziff, Larzer. *Return Passages: Great American Travel Writing, 1780–1910*. New Haven: Yale University Press, 2000.

Index

CPSIA information can be obtained
at www.ICGtesting.com
Printed in the USA
FSHW010926030319
56065FS

9 780226 871837